ARCTIC OCEAN
NUMBERED COUNTRIES
1. NETHERLANDS
2. BELGIUM
3. CZECH REPUBLIC
4. AUSTRIA
5. SWITZERLAND
6. SLOVENIA
7. CROATIA
8. SERBIA & MONTENEGRO
9. ALBANIA
10. MACEDONIA
11. GREECE
12. BULGARIA
13. ROMANIA
14. HUNGARY
15. SLOVAKIA
ICELAND
FINLAND
R U S S I A
SWEDEN
NORWAY
ESTONIA
DENMARK
LATVIA
LITHUANIA
BELARUS
UNITED KINGDOM
IRELAND
POLAND
GERMANY
UKRAINE
MOLDOVA
FRANCE
PORTUGAL
SPAIN
ITALY
GIBRALTAR (UK)
MALTA
TUNISIA
GEORGIA
ARMENIA
AZERBAIJAN
TURKEY
SYRIA
LEBANON
ISRAEL
IRAQ
IRAN
KUWAIT
BAHRAIN
QATAR
UNITED ARAB EMIRATES
SAUDI ARABIA
OMAN
YEMEN
KAZAKHSTAN
UZBEKISTAN
TURKMENISTAN
KYRGYZSTAN
TAJIKISTAN
AFGHANISTAN
PAKISTAN
MONGOLIA
CHINA
NEPAL
BHUTAN
BANGLADESH
INDIA
MYANMAR (BURMA)
LAOS
THAILAND
VIETNAM
CAMBODIA
SOUTH KOREA
JAPAN
TAIWAN
PHILIPPINES
SRI LANKA
MALDIVES
MALAYSIA
SINGAPORE
INDONESIA
EAST TIMOR
MOROCCO
WESTERN SAHARA
ALGERIA
LIBYA
EGYPT
MAURITANIA
MALI
NIGER
CHAD
SUDAN
ERITREA
DJIBOUTI
ETHIOPIA
SOMALIA
SENEGAL
THE GAMBIA
GUINEA-BISSAU
GUINEA
SIERRA LEONE
LIBERIA
CÔTE D'IVOIRE
BURKINA FASO
GHANA
TOGO
BENIN
NIGERIA
CAMEROON
CENTRAL AFRICAN REPUBLIC
EQUATORIAL GUINEA
GABON
CONGO
D.R.C.
UGANDA
KENYA
RWANDA
BURUNDI
TANZANIA
SEYCHELLES
CABINDA (Angola)
ANGOLA
MALAWI
COMOROS
ZAMBIA
MADAGASCAR
MAURITIUS
ZIMBABWE
MOZAMBIQUE
NAMIBIA
BOTSWANA
SWAZILAND
SOUTH AFRICA
LESOTHO
SOUTH ATLANTIC OCEAN
INDIAN OCEAN
NORTHERN MARIANA ISLANDS (USA)
MARSHALL ISLANDS
FEDERATED STATES OF MICRONESIA
PALAU
KIRIBATI
NAURU
PAPUA NEW GUINEA
SOLOMON ISLANDS
TUVALU
CORAL SEA ISLANDS TERRITORY (Aust.)
VANUATU
FIJI ISLANDS
NEW CALEDONIA (France)
AUSTRALIA
NEW ZEALAND
Km 800 400 0 0 200 400 Miles
0° 15° 30° 45° 60° 75°
VAN DER GRINTEN PROJECTION

TOP 100
BIRDING SITES
OF THE WORLD

TOP 100
BIRDING SITES
OF THE WORLD

DOMINIC COUZENS

University of California Press
Berkeley • Los Angeles

University of California Press, one of the most distinguished university presses in the United States, enriches lives around the world by advancing scholarship in the humanities, social sciences, and natural sciences. Its activities are supported by the UC Press Foundation and by philanthropic contributions from individuals and institutions. For more information, visit www.ucpress.edu.

University of California Press
Berkeley and Los Angeles, California

First published in 2008 by New Holland Publishers (UK) Ltd.
London • Cape Town • Sydney • Auckland
www.newhollandpublishers.com

ISBN 978-0-520-25932-4 (cloth : alk. paper)

Cataloging-in-Publication Data for this title is on file with the Library of Congress.

Commissioning Editor: Simon Papps
Editors: Simon Papps, Marianne Taylor, Nicole Whitton, and Liz Dittner
Designers: D&N Publishing and Alan Marshall
Production: Melanie Dowland
Publishing Director: Rosemary Wilkinson

Reproduction by Pica Digital PTE Ltd., Singapore
Printed and bound by Tien Wah Press, Singapore

Cartography by John Loubser
Special thanks to Chris Bradshaw, Mark Finn, Kathy Ombler, and Pippa Parker for checking various chapters

Photographs (a = above, b= below, l = left, r = right) by AGAMI (pages 7l, 10, 21, 24a, 27a, 28, 29, 31b, 38b, 41a, 52, 56a, 62, 67a, 69b, 70, 76b, 85a, 100a, 105a, 116b, 117, 118, 119, 122a, 122b, 130, 131, 144, 145b, 148a, 150a, 150b, 155, 157, 177b, 180, 202, 203, 204, 206a, 208b, 216a, 220, 221, 222b, 228b, 233a, 233b, 236, 239b, 244b, 259b, 261, 262b, 293a, 308, 309), Roger Ahlman (209, 219, 230), Soner Bekir (73a, 73b), Nigel Blake (7r, 17a, 17b, 120, 132, 133a), Nik Borrow (128, 129a, 129b, 133b, 134a, 134b, 137, 138, 139a, 139b, 143, 154, 156), Howard Bottrell (14, 24b, 27b, 30b, 32, 35), Andy Bunting (88), Bill Coster (8, 23, 25b, 33b, 49, 55a, 114, 170l, 170r, 177a, 178, 184a, 184b, 186b, 201a, 206b, 208a, 214a, 214b, 215, 216b, 232, 237a, 267, 268a, 271a, 271b, 272, 285b, 292b, 297, 300, 302a, 302b, 303, 304a, 304b, 305a, 313b, 314, 315), Gerald Cubitt (79, 82, 85b, 86, 146, 147, 148b, 159, 161, 165b, 167, 169l, 169r, 182b, 183, 185, 192b, 217, 218a, 218b, 228a, 229), Bob Gibbons (25a), Phil Gregory (172, 187, 188a, 188b, 189), Jens Søgaard Hansen (6l, 108b, 109, 110a, 110b), Simon Harrap (96l, 96r), Jon Hornbuckle (89, 90, 94, 136, 153, 196b, 197b, 211a, 211b, 238, 246a), Murray Lord (9), Ian Merrill (97a, 115, 193, 194, 195a, 195b, 223, 224a, 286), Pete Morris (91, 92a, 92b, 93, 107, 149, 160a, 162, 163, 164, 165a, 166, 168, 175, 176, 179, 212b, 213, 225, 226, 234b, 235, 252, 262a, 263), Rod Morris (190, 192a), Simon Papps (186a, 191, 196a, 239a, 269a), Jari Peltomäki (26, 37, 43a, 43b, 45, 50a, 51, 71a, 112b, 116a), René Pop (20, 36, 61, 67b, 68, 76a, 77, 127, 270, 284, 289b, 290, 291a, 305b, 313a), George Reszeter (33a), Don Roberson (158a, 268b), Alan Tate (15b, 101r, 112a, 197a, 224b, 237b, 240b, 269b, 288b, 301b), Adrian Thysse (103, 104), David Tipling (2, 3-4, 12-13, 16, 22, 34, 38a, 39a, 40, 41b, 46, 47b, 54, 58, 59b, 60, 66b, 72, 74, 105b, 108a, 125, 140a, 140b, 141, 142, 145a, 198, 200, 227, 231, 242, 245, 246b, 247a, 247b, 248, 266, 287, 289a, 291b, 307), Markus Varesvuo (6r, 15a, 42, 44a, 44b, 47a, 48, 50b, 53l, 53r, 55b, 56b, 57, 59a, 64, 65, 66a, 69a, 75, 78, 111, 123, 124), VIREO (30a, 83, 84, 87, 95, 102a, 102b, 135, 152, 158b, 160b, 171, 174, 181, 182a, 212a, 234a, 240a, 241, 244a, 249a, 249b, 250a, 250b, 251, 253a, 253b, 255, 256, 257, 258, 259a, 260, 273a, 273b, 274, 275, 276a, 276b, 277, 278, 279l, 279r, 280a, 280b, 281, 282a, 282b, 283, 285a, 288a, 293l, 293r, 294, 295, 296, 298, 299a, 299b, 310, 311a, 311b, 312), Gehan de Silva Wijeyeratne (80a, 80b, 81, 97b, 98, 99), and Steve Young (18-19, 71b).

Page 1: Scarlet-headed Blackbird in Brazil
Page 2: Osprey in Finland
Right: Lesser Flamingos in Kenya
Pages 6-7 (left to right): Ibisbill in China, Spotted Redshank in Estonia, King Penguins on South Georgia and Giant Kingfisher in The Gambia

CONTENTS

Right: Andean Cock-of-the-rock is one of more than 1,000 bird species so far logged at the biodiversity hot-spot of Manu in Peru.

Introduction

This book is a celebration of the world of birds, from the vast flocks of flamingos in the Great Rift Valley and the seabird colonies of Britain to the 'river of raptors' in Mexico and New Guinea's birds-of-paradise. Through the description of 100 of the world's top birding sites it gives an overall view of the variety and abundance of birds on this planet, something that I hope will delight and inspire readers. It is, therefore, essentially an aspirational book. It should not, as such, be taken as a site guide from which you can plot your way to see a huge number of species, although you can use it for this purpose. Instead, it is an introduction to what birds are found where on Earth, and to where one might have the greatest experiences looking for them.

The 100 sites are fairly evenly divided between the continents in order to offer a fair spread of sites around the world. Each site has its own mini-chapter, with a description of the place, the habitat it encompasses, what birds occur there and, overall, why it is included in the book. Hopefully, the uniqueness of each place will come across, and the descriptions of what is found there will go well beyond a banal list of what has been recorded. I have been sparing in details about how to get there and where to stay because, as mentioned above, this is not a 'where to watch' guide in the usual sense. There are plenty of books that tell you that sort of information, and most of these are listed in the bibliography.

The book will stand or fall on the places it includes, and I expect to stir up some arguments and emotions by presenting this finished product to you. You will doubtless be appalled that certain of your favourites are excluded, and you will be even more aggravated by what is included instead – perfectly livid, indeed, when you realise that the sites are given a World Ranking. So some explanation about the construction of the contents might be useful here. A preliminary list of sites was initially selected by a small editorial team at New Holland Publishers, so I hope that they will share the blame with me. Incidentally, I acknowledge that even with 100 sites to choose from, there are well in excess of another 100 for which a good case could be made for their inclusion, and happily there are simply thousands of places on this earth which can, at one time or another, enthrall birders and ecotourists with the spectacle or variety they offer.

Nevertheless, a choice had to be made, and certain considerations soon became apparent. For example, although a huge list of species is instructive, it cannot be anything but a rough guide of how good a place is. Long lists of species recorded may reflect more about the attention given to them by birders than their intrinsic richness. Another point is that, if high lists were the only guide, this book would contain nothing but tropical forests, the majority of which would be in South America.

Long-tailed Ground-Roller in Madagascar's Spiny Forest.

■ *Right: Stellers Jay in the western USA.*

So, once freed from the burden of working solely from numbers, other aspects of a place become important. Among these are: the sheer quality of birds to be found there (how rare, how beautiful, if they occur in spectacular numbers, and so on); how important the site is on a world scale for the conservation of the relevant habitat and/or species; the intrinsic beauty of the site; the degree to which a site is good for birds and birding all year round; the history and both ornithological and environmental significance of the site; the ease and comfort of birding the site, where relevant; the possibility of delighting in other features, such as other animals or natural or archaeological spectaculars; and yes, perhaps, the fame and star quality of the place. On the negative side, where a site has become unwise or dangerous to visit in recent times (Afghanistan or Kashmir, for example), it had been deliberately left out.

In the end, despite all of this, the selection of sites was, of necessity and expedience, largely a matter of personal experience. To begin with, we picked the sites we knew ourselves, or had heard about. During the research for the book, of course, many more places came to light, and a good number of these were considered sufficiently good to be included. However, there is no doubt that many places with at least a good a claim as those that are included have been overlooked, and I apologize for this. If your favourite has been missed out, or if you think that the selections for a particular region are flawed, please do let us know.

Finally, it is important to state that all selections were made free from commercial pressure. In this book there are some sites that are profit-making concerns, but we have chosen to include them for ornithological reasons only; in some cases, it would simply be disingenuous not to. None will have known for sure that they have been included in this book prior to publication, and none paid for the privilege. Nobody offered or gave free accommodation during the research, either.

I hope, therefore, that you will enjoy this offering of the 100 sites which I believe offer the best birding experiences in the world.

Dominic Couzens, Dorset, UK, May 2008

NOMENCLATURE

The English names used in this book follow *Birds of the World: Recommended English Names* by Frank Gill and Minturn Wright (2006, Christopher Helm, London), with the following exceptions (in species order). Names with an asterisk refer to subspecies mentioned in the text where the name differs from that in Gill and Wright. A double asterisk refers to a split where what was once regarded as a subspecies has been afforded full species status. The new scientific name of the split species is given in the second line:

Name used	Scientific name	Gill and Wright name
Orange-footed Scrubfowl	*Megapodius reinwardt*	Scrubfowl
Caucasian Black Grouse	*Lyrurus mlokosiewiczi*	Caucasian Grouse
Willow Grouse	*Lagopus lagopus*	Willow Ptarmigan
Common Pheasant**	*Phasianus colchicus*	Common Pheasant
Japanese Green Pheasant**	*Phasianus versicolor*	Common Pheasant
Tundra Swan*	*Cygnus columbianus columbianus*	Tundra Swan
Bewick's Swan*	*Cygnus columbianus bewickii*	Tundra Swan
Brent Goose	*Branta bernicla*	Brant Goose
Black Brant*	*Branta bernicla nigricans*	Brant Goose
Common Merganser*	*Mergus merganser americanus*	Common Merganser
Goosander*	*Mergus merganser merganser*	Common Merganser
Common Scoter	*Melanitta nigra*	Black Scoter
Black Scoter	*Melanitta americana*	American Scoter
Red-throated Diver	*Gavia stellata*	Red-throated Loon
Black-throated Diver	*Gavia arctica*	Black-throated Loon
Pacific Diver	*Gavia pacifica*	Pacific Loon
Great Northern Diver	*Gavia immer*	Great Northern Loon
White-billed Diver	*Gavia adamsii*	Yellow-billed Loon
Madeiran Storm Petrel	*Oceanodroma castro*	Band-rumped Storm Petrel
Slavonian Grebe	*Podiceps auritus*	Horned Grebe
Great White Egret*	*Ardea alba*	Great Egret
Great Egret*	*Ardea alba egretta*	Great Egret
Intermediate Egret	*Egretta intermedia*	Yellow-billed Egret
American Black Vulture	*Coragyps atratus*	Black Vulture
Gyr Falcon	*Falco rusticolus*	Gyrfalcon
Lammergeier	*Gypaetus barbatus*	Bearded Vulture
Black Vulture	*Aegypius monachus*	Cinereous Vulture
Short-toed Eagle	*Circaetus gallicus*	Short-toed Snake Eagle
Northern Harrier*	*Circus cyaneus hudsonicus*	Northern Harrier
Hen Harrier*	*Circus cyaneus cyaneus*	Northern Harrier
Rough-legged Buzzard	*Buteo lagopus*	Roughleg
Rough-legged Hawk*	*Buteo lagopus sanctijohannis*	Roughleg
Eastern Imperial Eagle	*Aquila heliaca*	Asian Imperial Eagle
Purple Swamp-hen	*Porphyrio porphyrio*	Purple Swamphen
Small Buttonquail	*Turnix sylvaticus*	Kurrichane Buttonquail
Grey Phalarope	*Phalaropus fulicarius*	Red Phalarope
Common Gull*	*Larus canus canus*	Mew Gull
Herring Gull**	*Larus argentatus*	Herring Gull
American Herring Gull**	*Larus smithsonianus*	Herring Gull
Caspian Gull**	*Larus cachinnans*	Yellow-legged Gull
Yellow-legged Gull**	*Larus michahellis*	Yellow-legged Gull
Lesser Black-backed Gull**	*Larus fuscus*	Lesser Black-backed Gull
Heuglin's Gull**	*Larus heuglini*	Lesser Black-backed Gull
Black-headed Gull	*Larus ridibundus*	Common Black-headed Gull
White-winged Black Tern	*Chlidonias leucopterus*	White-winged Tern
White Tern	*Gygis alba*	Angel Tern
Arctic Skua	*Stercorarius parasiticus*	Parasitic Jaeger
Long-tailed Skua	*Stercorarius longicaudus*	Long-tailed Jaeger
Brünnich's Guillemot	*Uria lomvia*	Thick-billed Murre
Common Guillemot	*Uria aalge*	Common Murre
Eurasian Eagle Owl	*Bubo bubo*	Eurasian Eagle-Owl
Northern Hawk Owl	*Surnia ulula*	Northern Hawk-Owl
Tengmalm's Owl	*Aegolius funereus*	Boreal Owl
Short-legged Ground-roller	*Brachypteracias leptosomus*	Short-legged Ground Roller
Scaly Ground-roller	*Geobiastes squamiger*	Scaly Ground Roller
Pitta-like Ground-roller	*Atelornis pittoides*	Pitta-like Ground Roller
Rufous-headed Ground-roller	*Atelornis crossleyi*	Rufous-headed Ground Roller
Long-tailed Ground-roller	*Uratelornis chimaera*	Long-tailed Ground Roller
Cuckoo-roller	*Leptosomus discolor*	Cuckoo Roller
Steere's Pitta	*Pitta steerii*	Azure-breasted Pitta
Rufous Scrub-bird	*Atrichornis rufescens*	Rufous Scrubbird
Iberian Azure-winged Magpie	*Cyanopica cooki*	Iberian Magpie
Crested Tit	*Lophophanes cristatus*	European Crested Tit
Siberian Tit	*Poecile cincta*	Grey-headed Chickadee
Shore Lark	*Eremophila alpestris*	Horned Lark
Japanese Bush Warbler**	*Cettia diphone*	Japanese Bush Warbler
Manchurian Bush Warbler**	*Cettia canturians*	Japanese Bush Warbler
Wren	*Troglodytes troglodytes*	Winter Wren
Stripe-breasted Rhabdornis	*Rhabdornis inornatus*	Stripe-breasted Creeper
Sri Lanka Scaly Thrush	*Zoothera imbricata*	Sri Lanka Thrush
Semi-collared Flycatcher	*Ficedula semitorquata*	Semicollared Flycatcher
Narcissus Flycatcher**	*Ficedula narcissina*	Narcissus Flycatcher
Elisa's Flycatcher**	*Ficedula elisae*	Narcissus Flycatcher
Lapland Bunting	*Calcarius lapponicus*	Lapland Longspur

USEFUL WEBSITES

This list is not exhaustive, but only includes those that were used significantly during writing and researching.

World Sites

www.birdlife.org.uk
www.fatbirder.com
www.splatzone.nl
www.surfbirds.com

Europe

www.birdingnorway.no (Varanger)
www.donanabirdtours.com (Coto Doñana)
www.finnature.fi (Oulu and Matsalu Bay)
www.kilda.org.uk (Outer Hebrides)
www.matsalu.ee (Matsalu Bay)
www.skof.se (Falsterbo)

Asia

www.drmartinwilliams.com (Mai Po and Beidaihe)
www.ecotours.ru (Ussuriland)
www.eilat-birds.org (Eilat)
www.hkecotours.com (Mai Po)
www.jetwingeco.com (Sinharaja)
www.kazakhstanbirdtours.com (Korgalzhyn)
www.orientalbirdclub.org
www.tommypedersen.com (Dubai)
www.wild-russia.org (Lena Delta)

Africa

www.africanbirdclub.org
www.birduganda.com (Bwindi)
www.gambiabirding.org (Gambia River)
www.gambiabirdguide.com (Gambia River)
www.natureseychelles.org (Seychelles)
www.sabirding (South Africa)

Australasia

www.alanswildlifetours.com.au (Queensland Wet Tropics)
www.birdingaustralia.com.au
www.birdsaustralia.com.au
www.cassowary-house.com.au (Queensland Wet Tropics)
www.oreillys.com.au (Lamington)
www.outback-australia.info (Strzelecki Track)
www.sossa-international.org (NSW Pelagics)
www.stewartisland.co.nz (Stewart Island)
www.wettropics.gov.au (Queensland Wet Tropics)

South America

www.birding-in-peru.com (Peru)
www.birdvenezuela.com (La Escalera)
www.inkanatura.com (Peru)
www.kolibriexpeditions.com (Peru)
www.manuwildlifecenter.com (Manu)
www.neotropicalbirdclub.org
www.tandayapa.com (Tandayapa)

Central America and the Caribbean

www.asawright.org (Asa Wright)
www.canopytower.com (Canopy Tower)
www.cct.or.cr (Monteverde)
www.guatemalabirding.com (Tikal)
www.pronaturaveracruz.org (Veracruz river of raptors)

North America

www.americanbirding.org
www.americanparknetwork.com
www.audubon.org
www.birder.com
www.birdinghawaii.co.uk (Alaka'i)
www.fws.gov
www.hawkmountain.org (Hawk Mountain)
www.houstonaudubon.org (High Island)
www.montereyseabirds.com (Monterey)
www.pc.gc.ca (Point Pelee)
www.sabo.org (Chiricahua Mountains)
www.shearwaterjourneys.com (Monterey)
www.texasbirding.net (High Island)

Europe

Europe has a phenomenal birding heritage, and the hobby has taken root as a national pastime in countries such as Britain, the Netherlands and Sweden. With only about 500 regularly occurring species, the continent cannot claim to have the most diverse avifauna in the world. However, few bird species have escaped detailed study, and Europe's habitats have been described and understood to an extraordinary extent.

Another important feature in Europe is how the relationship between people and birds has played out. Since there are almost no entirely natural and unspoilt habitats left in temperate Europe, monitoring the fortunes of the bird populations there is of great interest. Most of temperate Europe has been farmed, and the current trend towards the mechanization of the areas that had previously escaped this process (such as parts of eastern Europe) is a cause of great concern. Currently, Europe's most efficient farming areas are biodiversity deserts. The great forests that once covered the central sector have largely been felled, leading to severe declines in woodland species.

That is not to say that Europe lacks good birding areas – far from it. Many regions, from parts of the tundra and taiga to large temperate wetlands and vast intertidal reaches, are excellent, and the temperate climate means that the migration of birds can be spectacular in spring and autumn. One habitat that does deserve special mention is the Mediterranean scrubland known as maquis, which plays host to, among others, a delightful assemblage of Old World warblers from the genus *Sylvia*. Besides these, Europe is also rich in waders, wildfowl and seabirds such as auks and gulls.

Right: male Ruffs lekking in Finland.

North Norfolk

HABITAT Coast, mudflats, salt marsh, dunes, fresh water pools and marshes, shingle, scrub and woodland

KEY SPECIES [Dark-bellied] Brent and Pink-footed Geese, Red Knot, Dunlin, Pied Avocet, Bearded Reedling, Shore Lark, Twite

TIME OF YEAR Summer (June to July) is quietest; the rest of the year is superb

Below: every dawn and dusk in winter witnesses spectacular flights of Pink-footed Geese at sites along the Norfolk coast.

This magical corner of Britain must be one of the most popular birding areas in the entire world. In recent years, the tiny reserve of Titchwell Marsh alone has been receiving over 120,000 human visitors a year, exceeding the counts of all, or most of, the birds themselves. Britain has a rich tradition of birding and Norfolk, in most recent times, has been at the hub of this increasingly popular hobby.

North Norfolk certainly has plenty to attract birders, with delights all year round, together with an excellent infrastructure and easy advice and help virtually on tap. It hosts some scarce breeding birds, impressive numbers of waders and wildfowl in winter, while the two main migration seasons, April to May and August to November, bring an almost bewildering variety of birds. More than 360 species have been recorded between the estuary of The Wash to the west and the town of Sheringham to the east, and there is seldom a day in the year when some rarity is not in the vicinity.

The winter season brings big flocks of both geese and waders, especially in the western corner, near to the village of Snettisham. On a high tide a substantial number of the waders using the intertidal mudflats of The Wash move to the small complex of gravel pits and islands here for roosting and, on a particularly high tide in season, up to 50,000 Red Knots, 11,000 Dunlins and 6,000 Eurasian

■ *Right: Bearded Reedlings can be found in north Norfolk throughout the year. Good places to look include the reserves at Cley and Titchwell.*

Oystercatchers may be using this tiny area at the same time. The sight and sound of them packing on to the islands and spits and flying about is incredible; and indeed, even on a modest tide, the Red Knots still make an impressive spectacle on the flats themselves. These birds are famed for their gigantic gatherings, the flock moving as one amorphous mass, its edges moving back and forth like plumes of smoke.

Remarkably, that is not the only splendour of Snettisham. At dawn and dusk between October and February, a large flight of Pink-footed Geese also passes over the reserve as the birds commute from their roosting grounds on the salt marsh to the inland fields, where they graze on grain or potatoes. Indeed, this spectacle is repeated right along the coast as far as Holkham, some 30 km to the east. The geese fly high and their V-formations make handwriting in the skies, often to a backdrop of the rising or setting sun, while the merry *ang-ang, wick-wick* calls ring down on to spellbound birders. It is estimated that, at the very least, 70,000 Pink-footed Geese from Iceland winter in north-west Norfolk, along with nearly 10,000 Brent Geese from Russia, of the dark-bellied form *bernicla*.

Another of the attractions in north Norfolk is the presence of several scarce passerines that choose these wild, windswept coasts as regular wintering grounds. These include two species that breed primarily in the

■ *Right: The Wash is a good place to see high concentrations of waders – these are Red Knots.*

Above: Western Marsh Harriers are once again a familiar sight close to the Norfolk coast – the species recovered from the brink of extinction as a British breeding bird in the early 1970s.

high Arctic, Snow and Lapland Buntings, plus two birds that are mainly northern montane breeders in Europe, Twite and Shore Lark. All these species scour the dunes and salt marshes for seeds, and they can often be found together in flocks. The Snow Buntings are a particular favourite as they can often be seen in appreciable numbers (200 or more), and their flocks have the endearing habit of 'rolling' forwards, with birds at the back intermittently overflying the leaders to get in front.

The breeding birds of north Norfolk are almost as impressive as the winter visitors, and include several scarce species. The farmlands are the English stronghold of the fast-declining Grey Partridge, while other birds that are also being lost from the rural British scene are also present, including Barn Owl, Corn Bunting and Eurasian Tree Sparrow. Also important are the reedswamp birds, with Bearded Reedling, Eurasian Bittern and Western Marsh Harrier all being widespread, if not common. Pied Avocets, close to the northern edge of their range, breed on some of the lagoons within nature reserves – some of which have been specially constructed for the purpose. Terns also occur, especially on offshore Scolt Head Island and on Blakeney Point; Sandwich Terns can number nearly 4,000 pairs altogether.

For many an experienced birder, however, it is really the migration season that gives Norfolk its almost magnetic attraction. Norfolk is one of the closest parts of the United Kingdom to the European continent, and it only takes some kind of easterly blow, with a touch of drizzle, to dump drift migrants unexpectedly on to these coasts. On a good autumn day the many bushes and trees may be alive with migrants, especially Willow and Garden Warblers, Common Redstarts and European Pied Flycatchers, and a thorough searching will sometimes reveal a rarity such as Eurasian Wryneck, Red-breasted Flycatcher or Barred Warbler. As the season progresses, Yellow-browed Warblers in October and then Pallas's Leaf Warblers in November brighten the scene. Spring is also very good, with many of the same species as in early autumn, seen in their breeding finery. If conditions are right there might be a Bluethroat from the east or a Red-rumped Swallow or European Serin from the south.

Good numbers of waders are present throughout the year in the many pools and lagoons, and on a good day in September it is possible to see well over 20 species, including Curlew Sandpiper and both Little and Temminck's Stints. The many hides on nature reserves such as Cley and Titchwell can allow exceptional views of such birds at close quarters.

The sea-watching scene is just as lively as the comings and goings on land. North Norfolk is excellent for wintering seaduck such as Common and Velvet Scoter, Long-tailed Duck and Common Eider, while other inshore species include Red-throated Diver and Slavonian and Red-necked Grebes. Meanwhile, strong onshore winds in autumn bring skuas in good numbers, together with a few Manx and Sooty Shearwaters. Later on in the year a northerly blow can deliver good numbers of Little Auks from the Arctic.

Not surprisingly, the rarity list is long and includes some jaw-dropping encounters, some of which have gone down in the birding folklore that surrounds this place. For example, one spring both a Laughing Gull and a Franklin's Gull turned up on the very same day, more or less standing next to each other, while many British firsts, including a bewilderingly unlikely Rock Sparrow, have graced this coastline. This is the modern centre of British birding, and its status is fully deserved.

The Outer Hebrides

HABITAT Rocky islands, grassland, marsh, coast, bog, moorland

KEY SPECIES Seabirds including breeding European and Leach's Storm Petrels, Northern Gannet and Atlantic Puffin and impressive skua passage, Corn Crake, breeding waders and wildfowl, rarities

TIME OF YEAR Spring and early summer (May to July) is best

Right: the moorlands on Lewis and Harris hold large populations of European Golden Plover.

Few parts of the world are more densely populated with birdwatchers than Britain. Despite this, the indigenous birdlife is not especially rich, and more than half the list of species for the whole country consists of rarities visiting from elsewhere. However, there are two outstanding exceptions to this generalization – breeding waders and, most outstandingly, the seabirds. Nowhere in Britain are these two groups of birds so prevalent, and found in more breathtaking scenery, than in the archipelago of the Outer Hebrides (which is also known as the Western Isles).

The Outer Hebrides lie, on average, about 50 km off the north-west coast of Scotland. The main archipelago measures about 200 km from north to south, while a great deal of bird interest also lies in a cluster of outlying island groups, such as St Kilda, 64 km further out into the wild North Atlantic. On the main island chain

Right: now exceedingly rare elsewhere in Britain due to modern farming practices, the Corn Crake is a relatively common in the Outer Hebrides and the population is increasing there thanks to conservation measures.

■ *Above: Leach's Storm Petrels are found in the turbulent waters around the Outer Hebrides during the breeding season; they wander the open ocean for the rest of the year.*

there is an intriguing grading of habitats from west to east: the west coast is rimmed with long beaches of shell-sand, often backed by dunes. Just inland of this is a unique habitat known as machair, which is grassland growing on a mixture of shell-sand and peat. The dry areas of machair are farmed in a traditional manner, while the wet areas provide low-lying rich pools. Further inland still, the land becomes dominated by peat bogs and moorland dotted with lakes until it reaches the rugged, rocky coast on the eastern side, where every few kilometres the coast indents to form an inland finger of sea, known locally as a sea-loch. It is estimated that, taking both saltwater and freshwater bodies into account, there are some 6,000 lakes in the Outer Hebrides. On the whole, the land is low-lying, and the highest point in the whole archipelago is only 799 m above sea-level.

The peat bogs and machair form superb breeding habitat for waders, and the densities of breeding birds here are among the highest in Europe. As a general rule, the northern islands of Lewis and Harris provide the best moorland habitat, with large populations of European Golden Plovers, Common Greenshanks and Dunlins, while the southern trio of North and South Uist and Benbecula provide the cream of the machair, with high densities of Common Snipe, Common Redshanks and Common Ringed Plovers. Every year a few Red-necked

Phalaropes, extremely rare breeders this far south, turn up and sometimes they nest successfully.

In such a watery habitat it is perhaps not surprising that wildfowl also feature strongly. The Outer Hebrides maintain an impressive population (300 pairs) of entirely wild Greylag Geese (the rest of the British breeding population consists of feral birds), and there are also significant numbers of Mute Swans, Common Eiders and Red-breasted Mergansers. The lakes provide good habitat for both Red-throated and Black-throated Divers, making this a perfect place to compare their breeding ecology. Black-throated Divers breed by large lakes that provide fish on-site, while Red-throated Divers nest on often minute, sterile pools and they commute back and forth from the sea to deliver fish to their young.

While a few seabirds breed on the cliffs on the north coast of Lewis, the really impressive colonies occur far offshore. Some of these are important nationally and internationally. Sula Sgeir, off the northern tip of the main archipelago, hosts 10,000 pairs of Northern Gannets, and is the only place in Britain where young Gannets are harvested for food. Each year, the hunters spend two weeks on this remote rock and collect up to 2,000 youngsters. In the last few seasons a misplaced Black-browed Albatross has also graced these inaccessible cliffs. Meanwhile, the fearsomely isolated North Rona, 71 km out in the Atlantic, hosts a major population of European Storm Petrels. The Shiant Islands hold a big colony of approximately 76,000 Atlantic Puffins (and also, curiously, Britain's only population of Black Rats) while the Flannan Islands, 32 km west of Lewis, provide breeding habitat for both European and Leach's Storm Petrels (several thousand pairs of each).

However, it is St Kilda, 64 km off North Uist, that is by far the most famous for its seabirds, and no wonder: half a million pairs of various species breed there. A world heritage site with a rich history, this cluster of four small islands is all that remains of a long-extinct volcano. Looming above the stormy waters of the North Atlantic, where the heavy swell blights the lives of seabird-counters forced to use a boat, St Kilda was bypassed by the erosion of the last ice-age, and stands proud as a towering lump teeming with cliff-dwelling birds. Indeed, the cliffs on the island of Hirta, the largest island, are a sheer drop of 430 m, the highest in Britain. Remarkably, this remote outpost has a long history of human settlement. People are thought to have made it here 5,000 years ago, and to have introduced the islands' unique breed of sheep. These Soay Sheep are still present today and are undergoing intensive genetic study because of their isolation. What is certain is that the islands were occupied for some 2,000 years, until being abandoned finally in the 1930s, leaving their ruins to breeding Common Starlings of the local race, and to the unique 'St Kilda Wren', a rare large island form of the Wren.

The numbers of seabirds are indicative of the huge importance of the Outer Hebrides. These islands hold the world's largest colony of Northern Gannets (at least 60,000 pairs – a quarter of the entire population), some 90 per cent of Europe's Leach's Storm Petrels (49,000 pairs), about 250,000 pairs of Atlantic Puffins (half of the British population), 62,000 pairs of Northern Fulmars, 22,000 pairs of Common Guillemots and 150 pairs of Great Skuas. Many of these numbers are necessarily approximate, because of the extreme difficulty of counting and the high numbers involved.

Another seabird spectacular of the Outer Hebrides is of a quite different kind – an impressive regular return passage of skuas, mainly in May. The numbers of Long-tailed and Pomarine Skuas may sometimes run into the thousands during the short season.

The Camargue

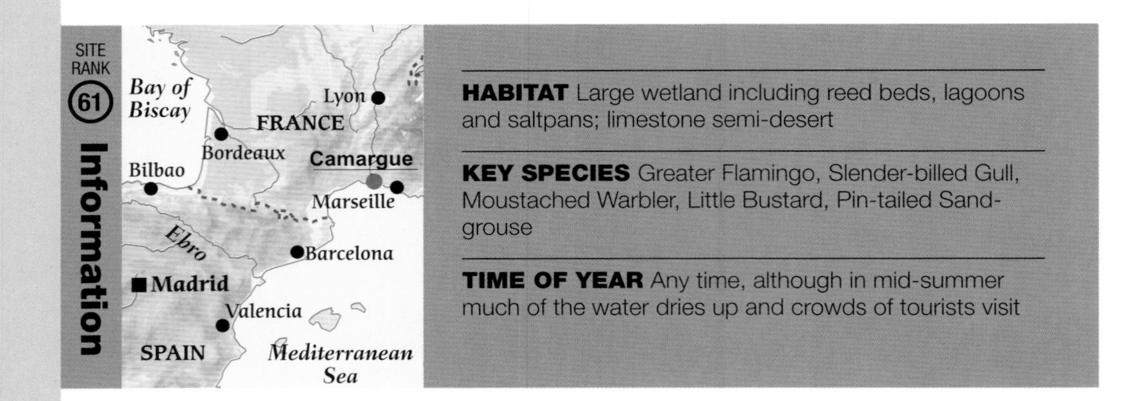

HABITAT Large wetland including reed beds, lagoons and saltpans; limestone semi-desert

KEY SPECIES Greater Flamingo, Slender-billed Gull, Moustached Warbler, Little Bustard, Pin-tailed Sandgrouse

TIME OF YEAR Any time, although in mid-summer much of the water dries up and crowds of tourists visit

The Camargue is a vast plain in the south of France. It is located between the Mediterranean Sea and the two arms of the River Rhone delta, the largest river delta in western Europe. It provides a large area (750 sq km) of fine wetland habitat for birds, and it is also famous for its free-ranging white horses and black bulls. It is a romantic and scenic region, steeped in history (the main town, Arles, was originally Roman), and it is dotted with marshes, lagoons, agricultural areas, beaches and, to the east, a large limestone semi-desert known as La Crau. Although much of its original area has been converted to agriculture, the Camargue is still ranked as one of Europe's finest wildlife sites.

One of the star attractions for birders is the very large colony of Greater Flamingos, mainly found on the Étang de Fangassier, a large saltpan in the central part of the Camargue, close to the sea. In the late 1960s the flamingos were at risk from the erosion of their nesting islands, so bulldozers were sent in to build new platforms for the birds to use. Since that time the population has more than doubled to nearly 15,000 pairs, and the flamingos are regarded as a major tourist attraction. They occur year-round, although only a quarter of the population remains

Below: Greater Flamingo is the Camargue's flagship bird species and almost 15,000 pairs breed.

Above: a boldly striped neck identifies an adult Purple Heron; this species nests on the ground in dense reedbeds.

during the winter, the rest retreating to Spain and northwest Africa. In recent years these flamingos have been supplementing their normal diet of small invertebrates (from which their pink coloration is derived) with a new fad food, rice, which is grown locally.

Sharing the shallow saline waters with the flamingos is another pink-flushed bird, the Slender-billed Gull. Indeed, this elegant species is often found feeding at the feet of taller birds, fielding shrimps fleeing oddly-shaped bills. This specialized, long-legged gull is one of the few of its family that can run after prey through the water, sometimes gathering in lines to herd fish into the shallows before diving into the panicking mêlée. This scarce gull began breeding in the Camargue as recently as 1993, and there are now almost 1,000 pairs.

The saline pools suit other specialized birds, too. The Pied Avocet and Black-winged Stilt get their long legs (blue and pink respectively) wet here, either scything their bills through the water (avocet) or picking delicately from the water surface (stilt). Both live up to the name 'wader', being able to use deep, saline water without swimming (although the avocet will swim if necessary). The Kentish Plover, on the other hand, keeps its feet dry, feeding by sight along the sandy edges of the saltpans and lagoons. It also nests on beaches, and benefits from being able to run about without the hindrance of stones or vegetation.

The very best time to visit the Camargue is perhaps during the winter, when the seasonal rains fill up the many ponds and lakes (which are dry between March and September) giving the whole place a lush, verdant look. Wildfowl are everywhere, and some impressive numbers have been counted in recent years: 13,000 Gadwall, 23,000 Eurasian Teal and 2,000 Red-crested

Above: Red-crested Pochard is resident in the Camargue; it both dabbles and dives to find food.

Pochard, for example, as well as 30,000 Eurasian Coot. This is the season when there can be a notable influx of raptors, including quite a number of rarities or oddities, such as Greater Spotted Eagle, Long-legged Buzzard and Booted Eagle (usually a summer visitor to Europe). The winter period also brings a regular small flock of Pine Buntings, which are usually very rare this far west.

By the spring, many of the breeding wetland birds are confined to the permanent lakes, such as the huge Étang de Vaccarès. This is a great place to see many of Europe's widespread marshland birds, including Purple and Squacco Herons, Western Marsh Harrier, Great Reed Warbler and Bearded Reedling. One of the few warblers to winter here is the Moustached Warbler, a bird that is able to pick smaller edible items off marshland vegetation than its competitors can – items that are more reliable year-round food than larger invertebrates. This bird sometimes has eggs in the nest by March.

On the eastern edge of the Camargue National Park, La Crau presents a picture very different from this abundance. It is as flat as the rest of the Camargue, but permanently dry, the ground mainly dotted with pebbles and herbs; a few sorry bushes survive here and there. The old delta of the nearby River Durance, it is France's only semi-desert area, and holds several species that are rare elsewhere in the country, including a population of several hundred Little Bustards and about 150 Pin-tailed Sandgrouse. These two species often gather together in flocks, since the sandgrouse use the bustards as lookouts. Birders should also look out for European Roller and also the Great Spotted Cuckoo, which uses the local Eurasian Magpie as host. However, both cuckoo and roller can be very difficult to find.

In contrast to the Camargue, most of La Crau is in private hands and therefore has no formal protection. Much is grazed, and the amount of habitat available to the special birds diminishes year by year through encroachment by agriculture and irrigation. The future of the sandgrouse, in particular, is threatened, because its population is isolated from others further south in Spain.

Organbidexka Col Libre

HABITAT Low mountain pass (1,440 m), forest

KEY SPECIES Raptors, especially Red and Black Kites and European Honey Buzzard; Black and White Storks, Common Wood Pigeon, White-backed Woodpecker, Lammergeier

TIME OF YEAR Mainly autumn migration July to November, but with light spring passage and some interesting resident species

The mountain chain of the Pyrenees, dividing France from Spain in western Europe, is a major barrier to post-breeding birds migrating to their wintering grounds in Africa. On the whole, travelling birds try to avoid high mountain crossings with their potential turbulence, cold and poor weather, so the Pyrenees present a problem to many millions of south-bound migrants. In the west of the range, however, the mountains are very much lower than their counterparts further east, and a series of mountain passes, or cols, present ideal short-cuts for the birds to follow on their way south into Spain. The most famous of these, with an interesting history as well as a good birding pedigree, is Organbidexka, in the Haute-Soule region of Aquitaine.

Below: the Lammergeier is scarce in Europe but birds are seen regularly at Organbidexka.

Long before birding became a widespread pastime, the locals in this rugged, proud region were well aware of a massive movement of Common Wood Pigeons and Stock Doves through their cols in October and November every year. Indeed, they took advantage, setting up shooting points along prominent parts of the pigeons' route. Somewhat resourcefully, they noticed that the birds could be tempted down to land; if the hunters flung small white discs, known as zimbelas, into the air, the pigeons mysteriously landed and could be shot or caught in nets. This pastime became ingrained into Basque culture as 'la Chasse de la Palombe' (the hunt of the Wood Pigeon).

Over time, what was both a sport and a means of easy food began to take a toll on the migrants. The hunting of birds is very popular in parts of France and Spain, and unfortunately by modern times the many north–south facing cols became littered with small bunkers for shooting; hard to believe in such a sophisticated and modern country as France. Thousands of pigeons and other migrants, including birds of prey, were also shot indiscriminately and, to the shame of the countries involved, this slaughter continues today.

However, in 1979 a small band of ornithologists and conservationists became sickened by the carnage and began, in the face of local opposition from hunters, to rent their own bunker at Organbidexka. They declared it a 'col libre' (hunting-free zone) and began to monitor not just the migration itself, but also hunting activities. Just to give an example of what they and the birds are up against, they recorded a total of 25,360 gunshots in a single day in 1982. Nowadays, some 30 years later, the col libre is still in existence, and indeed is now the site of a bird observatory. Every year, between August and November, the watch point is staffed by enthusiasts who both monitor the migration and introduce members of the public to the wonders of birds and their movements. In recent years they have added more Pyrenean passes to their list of watch points, including the Col de Lizarrieta, near Ascain, and Lindux, south of St-Jean-Pied-de-Port.

Despite its importance for pigeons, of which some 100,000 pass between late September and November, Organbidexka has in recent years become world famous for its raptor migration. The pass is a bottleneck for raptors avoiding the sea to the west and the higher peaks to the east, while being mountainous enough to enable birds to use updrafts for their journeys. In general,

Right: it was the huge numbers of passing Common Wood Pigeons that originally drew the attention of hunters, and then birders, to Organbidexka.

Below: September is the primary month for Montagu's Harrier passage.

mountain updrafts only exist some 300–400 m above the land, so the col is famous for affording quite exceptional low views of many of the passing birds.

The flight begins in mid-August, with the first Black Kites and European Honey Buzzards. These species are the two most numerous here, with averages of 17,209 and 12,354 birds passing each season, often in large flocks. By September the variety increases, with Honey Buzzards increasing and Black Kites slowing to a trickle, replaced by Montagu's and Western Marsh Harriers (86 and 207 per season respectively), Osprey (138) and Short-toed Eagle (131). There is also a small passage of White and Black Storks, as well as innumerable Barn Swallows. By October the raptor mix has altered slightly, and includes what must be one of the largest Red Kite flights in the world (1,721). In this month, Hen Harriers (69), Eurasian Sparrowhawk (324), Common Buzzard (91), Eurasian Hobby (41), Merlin (27) and Common Kestrel (57) pass by in numbers, while remarkably for this far north, up to about 50 Booted Eagles may also be recorded. October is also excellent for the visible migration of passerines such as larks and pipits, while November signals something of a change, when up to 15,000 Common Cranes may pass over, in the last migratory flight of the year. In all, some 31,000 migrant raptors use the narrow migration corridor along Organbidexka during the autumn.

Apart from the regular species, it is inevitable that the watch point should record a few oddities, and over the years these have included Egyptian Vulture, Eleonora's Falcon and, bizarrely, one Spanish Imperial Eagle. However, at least 95 per cent of all the raptors seen are European Honey Buzzards and Black Kites. The area also offers species such as Lammergeier and White-backed Woodpecker, which breed nearby in the dense forests.

But perhaps the best reason to go to Organbidexka is not so much as a birder, but as a human being enjoying the phenomenon of migration. Your presence will add another note of defiance to those whose only interest is in senseless shooting.

Lac du Der-Chantecoq

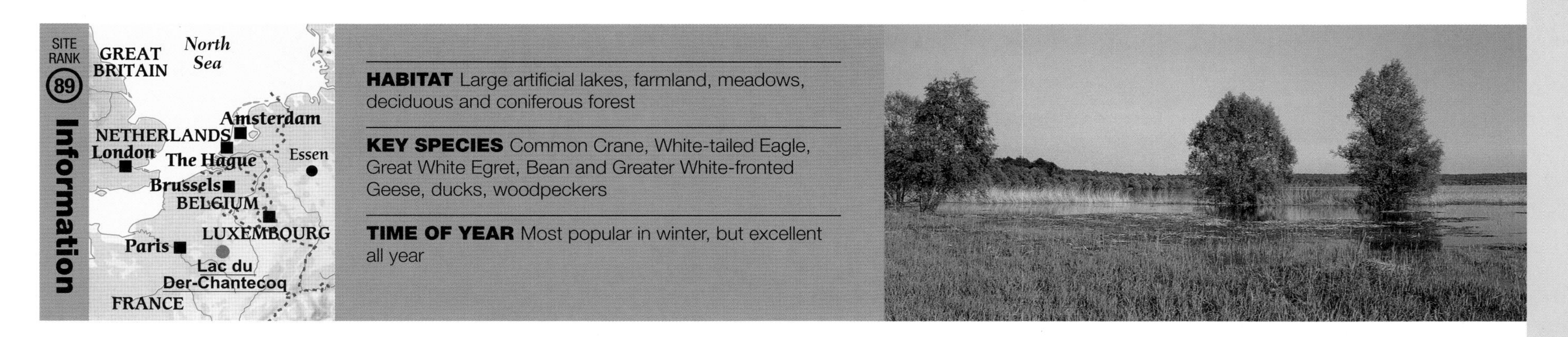

HABITAT Large artificial lakes, farmland, meadows, deciduous and coniferous forest

KEY SPECIES Common Crane, White-tailed Eagle, Great White Egret, Bean and Greater White-fronted Geese, ducks, woodpeckers

TIME OF YEAR Most popular in winter, but excellent all year

Above right: a quiet corner of the Lac d'Orient.

Very few of the world's best sites for birds are entirely created by human hand, but a complex of large reservoirs in the Champagne region of central France, about 190 km south-east of Paris, have this unusual distinction. The Lac d'Orient (23 sq km) was dug in 1966 in order to store floodwater from the River Seine, and was followed in 1990 by the 20-sq-km Lac du Temple just next door; a third reservoir, the 5-sq-km Lac d'Amance, which was also constructed in 1990, regulates the River Aube. Meanwhile, a few tens of kilometres to the north, the largest lake of the lot, the 48-sq-km Lac du Der-Chantecoq, was constructed in 1974 to hold back water from yet another river, the Marne, in order to prevent spring flooding in Paris. All in all, this massive creation of new freshwater habitat has completely altered the landscape of this sparsely populated, rural part of France, including the drowning of three villages, one of which gives its name, Chantecoq, to the largest lake.

It is not only the landscape that has been altered. Where once there were fields, there is now habitat for ducks, herons and cormorants. Where the ground was dry, there is now habitat for waders and gulls. Tens of thousands of waterbirds use the site in winter and on passage, while in the summer the lakeside marshes provide habitat for breeding species such as Eurasian and Little Bitterns, Gadwall and Purple Heron. Together with its surrounding farmland and rich deciduous and coniferous forest, the area as a whole has become one of the most important birding sites in all of France.

However, no single species has had more impact on birding in this area than the Common Crane. Cranes

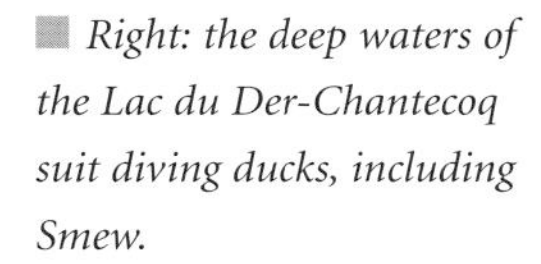

Right: the deep waters of the Lac du Der-Chantecoq suit diving ducks, including Smew.

Above: the star bird of the region is the Common Crane; November and February produce the best numbers as the wintering population is swollen by birds passing through.

must always have overflown or rested in the area on migration, but these days, with excellent feeding and roosting conditions available to them, they descend on the area in enormous numbers. It is estimated that almost the entire breeding population of cranes from Scandinavia and nearby Russia stops here on its way to Spain, some 60,000 birds in all. They don't all come at once, especially in the autumn when the migration period is fairly protracted, but on their return migration in the spring, especially in late February, the birds concentrate in impressive numbers, and counts on the roosts on the islands of the Lac du Der-Chantecoq have recorded 25,000 birds. In addition, a few thousand birds remain throughout the winter.

Cranes are marvellously exciting birds, flying with regal, slow wing-beats, moving in neat formations and uttering their extraordinarily loud, atmospheric calls. The far-carrying clanging is made by the elongated trachea, which actually winds around the bird's sternum and fuses with it to make a series of plates that vibrate and amplify the sound. The sight and sound of these cranes can be profoundly impressive, and on most evenings during the winter there will be birders perched on the western embankment of the lake, waiting for the inbound flight.

Not everybody in the area is a huge crane fan, especially those local farmers who lose large amounts of grain or potatoes to the birds every year. To combat this, a local farm has been acquired, the Ferme des Grues, where the birds can be protected and fed, and where they are entirely welcome; the problem is thus ameliorated, if not entirely solved.

Plenty of other waterbirds can be seen while you are enjoying the sight of the cranes, including a fairly new arrival into the area, the Great White Egret. These stately herons were once very rare in this part of France, but at least 30 to 40 now winter annually. In addition, there are plenty of ducks, and many hundreds of the usual common European species can be seen, such as Eurasian Wigeon, Gadwall and Eurasian Teal. Some parts of the lake are deep, so diving ducks turn up regularly, including such scarcities as Smew and Velvet Scoter. Alongside these, there are always some divers and grebes about and the Red-necked Grebe, in particular, is something of a speciality.

The profusion of wildfowl, and waders too (up to 10,000 Northern Lapwings may be present in winter), is the main attraction for another five-star species. This is the only area in France where White-tailed Eagle regularly winters, and in a good year between three and five individuals may be present. Although surprisingly hard to find in this area, the eagles are always stylish and spectacular, and the pandemonium they can cause to the assembled throng of waterbirds is in itself an impressive performance.

No visitor to the region should overlook the surrounding countryside. The fields and hedgerows are good for species such as Brambling and Great Grey Shrike, while the mature woodlands, especially those around the three southerly lakes, are an excellent draw in themselves. A bit of diligent searching is likely to bag at least five species of woodpecker – Great, Lesser and Middle Spotted, plus European Green and Black, while to see the very scarce Grey-headed requires considerable luck. Other forest birds include Firecrest, Northern Goshawk, Short-toed Treecreeper, European Crested Tit and Hawfinch, the latter being a regular visitor to feeding trays in the area, such as at the campground at Larzicourt, on the north of the Lac du Der-Chantecoq.

Altogether, 273 species have been recorded in the region as a whole which, considering that this is more than 250 km inland in a temperate area, is a hugely impressive total. And happily, for once, one can reliably report that, without the influence of humankind, the list would never have begun to approach such dizzy heights.

Waddensee

SITE RANK 56

Information

GREAT BRITAIN
North Sea
Waddensee
Amsterdam
NETHERLANDS
London
The Hague
Essen
Brussels
BELGIUM
LUXEMBOURG
Paris
FRANCE

HABITAT A long stretch of coast, including intertidal mudflats, salt marsh, dunes, barrier islands, freshwater marshes, grazing meadows and scrub

KEY SPECIES Huge numbers of wildfowl and waders, including Common Shelduck and Pied Avocet, gulls, terns including Gull-billed, raptors

TIME OF YEAR All year round

Above right: the band of intertidal ooze running the length of the Waddensee is, on average, 10 km wide.

Below: the shallow, saline conditions of the Waddensee are perfect habitat for Pied Avocets.

It seems hard to believe that three of Europe's most densely populated and industrialized countries could contain within their borders anything that could be described as 'wilderness'. And yet, incredibly, stretching along the North Sea coast from Den Helder in the Netherlands to the busy port of Esbjerg in Denmark is the Waddensee, 9,000 sq km of lonely, windswept, wild coastline where the footsteps of people cannot be seen and where people have not yet come to abuse and destroy the environment. No human can live here, on the roughly 10-km-wide skirt of perilous intertidal mudflats and salt marshes that rims this 500-km stretch of coast and, miraculously, the greedy hands of developers have been largely kept at bay.

The Waddensee is of crucial importance to Europe's birds. In the German section alone, it is thought that

some 1.5 million wildfowl and up to 4 million waders, gulls and terns use the rich feeding grounds for at least some of the year, and if it were not for the food-rich, sheltered and safe waters of the Waddensee, they would be dispersed and depleted to an extent that is hard to imagine. Almost all the Common Shelduck in western Europe congregate on the Waddensee between July and September in order to moult. Hundreds of thousands of waders use the estuarine ooze for refuelling on migration flights, or for sustaining themselves during the winter months. Without the Waddensee they would struggle to find the resources they need.

The sheer numbers of birds using the site are staggering. Counts of waders using the German section as a staging area on their migration include 463,000 Eurasian Oystercatchers, 400,000 Red Knots, 628,000 Dunlin, 192,000 Bar-tailed Godwits and 97,000 Eurasian Curlews. The Schleswig-Holstein Waddensee National Park, just one section of this, holds winter populations of 26,000 Barnacle Geese, 150,000 Common Eiders, while 133,000 Brent Geese, 104,000 Northern Pintails, 19,000 Curlew Sandpipers and 15,000 Sandwich Terns pass through on migration. In the Danish section, the Vadehavet area sees 35,000 Common Scoters, 10,000 Pink-footed Geese, 43,600 European Golden Plovers and 365,000 Dunlin pass by, while in the Netherlands a highly impressive total of 15,000 Pied Avocets has been counted on passage, along with 29,000 Grey Plovers. Not surprisingly, since the Waddensee is a huge site and the birds are constantly moving, it is difficult, if not impossible, to get complete counts for the whole area. However, for Common Shelduck the situation is slightly different because most of them congregate in the Schleswig-Holstein National Park. Here 150,000 of these bold black, white, bottle-green and chestnut ducks while away the late summer, feeding in these rich, shallow waters whilst they exchange old flight feathers for new ones. Interestingly, this gathering does not include the youngsters, who remain in the breeding areas, initially under the care of a small number of selected adults.

Birding the Waddensee is, not surprisingly, very rewarding, and not just for big numbers. Right across from the Netherlands to Denmark there are superb locations hosting an excellent variety of species throughout the year. In Denmark, for example, at the very northern tip of the Waddensee, the sandy peninsula of Skallingen is the location for one of Denmark's most famous bird observatories, Blåvandshuk. A superb migration watch point, it attracts ducks, geese and waders like much of the rest of the area, but also, at times, seabirds including skuas and even storm petrels. Further south, a group of barrier islands lie along the coast right down to the German border, including several that are excellent for birds. Romo, like many of the sites on the Waddensee, has freshwater marshes as well as saline habitats, and here birds such as Red-necked Grebe, Eurasian Bittern, Spotted Crake, Bearded Reedling and Eurasian Penduline Tit breed. In the marshes, heathland and dunes a trio of harriers, Hen, Montagu's and Western Marsh, all breed, and gorge themselves on the plenty. In winter, Rough-legged Buzzards and even White-tailed Eagles patrol the general area.

Another good barrier island is Fano, to the north. About 20,000 ducks, geese and waders use this 16-km-

■ *Above: one small German section of the Waddensee alone holds 104,000 wintering Northern Pintails.*

■ *Below: huge flocks of Bar-tailed Godwits use the site as a refuelling stop before they continue their northbound migration.*

long island as a staging post on migration. In addition, it hosts a number of species that are especially attracted to the peculiar nature of the Waddensee, and are thus widespread throughout the area. These include the Kentish Plover, which is drawn to flat, sandy beaches, and the Pied Avocet, which relishes the shallow, saline water. Throughout the Waddensee, low shingle and beaches provide excellent breeding habitat for terns, including Common, Arctic, Sandwich, Little and Gull-billed. Fano is also a noted area for visible migration, not just for waterbirds but also for passerines such as pipits and finches.

At the other end of the Waddensee, in the Netherlands, there is another set of barrier islands, offering similar habitats to those in Denmark and Germany. The most famous of these is Texel, which holds a superb variety of breeding birds, including Eurasian Spoonbill and Black-tailed Godwit, and even provides a little woodland for such species as Short-toed Treecreeper, Hawfinch and Long-eared Owl. It is a superb place for observing migration, including passerines. Visitors include 'regular' rarities such as Yellow-browed and Pallas's Leaf Warblers.

However, Texel is far removed from the core part of the Waddensee. In truth, the value in this remarkable area is in its lonely flats, beaches and dunes, where the birds can feed for days, or even months at a time, without being seen or disturbed by a single human being.

Coto Doñana

HABITAT Seasonally flooded marshes (marismas), salt flats, woodland, scrub, dunes, beach

KEY SPECIES Herons and egrets, Red-knobbed Coot, Marbled and Ferruginous Ducks, Purple Swamp-hen, Spanish Imperial Eagle

TIME OF YEAR All year, although July and August are poor for waterbirds

Right: about seven pairs of the endangered Spanish Imperial Eagle breed each year in the Coto Doñana.

This wild, unspoilt corner of south-west Spain, at the mouth of the Rio Guadalquivir, can be said to have been a nature reserve for at least the last 400 years. Back in the 1600s it was in the hands of the Dukes of Medina Sidonia, who enclosed it as a private hunting area and discouraged any kind of settlement or development. Every so often they would invite the incumbent monarchs for a seventeenth-century knees-up, and royalty would come with their massive entourage (reputedly up to 12,000 people) and shoot a few head of game and compliment the current Duke on his fine estate. Thus, the area remained safe in the hands of generations of complimented Dukes, and escaped the bulldozed fate of much of the rest of southern Spain. These days it still survives, almost intact, as a huge

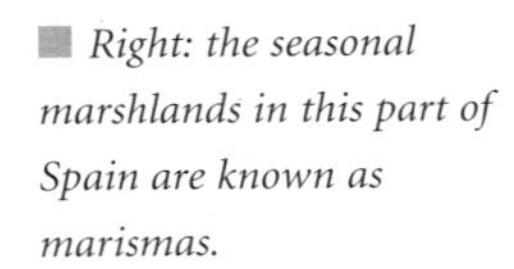

Right: the seasonal marshlands in this part of Spain are known as marismas.

national park covering an area of 1,300 sq km.

Even in 1953, this was still a remote area. In that year a team of naturalists, led by the late distinguished British ornithologist and conservationist Guy Mountfort, visited Doñana and carried out scientific research, including counts and descriptions of the birdlife. The resulting account of this and subsequent expeditions, *Portrait of a Wilderness*, with its evocative descriptions of a lost world not so far away from enraptured readers, where the sun shone and the only mode of transport was on horseback, became something of a bestseller. This entrenched the Coto Doñana's reputation as one of the most exciting places in Europe for birdwatching, and gave it something of an aura. Happily, and emphatically, in the 50 years since, it has maintained its reputation, and is almost as good as it ever was.

The Coto Doñana is what you might call a failed delta. In sluggish old age, many a river breaks into dozens of small channels as it lurches towards the sea. However, owing to the action of ocean and wind, all but one of the Guadalquivir's routes to the sea have been blocked by a large system of sand-dunes that, over the centuries, has built up on the coastal side. Thus, these days, there is but one river mouth, and a vast area of flat land hemmed in by the dunes is seasonally flooded by winter rains, usually to not much more than 1 m in depth. The resulting marshlands are known as marismas, and provide an enormous area of bird habitat. Together with the dunes, salt-flats, patches of Mediterranean-type scrub formed by various aromatic plants, and with woodland on the higher ground, these marshlands provide a patchwork of rich habitats attracting an astonishing variety of birds. Up to 2007, 380 species have been recorded, of which a staggering 150 regularly breed.

With so much shallow water about, it is hardly surprising that the heron family does very well here. Several species maintain populations in the hundreds of pairs, including the Purple Heron (minimum 300 pairs), Little and Cattle Egrets and Black-crowned Night Heron. Squacco Heron and Little Bittern are present in lower numbers, but can be easier to see here than in many other parts of Europe. Most of these species breed in mixed colonies, often in the cork oaks in the woodlands, but Purple Heron and Little Bittern are solitary breeders. In recent years the Great White Egret, so mysteriously rare in Europe as a whole, has become a familiar sight in the area.

The herons don't hold the monopoly on long-legged wading birds in Doñana. An increasing population of Glossy Ibises is present, along with about 400 pairs of Eurasian Spoonbills, often mixed in with the breeding colonies of herons, while there are always some non-breeding Greater Flamingos about in the shallow, saline waters. White Storks are abundant, and the largest colony in Europe, 400 pairs, is found in the north of the reserve. Meanwhile, both Black-winged Stilt and Pied Avocet are a familiar sight on the saltpans or saline pools, with thousands of pairs breeding. It might be stretching the point to call the Purple Swamp-hen long-legged, but these 'big-

Below: the vast wetlands of the Coto Doñana are predictably superb for herons, including Squacco Heron.

nosed' iridescent blue lumps are increasingly common in Doñana, despite being rare and localized in Europe overall.

Several other great rarities for Europe occur in Doñana. One of these, the Red-knobbed Coot, is mainly a bird of sub-Saharan Africa, but occurs here in a relict population shared with the Maghreb. Once, finding this species required a patient check through thousands of Eurasian Coots, but researchers are now putting collars on individuals to make them easier to study and conserve. The secretive Marbled Duck, not found in Europe outside Spain and remarkable for its sheer sluggishness, still occurs in very small numbers. However, time has probably already run out for the Small Buttonquail, which is also known by its old name of Andalusian Hemipode. The last Doñana record of this species was in 1999, so it is now probably extinct here. The Spanish Imperial Eagle is faring somewhat better, with a stable population of about seven pairs. In recent years, a shortage of rabbits has made life difficult for these specialists, but measures are being put in place to increase the population of the eagle's favourite food, and its future seems reasonably secure.

If you visit Doñana in the winter you may notice – if you are not distracted by the 60,000 Greylag Geese, 100,000 Eurasian Teal and host of other wildfowl – that quite a few familiar faces remain behind when they should, by all accounts, be deep into Africa. The most obvious is probably the Barn Swallow, but other migrants such as White Stork, Black Kite, Lesser Kestrel, Whiskered Tern and Common House Martin regularly stay put and could all be considered resident. Clearly, impending climate change could soon add to this list, with birds altering their migratory journeys as the need for transcontinental movement is removed. Doñana also attracts some astonishing rarities from Africa, and the trickle of these could perhaps turn into a more substantial feature in years to come. Among the more interesting are Pink-backed Pelican, African Spoonbill, Yellow-billed Stork and Marabou, while the Lesser Flamingo, recorded here several times recently, actually bred in Spain in the summer of 2007, although not here. Who knows what might colonize next?

The fabulous Coto Doñana has been one of Europe's top wetland sites for a great many years, and excites continuing affection and delight among birders throughout Europe. It is thus scandalous that, on the whole, the Spanish government, which owns most of the land, has generally treated Doñana with contemptible disregard. One might think that it would recognize the site as the jewel it is and yet Doñana faces constant threats from development and other forms of abuse. In 1998 a chemical spill upstream nearly reached Doñana, and would have done but for some last-minute building of dykes that only just averted wide-scale disastrous damage. The needs of tourism and development around the site lower the water-table, and chemicals leach into the Guadalquivir from nearby agri-business. The attitude of the Spanish government threatens not just the wildlife, but a special part of Spanish history, too.

Right: it usually takes sharp eyes and patience to obtain a good view of the secretive Little Bittern.

Extremadura

HABITAT Steppe, rolling farmland, open woodland, scrub, rugged mountainsides, small towns

KEY SPECIES Great and Little Bustards, Pin-tailed and Black-bellied Sandgrouse, Black Vulture, Spanish Imperial Eagle, Iberian Azure-winged Magpie

TIME OF YEAR All year, although April to June is best for the breeding birds

Above right: a good time to catch up with Pin-tailed Sandgrouse is early morning or evening when they commute to a drinking source.

Lying some 250 km to the west of Madrid, the region of Extremadura could hardly be more different from the bustling Spanish capital. This is a sparsely populated, peaceful area with rolling hills, scattered farmsteads, Mediterranean-type scrub and steppe, its wide vistas and undisturbed landscapes in many ways presenting a throwback to a time when most people worked the land, and high-intensity agriculture had not yet entered the vocabulary. It is a refreshing place to visit, and an inspiring region indeed to go birdwatching.

In terms of ornithology and conservation, Extremadura's most important habitat is its steppe, which is dotted about the region, with major patches close to Merida, Caceres and Trujillo. This rolling grassland holds one of the most significant populations of Great and Little Bustards in the world, never mind Europe. Over 10,000 pairs of the latter occur and, until recently, were known to gather in flocks of more than 1,000 individuals, while Great Bustards still number 6,900 birds (counted 2006). So Extremadura ranks as one of the easiest places in the world to see these shy terrestrial birds. All you need

Right: the globally endangered Lesser Kestrel breeds colonially on some of the region's historic buildings.

Above: a population of nearly 7,000 Great Bustards patrols the steppes of Extremadura.

to do is to stop on one of the many minor roads and scan likely areas. Displaying Great Bustards, which seem to ruffle up their plumage into a lather, often resemble sheep from a distance, so don't fall into the trap of wondering where all these sought-after birds are among the livestock. Large males hold court and dominate copulations with the local females. Male Little Bustards, meanwhile, sometimes combine several displays at once, stamping their feet, leaping into the air while flashing white wings and making a flatulent snort, all at the same time.

Often accompanying flocks of the relatively tall Little Bustard, especially in winter, are Pin-tailed Sandgrouse. Extremadura is the best place in Europe for these, holding many hundreds of pairs. A much tougher proposition is the Black-bellied Sandgrouse, a rarer species mainly occurring at higher elevations (above 1,300 m), and difficult to find even there. Sandgrouse, which are herbivores, often delay their breeding season here until June, when the first glut of seeds appears in the grasslands.

Other breeding species of the steppe include Common Quail, Montagu's Harrier and Eurasian Stone-curlew, while the air rings in spring with the songs of Greater Short-toed and Calandra Larks, the latter sporting conspicuous black undersides to the wing. Meanwhile, in the winter, many thousands of Common Cranes migrate from northern Europe to spend the winter on nearby agricultural land.

Another important habitat of Extremadura is known as dehesa, a type of open woodland with scattered cork oaks, home to a highly distinctive set of birds. Its signature species is the Black-winged Kite, a very widespread

■ *Above: Great Spotted Cuckoos are brood parasites; they spell trouble for magpies, which are the main hosts.*

Old World species that is abundant in Africa but very rare in Europe. With its large, forward-facing eyes, this species hunts over tall grass by slowly quartering and hovering, and nests in cork oaks and other low trees. Alongside this species are birds such as European Roller, which relies on ground very rich in large invertebrates, Southern Grey and Woodchat Shrikes, Great Spotted Cuckoo and the localized Iberian Azure-winged Magpie. The cuckoo is a brood parasite of both Eurasian and Iberian Azure-winged Magpies. The latter often breed in loose colonies, and can be seen passing through groves of trees one after the other, each member uttering its shrill calls. Until recently this species was considered conspecific with the Azure-winged Magpie of the Far East, and there were suggestions that, in view of the extraordinarily disjunct distribution, the Iberian birds might have been introduced. Recent sub-fossil remains, however, have confirmed that the presence of these subtle blue, pink and black birds is entirely 'natural'.

One of the jewels in the crown of the region of Extremadura is the magnificent Monfragüe National Park, about 60 km to the north of Trujillo. This 1,550-sq-km, rich mixture of broad-leaved forest, scrub and rugged gorges at the confluence of the Rio Tejo and Rio Tietar should be considered as one of the very best sites in Europe for raptors, with some 16 breeding species, including probably the highest breeding populations in the world of Black Vulture (200 pairs) and Spanish Imperial Eagle (at least 10 pairs, some of which nest on pylons). Egyptian Vultures breed along the gorges, while there is a healthy colony of Griffon Vultures, some of which can be seen at the famous landmark of Penafalcon, a huge rocky outcrop towering over the road into the park. Along with the Spanish Imperial Eagles, Golden, Booted, Bonelli's and Short-toed Eagles all breed, as do Red and Black Kites. On a good day, it is perfectly possible to see every one of these.

The rocky areas such as Penafalcon also hold a range of other exciting species. For many years a pair of Black Storks has bred on the rock, much to the delight of visiting birders, while the supporting cast includes Black-eared Wheatear, Alpine Swift, Red-billed Chough and Blue Rock Thrush, among others. In recent years the very rare (in Europe) White-rumped Swift has also appeared in small numbers, although it usually turns up after most birdwatchers have left the area after the end of May.

Another interesting raptor that occurs in Extremadura is the now globally threatened Lesser Kestrel. This is a species of unusual habits, tending to select towns and buildings in which to nest in colonies, rather than wilder habitats. From its urban base it commutes out over fields and rivers to hunt insects and, occasionally, lizards. Several of the towns in Extremadura have colonies of these birds, including Caceres and Trujillo, and they often share the rooftops with colonies of White Storks and Pallid Swifts.

It seems that even the towns and villages of Extremadura are great for birds.

Po Delta

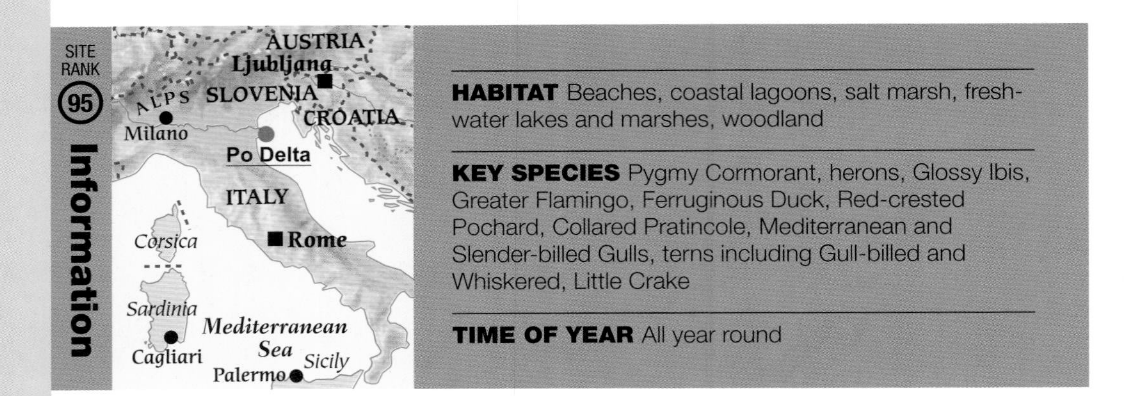

SITE RANK 95 **Information**

HABITAT Beaches, coastal lagoons, salt marsh, freshwater lakes and marshes, woodland

KEY SPECIES Pygmy Cormorant, herons, Glossy Ibis, Greater Flamingo, Ferruginous Duck, Red-crested Pochard, Collared Pratincole, Mediterranean and Slender-billed Gulls, terns including Gull-billed and Whiskered, Little Crake

TIME OF YEAR All year round

Right: Black-winged Stilts abound on the delta's saltpans, allowing visitors to enjoy their delightful, elegant mating display.

Italy's Po Delta region is one of Europe's best-kept birdwatching secrets. Situated in the north-east of the country, on the Adriatic Coast, it is of vital importance for all sorts of breeding, passage and wintering birds, so it is a classic site throughout the year. Over 50,000 waterbirds use the site in winter, while several species of gulls and terns reach into the thousands of breeding pairs. Irresistibly for those with wider interests, the delta also happens to span the gap between two of Italy's most fabulous cultural centres, world famous Venice to the north, and the smaller, picturesque and underrated Ravenna to the south. Thus it is a good destination to visit in order to share pastimes, and the birding half will most certainly not be disappointed.

Although a great deal of the delta has been built over and strung with canals and dykes over the years, there is still plenty of superb 'wild' habitat left, especially for wetland birds. In 1988 the area was declared a regional park, covering some 591 sq km, and since that time has been adapted for all sorts of tourist interests, including birdwatching. There are dozens of excellent walks and loops that are easily followed into the best areas, many fitted with hides or raised platforms, making the birdwatching easy. In Italy, where birds have been traditionally shot rather than watched, the facilities and provisions for birding represent a significant shift in people's attitudes; the sport of hunting is no longer widely supported, although not many of the locals yet go birding themselves. Inspired by places like the Po Delta, this could quickly change.

At first sight this can be a bewilderingly big area to work, but there are plenty of reserves within the delta area where birders can focus their attention. In the northern stretch, for example, nearest to Venice, a large arm of the sea, the Sacca di Gor, is a superb place for resting and roosting gulls, including Slender-billed and Mediterranean Gull and Gull-billed Tern. Nearby is the large brackish lagoon of Valle Bertuzzi, where Black-winged Stilts and Mediterranean Gulls nest on the salt flats, and the adjacent reedbeds hold such species as Purple Heron and Great White Egret, which actually occur widely. In all, the Po Delta is by far the most important breeding site in Italy for herons, and one of the most important in Europe.

Further south there is another gem of a place, the Valli di Cammachio, a huge complex of lagoons (the largest in Italy) and nearby salt marshes that simply teem with birds. Italy's largest colony of Greater Flamingos (nearly

■ *Right: Punta Alburete hosts Italy's only colony of Pygmy Cormorants.*

1,000 pairs) is found here, along with its only colony of Eurasian Spoonbills. Other breeding species include Kentish Plovers (maximum 100 pairs), Mediterranean Gulls (nearly 2,000 pairs), Gull-billed, Common, Sandwich and Little Terns and Collared Pratincoles. At Boscoforte, on a small peninsula that juts out into the giant lagoon, there have been nesting attempts by the very rare Lesser Crested Tern among the more numerous species. In the many reedbeds surrounding the lagoons are significant numbers of Eurasian Bittern, Purple Heron, Western Marsh Harrier and Great Reed Warbler, while the breeding wildfowl of the area include Common Shelduck, Gadwall and Garganey. Montagu's Harriers nest in the surrounding fields.

The Worldwide Fund for Nature reserve at Punte Alburete represents something of a change of scene. In contrast to all the open lagoons and salt marshes in the northern part of the delta, this reserve protects a small area of flooded forest made up of willow and ash trees home to, among other attractions, Italy's only colony of Pygmy Cormorants. The 2-km route around the reserve also takes in reedbeds and meadows, and overall this is one of the richest sites for birding in the whole delta. The roll-call of herons is especially impressive, with Purple, Black-crowned Night and Squacco Herons and Great White Egrets all breeding, along with Little and Eurasian Bitterns. There is also a colony of Glossy Ibises. Ducks are well represented, with scarcer species such as Ferruginous Duck and Red-crested Pochard usually present in the freshwater.

South again and the saltwater, or at least brackish, theme returns, with the lagoons at Pialassa della Baiona and Pololonga supporting one of the best gull breeding sites in the delta: Black-headed, Mediterranean and Slender-billed Gulls nest here on specially constructed islands, along with Common, Little and Gull-billed Terns. The same species also occur not far away at the southern edge of the delta, at the Cervia Saltpans, together with large numbers of Yellow-legged Gulls (some 16,000 pairs have been counted). These huge saltpans, which date back at least to the time of the Romans, are still working today; and while the extreme salinity restricts the number of species using the waters, the pans are much appreciated by Pied Avocets and Black-winged Stilts, while Kentish Plovers breed on the bare surrounding sand. Outside the breeding season, the saltpans attract Greater Flamingos and large numbers of waders, including Dunlin and Ruff, and wildfowl such as Eurasian Wigeon.

Finally, some 30 km inland near the town of Argenta, but still within the regional park, the scene changes once again. The delightful Valle Santa reserve encompasses a freshwater lagoon skirted by wet meadows. Both Spotted and Little Crake breed in the reedbeds here, while Whiskered Terns nest on the water-lilies and Eurasian Penduline Tits frequent patches of willow scrub. Very different in character from most of the rest of the delta, it demonstrates the exciting and diverse range of habitats that occur in this part of what is, ornithologically, largely a neglected country.

Varanger Peninsula

HABITAT Arctic tundra, cliffs, coast (mainly ice-free), scrub

KEY SPECIES Steller's and King Eiders, White-billed Diver, Brünnich's Guillemot, Gyr Falcon

TIME OF YEAR March to mid-June (peak May to June). Thereafter there are plenty of birds, but mosquitoes are a problem

Right: Varanger combines superb Arctic birding with good accommodation and logistics.

Opposite: the Dotterel is one of several northern-breeding waders in which the roles of the sexes are partly reversed. The colourful female (pictured) initiates display, while the male incubates the eggs and tends the young.

Perched at the northern tip of the European continent, the Varanger Peninsula would be an impressive place even without its birds. The countryside, with its barren, ice-covered rocks, patchwork tundra, colourful scattered houses and towering sea cliffs, has its very own jolting sort of bleakness, and the ever-changing weather, which results from the continuous tussle between the fierceness of the Arctic latitude (400 km inside the Arctic Circle) and the mollifying effect of the Gulf Stream, simply serves to provide endless different backdrops for this dramatic landscape. It is a great setting for some fantastic birding.

The area offers the chance to see some northern specialities that are hard to find elsewhere on the European continent. The most famous of them are the eiders, which come here from further east to take advantage of the unusual ice-free winter conditions in the Varangerfjord and around the peninsula (an effect of the Gulf Stream). Three species are easy to find: the Common Eider, a numerous bird of the north European seaboard, the King Eider, a deepwater duck of rough Arctic seas,

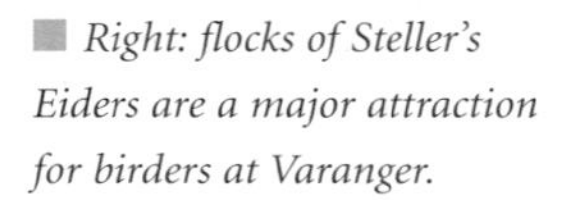

Right: flocks of Steller's Eiders are a major attraction for birders at Varanger.

and Steller's Eider, another Arctic species which prefers to feed in the shallows. A fourth species, the Spectacled Eider, is a real rarity, having been recorded on just three occasions.

Both King Eider and Steller's Eider use the area as a major non-breeding base. Up to 15,000 Steller's and 5,000 King Eiders arrive in October from their main breeding sites in Russia and can be seen all around the peninsula, the male Steller's with their glorious butter-scotch-brown underparts, black plumes and white head, and the drake Kings with their smart blue-grey heads, red-and-orange bills and bold black-and-white bodies. Both species remain in large flocks well into the spring, and a few, mainly second-year Steller's, remain throughout the summer. A good spot for the summering birds is offshore from the small settlement of Nesseby.

The richly productive waters of the Barents Sea provide the eiders with plentiful molluscs, crustaceans and echinoderms, and they share this bounty with large numbers of Common and Velvet Scoters, Greater Scaup, Common Goldeneyes and Long-tailed Ducks. The area is also an exceptionally rich fishery, and this attracts a slightly different guild of birds, chief among them the divers. These thick-necked, back-propelled fish-chasers are common in the ice-free waters and, as with the eiders, include in their number a mixture of the routine and rare. Both Black-throated and Red-throated Divers are plentiful and breed, while a small number of non-breeding Great Northern Divers may be seen in the summer. However, Varanger is most famous for its White-billed Divers. They are primarily winter visitors but, like the eiders, leave a few representatives behind in the summer. On a good day in spring, up to

300 passing birds have been counted from the headland at Hamningberg, on the north-east side of the peninsula, suggesting quite a substantial wintering population.

One of the fish species that spawns off these coasts is the Capelin, a great favourite of another of the region's specialities, the Brünnich's Guillemot. This Arctic auk, which has been known to dive 210 m below the surface to find food, breeds on the Varanger Peninsula and just a handful of other places on the European mainland, being found principally in island groups in the Arctic Ocean such as Svalbard. The Brünnich's Guillemot feeds on the Capelin in the early spring before settling on its breeding ledges by the end of April. A good place to find it is the island of Hornoya, just off the town of Vardo on the eastern extremity of the peninsula. Overall, this is a great location for seabirds, with breeding Common Guillemot, Razorbill, Black Guillemot and Atlantic Puffin keeping the Brünnich's Guillemot company. To complete the set of European auks, the Little Auk also occurs here in winter, and may breed occasionally. Other breeding birds include 25,000 pairs of Black-legged Kittiwake, plus European Shag and one of Europe's largest colonies of Herring and Great Black-backed Gulls. White-tailed Eagles, which are numerous at Varanger, plus Gyr Falcons, which are scarce, visit these colonies regularly during the summer in search of an easy meal. Arctic, Long-tailed and Pomarine Skuas may also cause their fair share of trouble.

In recent years a colony of Leach's Storm Petrels has also been found on Hornoya. It was probably overlooked in the past because the birds don't enter their breeding colonies until late summer, when most birdwatchers have left the area. This late breeding helps them avoid the attention of predators; in midsummer there is no night-time or twilight to hide the birds, so they delay until there is some darkness at night.

Besides the seabirds, the Varanger Peninsula has many other delights, not least its breeding waders. These include abundant Temminck's Stints, which may perch on top of street-lights in the towns, colourful Red-necked Phalaropes that spin in the pools, Spotted Redshanks, which breed in the bogs, Ruffs, European Golden Plovers and numerous Purple Sandpipers. The last named are famous for their remarkable 'rodent run' distraction display, used to lure marauding predators like Arctic Foxes away from their nests, in which they run through the vegetation and squeak like lemmings.

Away from the coast there are several excellent sites for a different suite of Arctic birds. In the south-west of the peninsula are some woodlands, where such species as Arctic and Common Redpolls, Siberian Tit and Siberian Jay occur. The hills and fells in the centre (especially Falkefjell at 548 m) are worth checking for Snowy Owl (rare), Long-tailed Skua, Eurasian Dotterel and Bluethroat, while scrubby areas offer such species as Red-throated Pipit. One of the great advantages of this marvellous area is that it is very easy to bird here, with good roads and good accommodation, so all of these different habitats are within comparatively easy reach.

Right: rare at Varanger, the Snowy Owl breeds inland from the coast.

Falsterbo

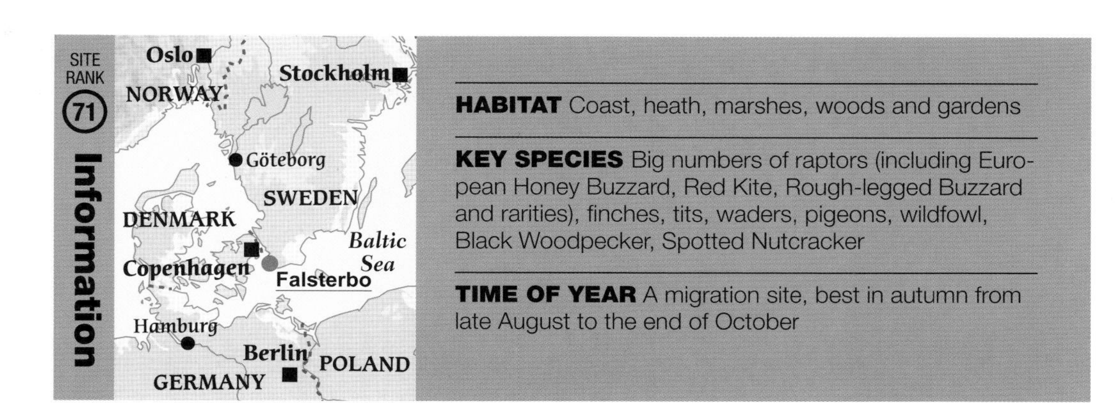

HABITAT Coast, heath, marshes, woods and gardens

KEY SPECIES Big numbers of raptors (including European Honey Buzzard, Red Kite, Rough-legged Buzzard and rarities), finches, tits, waders, pigeons, wildfowl, Black Woodpecker, Spotted Nutcracker

TIME OF YEAR A migration site, best in autumn from late August to the end of October

Above right: sunrise is the best time for visible migration at Nabben, at the southern tip of the Falsterbo peninsula.

It is estimated that about 500 million birds pass through southern Scandinavia each autumn, on their way to their wintering grounds in the milder parts of western Europe and beyond. For many, this southward journey leads them to the shores of the Baltic Sea where, in their need to avoid the potential risks of a long sea crossing, they are funnelled south and west along the coast to the very south-western tip of Sweden, where the Danish island of Sjaelland, and safety, lies only 20 km away. At this point, jutting out into the Baltic, lies the peninsula of Falsterbo, one of the finest places in the world to watch bird migration, and the site of a bird observatory since 1955. Between late August and the end of October, untold millions of birds take off from this small spit of land and leave the Scandinavian mainland behind.

Although the overall variety of bird species is superb, what has really made Falsterbo's reputation is the numbers of raptors that pass through here. Almost 30 species have been recorded altogether, which is impressive enough, but it is the sheer potential numbers that coax visiting birders to come here from all over Europe to witness the spectacle. Up to 14,000 individual raptors have been counted passing over in a single day. Daily totals of certain species, such as Eurasian Sparrowhawk, routinely exceed 1,000, with similar totals for Common Buzzard and slightly fewer (600) for European Honey Buzzard. Needless to say, these birds make a spectacular sight,

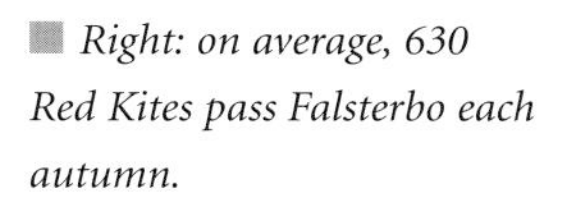
Right: on average, 630 Red Kites pass Falsterbo each autumn.

■ *Above: Black Woodpeckers regularly turn up at Falsterbo, but nobody knows where they come from, or where they go afterwards.*

especially in mid-morning during light south-westerly winds, when the air seems to be full of shapes soaring to gain height before they break out over the sea. Many of these fleeting visitors also live up to their predatory nature by attacking smaller birds *en route*, especially Eurasian Sparrowhawks and Merlins, which frequently follow travelling flocks of small passerines over the water.

The seasonal numbers of raptors illustrate the importance of Falsterbo most effectively. In the course of the whole autumn the average total counts for the commoner species are as follows: Eurasian Sparrowhawk 15,300, Common Buzzard 10,500, European Honey Buzzard 5,000, Rough-legged Buzzard 1,100, Western Marsh Harrier 680, Red Kite 630, Common Kestrel 400, Osprey 240, Hen Harrier 210 and Merlin 200. Other regular migrants that pass through in smaller numbers include White-tailed and Golden Eagles, Montagu's Harrier, Eurasian Hobby and Peregrine Falcon. Not surprisingly, rarities often get mixed up with all these birds. Both Greater and Lesser Spotted Eagles are recorded from time to time, and in recent years increasing numbers of Pallid Harriers have also been turning up.

As you might expect, each species has its own peak time of appearance. Western Marsh Harrier and European Honey Buzzard are commonest from late August to early September, while Red Kites peak in the third week of September and both Common and Rough-legged Buzzards hit their greatest passage in mid-October. On the whole, go for variety in mid-September, and for big numbers in October.

Falsterbo is far from being just a raptor site and, indeed, numbers of other species can be just as impressive. This corner of Scandinavia is arguably the best site in the whole of Europe for seeing the visible movement of smaller day-flying migrants. These include Common Starling, finches, pipits and pigeons. In contrast to such birds as warblers or flycatchers, these migrants begin their daily journeys at dawn, rather than at dusk, and their movements often continue until mid-morning. Daylight, of course, makes it possible for birdwatchers to witness them in action, and at times at Falsterbo flocks of birds can seem to simply deluge through. If the wind is south-westerly it is perfectly possible to see about 1,000 birds passing by every minute! In such conditions it is easy to be overwhelmed and lose concentration.

From September onwards, each dawn sees crowds of birdwatchers huddled at the very tip of Falsterbo itself, at Nabben. From here flocks of birds stream past from first light, often barely over head height, challenging the birders present to identify them from a few calls. Most of these shapes will be finches, of which the most numerous are Common Chaffinch and Brambling. These closely-related birds often flock together and, because of the somewhat subtle differences between them in flight, it is impossible to count them separately. Instead the term 'chaffling' has arisen here, for convenience. On a good migration day 10,000 'chafflings' may pass Falsterbo and, exceptionally, 500,000 have been seen in a 12-hour period. Flocks of these birds are a common sight from late September to the middle of October.

Many other species pass by as well, of course. The average seasonal totals for the more numerous visible migrants include 207,000 Common Wood Pigeons, 134,000 Common Starlings, 40,000 Western Yellow Wagtails, 32,000 Western Jackdaws, 30,000 European Greenfinches, 26,000 Common Linnets, 24,000 Eurasian Siskins, 23,000 Barn Swallows and 20,000 Tree Pipits. The seasonal total of 17,000 Blue Tits will also surprise the many European birders who are unaware how migratory this species can be. These small birds often seem to baulk at the shore at first, making several false starts before finally committing themselves to crossing the sea.

As with the birds of prey, a whole host of less common

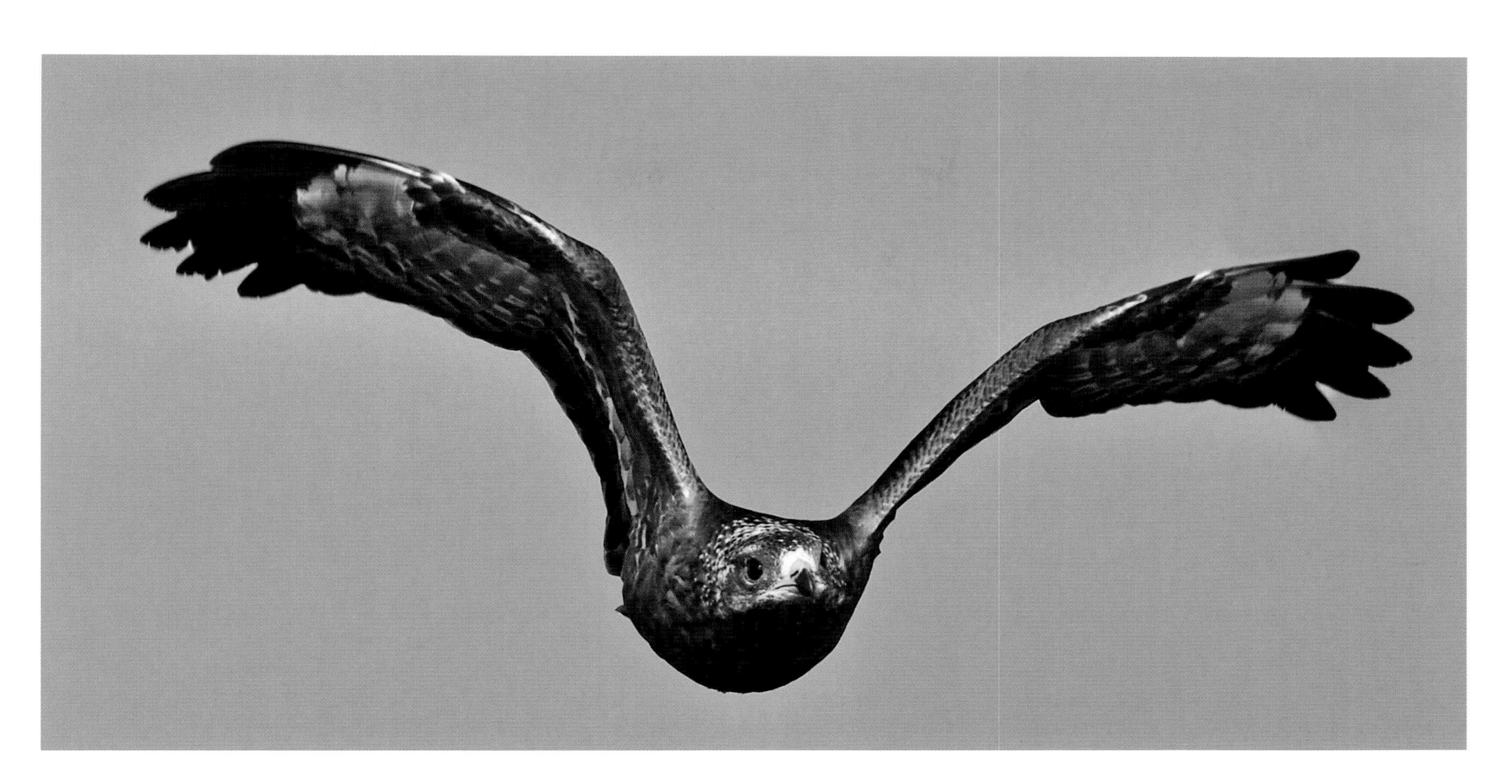

■ Above: a dark-morph juvenile Honey Buzzard; this is the third most numerous migrant raptor to pass through Falsterbo and its numbers peak early in the autumn.

species and rarities can also become caught up in these movements. Notable regulars include the Spotted Nutcracker, Bohemian Waxwing, Red-throated Pipit and Lapland Bunting. Even Black Woodpeckers are regularly seen in the lighthouse garden, although they are not resident and both their origin and destination are obscure.

Nevertheless, steps are continually taken at Falsterbo to try to understand the patterns of migration of the visitors. The first scientific studies and censuses were carried out here in the 1940s, and ringing has taken place ever since the 1950s. In conjunction with the University of Malmö, much pioneering work has been undertaken at Falsterbo into the biology of migration as a whole.

For birders, though, the science is an afterthought. For a few months in autumn, the entrancing and thrilling spectacle of mass migration in this small part of Sweden is all that counts.

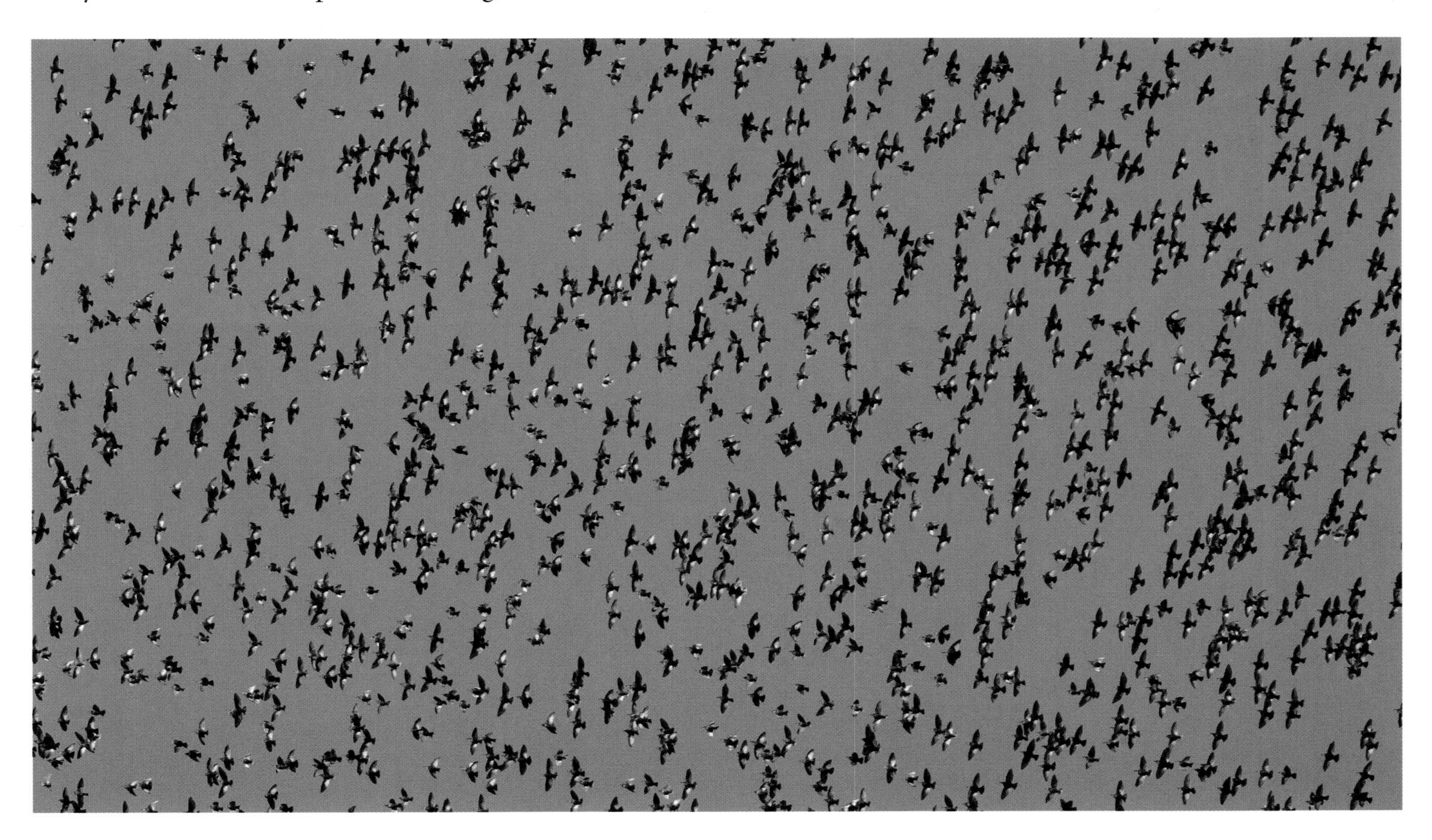

■ Right: Common Starling is an abundant migrant at Falsterbo; flocks travel by day and are known to orient themselves by using the sun's position in the sky.

Oulu

HABITAT Coniferous forests, lakes, bogs, marshes, coast

KEY SPECIES Owls including Great Grey, Ural, Eurasian Pygmy and Tengmalm's, grouse including Black and Hazel, and breeding waders such as Ruff and Spotted Redshank

TIME OF YEAR April to June is best for the owls

Right: male Ruff in breeding plumage. The species is numerous at Liminganlahti, south of the city of Oulu.

There is nothing particularly special about the coniferous forests fringing Oulu, the largest city in Northern Finland. They are exactly what you would expect to find at 65°N. What does make them unusual is their resident network of spies.

In most European taiga forests the birds are thinly dispersed over a wide area, are generally shy and can be almost impossible to find. However, since Oulu is a population centre, and birding is a popular hobby in Finland, even the most elusive species tend to be tracked down by local enthusiasts year after year. Thus it is possible for a visiting birder to tap into this information by hiring a guide, and thereby see a wide range of desirable species

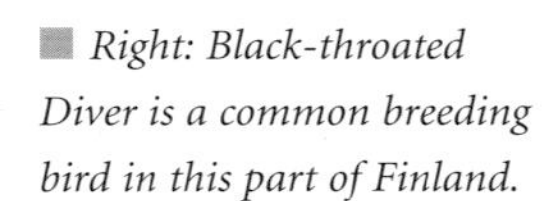

Right: Black-throated Diver is a common breeding bird in this part of Finland.

with minimal effort. It may not be everyone's idea of 'proper' birding, but those who have feasted their eyes on some of Europe's trickiest birds tend not to complain too loudly.

Top of the list for visitors is Oulu's superb collection of owls. In the spring it is possible to find eight species within about half an hour's drive of the city, and occasionally the whole set has been seen in a single night. Nowhere else in Europe offers such diversity of these magical birds, nor such convenience in finding them. Many of the nests are discovered in April and May, and by the time most birders arrive the adults are feeding youngsters, making for a great show.

The most famous owl at Oulu is the Great Grey Owl, a huge mottled, smoky-grey bird with yellow eyes and narrow concentric 'growth rings' on the facial disc; the Oulu area may be the best place in the world to see it. Although it looks large enough to carry a birdwatcher away, this bird is actually quite light underneath its thick plumage, and it specializes, as do so many of the northern owls, on voles. These creatures often make burrows under the snow, but such is the Great Grey's hearing that it can detect movements under the white blanket, and smash through the crust with its talons. These birds tend to nest in the structures abandoned by buzzards and other birds of prey.

The shy and secretive Ural Owl is another highlight. With its comparatively small, dark eyes and gentle expression, it looks placid, but that totally belies its fearsome nature. Woe betide any birders who approach too close to an occupied nest; they could be viciously attacked, and a few have been badly injured. Thus any encounter with this big bird carries a frisson of danger, something we birders are not really used to; several nests are staked out each year.

A Eurasian Pygmy Owl certainly couldn't do you much damage, as this pint-sized predator lives up to its name, being no larger than a Common Starling. It is, nevertheless, a ruthless predator capable of tackling prey species larger than itself which, unlike most of the other owls in the area, includes a substantial proportion of birds. This is one forest owl that a visitor to Oulu might run into without a guide, since it is common and will often perch high on top of a conifer, even in broad daylight, and call. A

Below: Oulu might be the best place in the world to see the magnificent Great Grey Owl.

Above: the Northern Hawk Owl is much easier to see when vole populations are high.

much trickier customer is the Tengmalm's Owl, which occupies the same forests. This species is strictly nocturnal, even during the short nights up here, and is often found in the darkest, densest forest interior. This richly patterned species often stares out from its nest hole when people visit a stake-out, glaring with wide-open yellow eyes that produce a slightly mad expression.

Not all the owls around Oulu are strict forest-dwellers. One of the commonest species, the Short-eared Owl, breeds in open areas and bogs, and nests on the ground. The Eurasian Eagle Owl, meanwhile, can be found in the insalubrious surroundings of the city dump, and also has a habit of perching on highway street lights. Whereas the Short-eared Owl is numerous, to see the Eurasian Eagle Owl requires being in the right place at the right time.

While all the above species can be found around Oulu every year, two other species are less reliable: the Long-eared Owl and the Northern Hawk Owl. Both are most numerous in good vole years. The latter is highly sought after, but as often as not birders need to travel 200 km north-east to the Kuusamo area to catch up with it. The Northern Hawk Owl is unusual for being entirely diurnal, and actually roosts at night. In common with the pygmy owl it often perches on the tops of tall trees.

Between owls, there are plenty of other birds to see. Indeed, Liminganlahti, to the south of Oulu, on the edge of Liminka Bay in the Gulf of Bothnia, is Finland's most important wetland. It was once home to Europe's only population of Yellow-breasted Buntings, but the numbers have crashed to nothing recently, perhaps as a consequence of over-hunting on its wintering grounds in southern Asia. Instead, visitors can admire Europe's most westerly population of the singularly odd Terek Sandpiper, as well as such species as Common Rosefinch, Greylag Goose, Black-throated Diver, Spotted Redshank and Western Marsh Harrier. It is also a terrific site for breeding Ruffs, with several hundred females breeding. These birds operate a highly unusual 'lek' breeding system, in which males compete among themselves to gain possession of a central territory, to which the females make a bee-line when they are seeking sperm for their eggs. All male Ruffs look slightly different, even to our eyes – a situation unique in the bird world – and part of the system is that males behave according to the colour code of their plumage. Black- and brown-ruffed males remain on a single lek all season, whereas white-marked birds ('satellites') may commute between leks to obtain the odd opportunistic copulation.

Watching the Ruffs – which display in complete silence – from one of the five observation towers in the Liminganlahti Nature Reserve is certainly a highlight of visiting Oulu. It can almost, but not quite, trump the owl spectacular.

Lake Myvatn

SITE RANK 79

Information

HABITAT Large shallow freshwater lake, river, moor and bog

KEY SPECIES Barrow's Goldeneye, Harlequin Duck, other ducks, Gyr Falcon, Rock Ptarmigan, Red-necked Phalarope

TIME OF YEAR Breeding season: late April to August. Avoid the months of June, July and August because of the swarms of biting insects

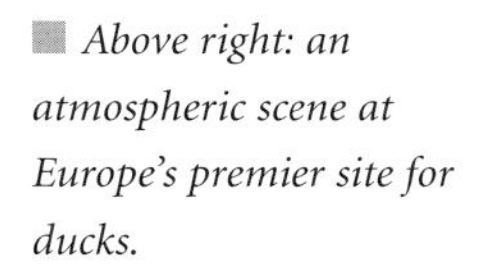

Above right: an atmospheric scene at Europe's premier site for ducks.

Below: the colourful Harlequin Duck is one of Lake Myvatn's top attractions; these three are drakes.

Many of us first notice birds when, as youngsters, our parents take us to the park to feed the ducks. A visit to Lake Myvatn, therefore, is rather like the grown-up fulfilment of those early experiences, for this place is an absolute paradise for ducks. More species (15) breed here than anywhere else in Europe, and probably the world.

Situated in the north-central part of Iceland, the Lake Myvatn area is a unique ecosystem. With an area of 37 sq km, the lake is Iceland's fourth largest, but it is highly unusual in being almost entirely spring-fed, rather than river-fed. Rains are absorbed quickly into the bedrock and re-emerge as mineral-rich springs all around the lake, together producing a net inflow of 35 cubic metres of water per second, most of which eventually exits from the west corner as the River Laxa. Fed by the sunlight, algae grow abundantly in these fertile waters, and provide food for midges and blackflies. These insects, indeed, are found in such vast abundance here that they have given the lake its name: Myvatn is Icelandic for 'Lake of Flies'.

■ *Above: like all members of its family, Red-necked Phalarope exhibits reverse sexual dimorphism, with the female being larger and more colourful than the male. Here a female reduces the lake's midge population by one.*

The flies provide ample sustenance for ducks (especially the young) and their presence is a major contributory factor in making this place so special for wildfowl. Two other aspects of the wetland are also important: the whole lake is extremely shallow, never more than 4 m deep (average 2 m), and it is topographically varied, with myriad inlets and islets. The depth allows the ducks to obtain food easily, and the topography provides many and varied breeding sites.

As for the ducks themselves, Iceland's position between the European and North American continents means that the cast list is drawn from both sides of the Atlantic, and two primarily New World species, the Harlequin Duck and Barrow's Goldeneye, breed nowhere else in Europe. The Harlequin Duck is widespread on the island, but Barrow's Goldeneye is almost confined to the Myvatn area. In North America this duck nests in holes in trees, but here in Iceland it would quickly become extinct if it had to depend on these, since no large trees grow here at all. Instead it utilizes holes of a different kind. The whole Myvatn area is highly active volcanically (there are hot springs near the lake), and females often fly out from the lake in groups to the lava fields to look for a suitable crater or burrow in which to nest, a habit unique to this region. The birds also use holes in buildings, and visiting birders are sometimes met by the incongruous sight of a female Barrow's Goldeneye peering down a rooftop chimney!

The other star species is the Harlequin Duck, a bird of exquisite colour and pattern which is a specialist in feeding in turbulent waters. These delightful ducks, top of most visitors' 'wanted' lists, are easily seen in the River Laxa, which has the highest breeding concentration in the world. In summer the Harlequin Duck specializes in foraging upon blackfly larvae, which tend to be most numerous within these highly oxygenated, fast-flowing waters; when the adult flies emerge in August, they provide food for the young.

The population of Harlequin Ducks in this area is about 250 pairs, while Barrow's Goldeneyes now numbers about the same, having been in sharp decline in recent years. After breeding, the Harlequins leave the area and indulge their passion for turbulent water along the coasts. Barrow's Goldeneyes, meanwhile, mainly sit out the winter at Myvatn in patches of unfrozen water.

Aside from these specialities, the most numerous duck at Myvatn is the humble Tufted Duck, a rather common European species. In 1970 this diving duck,

■ *Above: displaying drake Barrow's Goldeneye. This species nests in holes in Myvatn's volcanic lava fields, a habit unique to the region.*

which feeds on freshwater molluscs as well as the ubiquitous midge larvae, overtook the Greater Scaup in abundance at Myvatn, and now outstrips it four to one (6,000:1,500 males). The scaup takes fewer snails than the Tufted Duck and more crustaceans. These species are closely related, and when the females lead their broods on to the lake (late July onwards), the ducklings sometimes become intermixed.

Myvatn attracts a number of other ducks to its water. Among the diving ducks the most interesting are the Common Scoters (approximately 350 males), which are rare elsewhere in Iceland; for food, they favour crustaceans over midge larvae. Long-tailed Ducks (150 males) have similar tastes. The Red-breasted Merganser (700 males in 2005) and Goosander (15 males), on the other hand, are both specialist fish catchers. One other diving duck, the Common Pochard, used to breed in very small numbers but has done so only sporadically since the 1950s. Among the surface feeding ducks the most numerous is the Eurasian Wigeon, with about 1,000 pairs, followed by the Gadwall (nearly 300), the Mallard (about 200), the Eurasian Teal (50–100) and the Northern Pintail (20–40). Northern Shovelers are sometimes seen in spring, and may breed regularly in very small numbers. All the dabbling ducks feed on the midges.

In addition to the ducks, Lake Myvatn offers an excellent supporting cast of other birds. One of these, the Great Northern Diver, is another of the New World brigade; along with Barrow's Goldeneye and Harlequin Duck it has its European headquarters in Iceland; a few pairs nest in the district. The Slavonian Grebe is common (about 600 pairs) and hard to miss, and the gloriously plumaged Red-necked Phalarope can be seen almost everywhere, spinning in the water as it picks midges from the surface. These waders are as sociable as they are approachable.

If you can tear yourself away from the lake for a while to visit the lava flows and moors, you are certain to stumble across the Rock Ptarmigan, which is common here. That is exactly what the Gyr Falcon is intending to do. This large and spectacular raptor is a major predator of ptarmigan in Iceland; indeed, it is thought that predation is such that the ptarmigans have a slightly different breeding system here than they do elsewhere. Instead of being monogamous, the birds seem to have no more than casual relationships; life, it seems, is too short for serious commitment.

Matsalu Bay

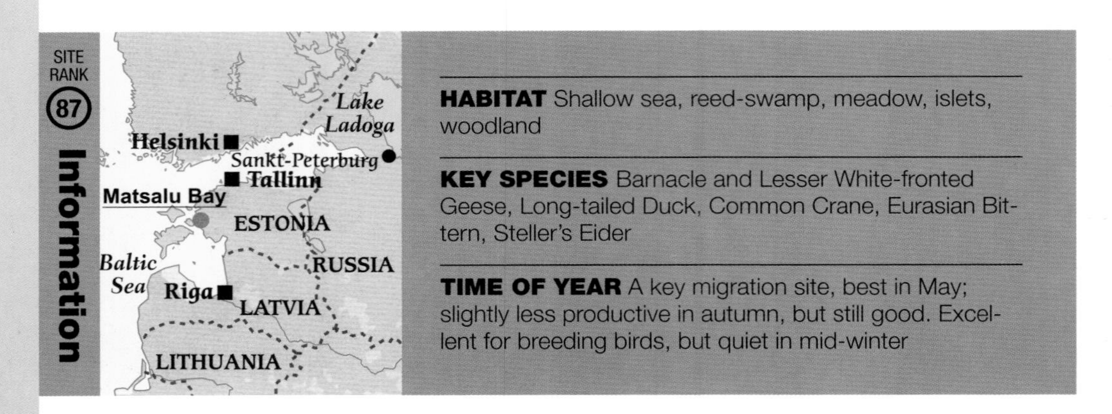

HABITAT Shallow sea, reed-swamp, meadow, islets, woodland

KEY SPECIES Barnacle and Lesser White-fronted Geese, Long-tailed Duck, Common Crane, Eurasian Bittern, Steller's Eider

TIME OF YEAR A key migration site, best in May; slightly less productive in autumn, but still good. Excellent for breeding birds, but quiet in mid-winter

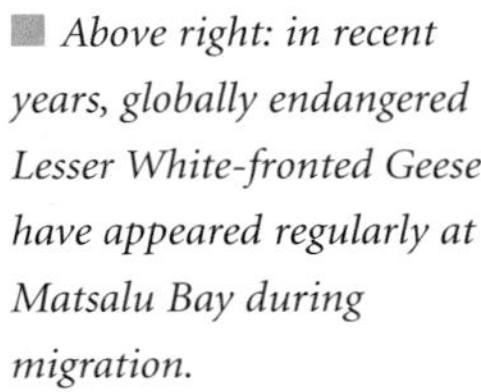

Above right: in recent years, globally endangered Lesser White-fronted Geese have appeared regularly at Matsalu Bay during migration.

Below: at least 20,000 Barnacle Geese pass through the area each spring.

Matsalu Bay is one of the most important wetland areas for birds in the whole of Europe. This is not so much because of its breeding habitats, although these are very rich; neither is the site famous for the birds that spend the winter here. Instead Matsalu is most celebrated as a staging site for migratory birds, equivalent, if you like, to a roadside service station. Birds turn up, stay for a short while to refuel, and then move on. Matsalu has a prime position on the East Atlantic Flyway, the major route between birds' wintering grounds in western Europe or Africa and the vastness of the Arctic tundra, and therefore the travellers can pass by in staggering numbers. It is estimated that at least one million waterbirds traverse the area each spring, peaking in the first half of May; about a third as many pass by in autumn, between August and October. On a good day the spectacle of mass migration here can be truly astounding.

This is a place of open landscapes and large flocks. The bay itself is an east-west facing inlet 18 km long and up to 5 km wide, surrounded by coastal fields and, further inland, by reed-swamps and floodplain meadows.

Above: in spring Long-tailed Ducks gather to court in the shallow waters of the bay.

There are also about 60 low-lying islets, some in the bay and some in the Moonsund area on the coastal side; on the inland fringes of the core reserve are also a few deciduous and mixed woodlands. Such diversity of habitat is an open invitation for a rich avifauna; besides the many migrants, there are about 170 species that breed on the nature reserve itself.

The waters of the actual bay are shallow, nowhere more than 5 m deep, providing superb feeding areas for ducks and divers. On a busy day in May the dull-grey waters can be dotted with swimming birds as far as the eye can see, with many more passing by in flying flocks, sometimes overhead. These may include simply thousands of Long-tailed Ducks, Common Scoters, Common Goldeneyes, Tufted Ducks, Common Pochards and Greater Scaup, together with hundreds of Red-throated and Black-throated Divers and smaller numbers of such birds as Goosander and Smew. Sooner or later the less expected species make an appearance: birds such as Velvet Scoter and, if your luck is in, perhaps a few Steller's Eiders, which are something of a feature at Matsalu. Just to add to an already impressive spectacle, many of these wildfowl cannot resist the opportunity, even on migration, to engage in bouts of courtship while resting in the bay, and the waters may seethe with excited activity, Long-tailed Ducks pointing their tails skyward, Common Goldeneyes head-throwing and Tufted Ducks peering. The loud yodelling calls of the numerous Long-tailed Ducks mix with the piping of scoters and the suggestive croons of the locally-breeding Common Eiders, making the whole experience of sea-watching here unforgettable.

Fringing the bay are coastal meadows where geese and swans graze, often as impressively numerous as the ducks. At least 20,000 Barnacle Geese pass through Matsalu Bay every spring, together with a similar combined number of Bewick's and Whooper Swans, about 10,000 Greylag Geese and several thousand Bean and Greater White-fronted Geese. It is well worth searching among these goose flocks for the occasional rarity: Red-breasted Geese have been known to mix in with the Barnacles, and in recent years the nature reserve has become a noted site for the perilously rare Lesser White-fronted Goose. This species can be hard to pick out among the distant flocks of Greater White-fronts, but in 2004, for example, an impressive gathering of 20 was present for a few days.

One of the more unusual delights of birding at Matsalu is the system of high viewing platforms set up to enable you to see for great distances over the surrounding landscape. One of these (Kloostri) is, quite literally, a watchtower, a relic of the time when Estonia was an outpost of the old Soviet Union; these days it is put to much better use! There are five main towers and they vary in height, from 6 m (Keemu) to a giddying 21 m (Suitsu), the latter overlooking a patch of forest. As an illustration of how good these towers can be, in May 1997 a group of birders recorded 128 species of birds from the 8-m-high Haeska tower in a period of just 24 hours.

Several towers have good views over reed-swamp, of which there are some 3,000 hectares in the reserve overall. The ecologically fussy Eurasian Bittern breeds in this habitat (the reedbeds must be *just right*), with about 15 booming males, and there are also high numbers of breeding Spotted Crake, Western Marsh Harrier, Great Reed and Savi's Warblers and, in more open areas, Black Tern. Inland from here are the alluvial meadows, rich grassland areas often overgrown with willow; these have significant, but currently declining populations of waders, which include a few Ruff and the occasional Great Snipe, a globally threatened species. Birds such as Corn Crake, Thrush Nightingale, Common Rosefinch and River Warbler also occur here.

Once autumn begins the scene is again set for impressive migratory movements, and one of the most distinctive differences from the spring is the appearance of large numbers of Common Cranes. These arrive from their northern breeding grounds from mid-August, peak in mid-September and have all but disappeared by early October. During their stay, numbers may build up to 20,000, making this the largest autumn gathering of Common Cranes in Europe. Autumn can also be a very good time for geese, and numbers may actually exceed those recorded in spring. Few of the cranes remain here for long; winters are harsh, and most of the birds evacuate, leaving Matsalu Bay, for a few months at least, a shadow of its former vibrant self.

Right: Black Terns, which feed mainly on insects picked from the surface of the water, breed on the marshes around Matsalu Bay.

Bialowieza Forest

HABITAT Old growth and managed forest; meadow

KEY SPECIES Woodpeckers including White-backed and Eurasian Three-toed, flycatchers including Collared and Red-breasted, raptors, Thrush Nightingale, River Warbler, Corn Crake

TIME OF YEAR Best in spring and early summer, particularly May and June, but interesting all year round

Right: White-backed Woodpecker is a specialist that relies on the presence of plenty of rotting timber.

Below: the Red-breasted Flycatcher has a sweet song and is common at Bialowieza, although its preference for inhabiting the canopy means that it can often be elusive.

This vast lowland forest, some 50 km from north to south and west to east, straddles the border between south-western Belarus, where it is known as *Belavezhskaya Pushcha*, and Poland, where it is known as *Puszcaz Bialowieska*. In many ways it is unique, being the largest remaining, relatively intact area of lowland deciduous/mixed forest in Europe, a remnant of the wild wood that once covered vast areas of the European Plain. Some parts have been untouched for more than a century, allowing the trees to grow to enormous heights (some over 50 m) and enabling the natural process of forest succession to run its course, with dead and decaying wood left in place. Not surprisingly, species diversity is extremely high, and includes nearly 170 species of breeding birds, 55 species of mammals, 900 species of vascular plants, 1,500 species of fungi and many thousands of invertebrates.

This is definitely not an area in which to rush about looking for birds; it is best to stand still and let the inhab-

Above: the Hawfinch feeds on seeds in winter, and can even crack open a cherry stone with its powerful bill.

itants show themselves one by one. Besides, the very special atmosphere in the strictly protected areas in spring and summer, standing in the dappled shade of the multi-layered forest and dwarfed by the towering trees, tends to instil hush in even the most fevered birder. Europeans are no longer accustomed to this kind of place, despite centuries of love-hate folklore about vast, mysterious forests.

Even a person with little knowledge of birds would expect woods to attract woodpeckers, and Bialowieza boasts a proud list of nine species. One of these, the White-backed Woodpecker, is a major beneficiary of the park's protected status, since it cannot sustain itself whenever forests are over-managed and no significant rotting timber is available. Moreover, the dominant woodland community of Bialowieza, which is made up of mixed stands of oak, lime and hornbeam, also suits the pernickety Middle Spotted Woodpecker, an oak specialist. The mighty Black Woodpecker and the Grey-headed Woodpecker could almost be described as common, and in the park's spruce forests the Eurasian Three-toed Woodpecker can also be found, although it tends to be quiet and elusive.

The presence of Eurasian Three-toed Woodpecker, together with such other northern species as Tengmalm's Owl, Eurasian Pygmy Owl and Spotted Nutcracker, reflects the park's position at the transition zone between European coniferous and European deciduous forest. Norway Spruce reaches its southern limit here and Sessile Oak is at its north-eastern extremity. This mixing allows for some 12 different forest communities to flourish in the park. The oak, lime and hornbeam forest is the most widespread, but other important communities include pine, spruce and oak and, in the wetter areas, alder and pine.

The deciduous sector is excellent for flycatchers, with four species breeding. The Spotted Flycatcher is common, the European Pied Flycatcher benefits from the presence of Sessile Oak, while the Collared Flycatcher and the Red-breasted Flycatcher are seen here as well as anywhere in Europe. The Red-breasted Flycatcher can be a very difficult species to locate, however, because it is a canopy feeder that shuns the open perches favoured by the others. Other common smaller birds of the deciduous areas include Wood Warbler, Hawfinch, Willow and Crested Tits and Eurasian Golden Oriole.

Aside from the forest itself, the reserve contains significant clearings and marshland areas on both sides of the border. Unsurprisingly, these sites hold a different suite of species from the forest proper. A walk in such an area is likely to reveal Thrush Nightingale, Common Rosefinch, Red-backed Shrike and Barred and River Warblers, while two rare and declining species also occur: the Corn Crake, which is still relatively numerous, and the Great Snipe, which occurs only sparingly on the Belarusian side. Black Storks rely on the juxtaposition of marsh and forest, feeding in the former and breeding in the latter, while Lesser Spotted Eagles are somewhat similar, requiring forest for breeding and open country over which they can forage.

As well as the Lesser Spotted Eagle, the area generally is excellent for raptors, and about 15 species breed here. As might be expected for a large, forested area, the wasp-eating European Honey Buzzard is numerous, although, as ever, highly elusive when breeding. Montagu's Harrier occurs on the open areas and the Short-toed Eagle can be seen soaring in search of the park's various species of snakes. The Booted Eagle is another large raptor breeding here, feeding mainly on birds which it captures with spectacular plunges from a great height.

Birders will need to choose which kinds of forest to go birding in. On the Polish side there is a strictly protected zone of about 50 sq km, in which there is no manage-

■ Above: birding is not always easy here, with the immensely tall trees and thick canopy.

■ Right: the Golden Oriole is common in the deciduous parts of the forest.

ment whatsoever, and entry is strictly by permit along a single 4-km route. There is also a much broader, managed zone, where visitors can walk relatively freely. The core zone is much more exciting and atmospheric, while the managed area outside is actually somewhat easier to work for birds. On the Belarusian side the park has three zones: a core area (157 sq km), a buffer zone and a transition zone, the latter allowing some farming and other commercial practices. Within its larger core zone, the Belarusian park authorities have several designated oaks of more than 400–600 years old, together with ashes and pines over 350 years old and spruces up to 250 years old. It is these that add to the magic of the whole place, quite apart from its variety of birds and other wildlife.

Incidentally, birders have a special park inhabitant to thank for the continued existence of Bialowieza: the European Bison, or Wisent. It was the presence of this huge mammal that originally marked the area out as special, and Bialowieza has spent much of its history as a hunting reserve for the rich and influential. A visit to the area would not be complete without at least one sighting of this four-legged icon.

High Tatras

HABITAT Forest and mountain

KEY SPECIES Black, Syrian, White-backed, Grey-headed and Eurasian Three-toed Woodpeckers, Tengmalm's, Eurasian Pygmy and Eurasian Eagle Owls, Spotted Nutcracker, Western Capercaillie, Alpine Accentor, Wallcreeper

TIME OF YEAR Best in spring and early summer, particularly from April to July

Some of the wildest country in Europe can be found within the bounds of this fine national park, with its peaks and valleys. Healthy populations of large wild mammals lurk here, such as Lynx and Brown Bear, as well as a superb range of mountain and forest birds, including many that are rare or restricted in the western half of Europe. Yet it is a convenient area to work; the Slovak part of the mountain range is comparatively compact, at only 741 sq km, and the area attracts large numbers of tourists for walking in summer and skiing in winter, so that there is plenty of accommodation nearby. As a bonus, there are 600 km of well-marked trails, reaching ten of the high peaks. For a bit of luxury, there are even cable-cars to take visitors into the Alpine Zone (above 1,500 m).

Nevertheless, in the forests of the lower slopes it is easy to imagine that you are far from civilization. Early in the season, in April for example, it is possible to walk for many hours without coming across another human visitor; take care, however, because the weather can easily turn a gentle hike into a fight for survival. In this comparative wilderness such birds as the Western Capercaillie thrive. Europe's largest grouse is a shy species, but there are estimated to be over 100 male birds in the national park (and many more females), and if you are very fortunate you may come across a small group of birds in display, strutting on the ground and making petulant jumps into the air, to a background of loud belches and odd popping noises like the release of champagne corks. The Hazel Grouse is another species of the deep forest, while the Black Grouse is often seen at the edge of clearings.

The very richness of these forests is illustrated by the fact that, within this relatively small area, no fewer than ten species of woodpecker occur, each with its own specific ecological requirements and favourite trees. One of the star birds of the High Tatras, for example, is the Eurasian Three-toed Woodpecker. Scarce over much of Europe, this species is locally common here, occurring in the older stands of spruce where dead wood is left to rot; it specializes in searching for the larvae and pupae of wood-boring beetles. The White-backed Woodpeck-

Opposite: one of the star birds of the High Tatras, the Eurasian Three-toed Woodpecker, is drawn to areas of dead wood, including recently burnt or felled areas.

Opposite top: Alpine Accentors play out their strange pairing games on the higher scree slopes.

Above: watch out for the Western Capercaillie – it is relatively common in the Tatras and males can be very aggressive in the spring.

er has a similar diet, but is drawn to deciduous forests rather than conifers. It is scarce, but can be found in stands of beech or aspen, its presence betrayed by deep gashes at the base of dead or rotting trees. Meanwhile, the local oak woods are favoured by the Middle Spotted Woodpecker, which gleans most of its food from the surface of the bark. Lesser Spotted Woodpeckers exist in stands of alder and other deciduous trees, while Black Woodpeckers favour mixed beech forest, and Great Spotted Woodpeckers tend to occur in the conifers at higher altitudes. Grey-headed Woodpeckers are an indicator of mature deciduous woodland, and the picture is completed by the Syrian Woodpecker, the European Green Woodpecker and the unusual Eurasian Wryneck, species favouring more open areas at low altitude.

Where there are woodpeckers there are often owls, too, which can be equally fussy about their habitat requirements. In the High Tatras, Eurasian Pygmy and Tengmalm's Owls inhabit spruce and spruce/pine forests, mainly above 900 m, the Eurasian Eagle Owl nests on rugged cliffs and gorges, and there are a few Ural Owls in the beech forests in the eastern part of the national park, in the Belianske Tatra.

This steep mountain range is a good place for the hiker to observe changes in habitat and bird life as the altitude increases. For example, on the lowest ground within the national park Corn Crakes breed in the meadows, and the woods support species such as Red-breasted Flycatcher, Firecrest and Lesser Spotted Eagle. Above 900 m the montane zone begins; here the woodland is mainly coniferous, and birds such as Western Capercaillie and Crested Tit are commoner than lower down. In the submontane zone, at 1,200 m, some Arolla pine grows and provides a niche for the Spotted Nutcracker. At 1,500 m the tree cover begins to be dominated by mountain pine, with dwarf shrubs growing underneath (the subalpine zone), where species such as Common Redpoll and Ring Ouzel eke out a living. Finally, above 1,850 m, the trees give way to a rugged, boulder-strewn Alpine tundra.

Above: the cliff-inhabiting Wallcreeper is a big prize for a skilful and patient birder; it can be notoriously difficult to locate this small bird on a vast rock-face.

There are not many species on these very high tops, but those present are charismatic and intriguing. The main predator up here is the Golden Eagle, which is relatively easy to see and quite unfazed by the often violent changes in the weather that commonly afflict these mountains, even in summer. Water Pipits breed in good numbers, favouring areas with short grass, and Rufous-tailed Rock Thrushes breed among the boulders. On scree slopes there are Alpine Accentors, which have recently been discovered to have an extraordinary mating system. A group of three to six males and three to five females live together, and any given individual may copulate with all the members of the opposite sex while trying to prevent the others from doing the same. To attempt to safeguard their paternity, male Alpine Accentors may mate more than 100 times a day.

One further species occurring at these high levels is the Wallcreeper, an extraordinary passerine with a long, curved bill and broad, butterfly-like wings of brilliant crimson, dotted with black and white. It feeds on invertebrates cleaned from tall rock faces or gathered by the sides of streams. The Wallcreeper is harder to find here than in some other mountain locations in Europe. But where, other than in these wild, unspoilt mountains, could there be a better setting for such a bird?

Danube Delta

HABITAT Freshwater wetland, fringed by marine habitats

KEY SPECIES Great White and Dalmatian Pelicans, Pygmy Cormorant, Glossy Ibis, herons, Red-breasted Goose, Little Crake, migrant waders, Moustached Warbler, Eurasian Penduline Tit

TIME OF YEAR All year round, but cold and difficult of access in the depths of winter

Above right: the Eurasian Hobby is common all over the delta and benefits from the area's great profusion of dragonflies.

There are many who consider this to be quite simply the finest site for birds in the whole of Europe, and it would be difficult to disagree. It certainly has some measurable superlatives; it is, for example, the largest continuous marshland in Europe, and has an authentic claim to the largest expanse of reedbeds in the entire world. Within its core area of 733 sq km there are extraordinary and scarcely rivalled numbers of breeding birds: 2,500 pairs of Pygmy Cormorant and Great White Pelican, 3,000 pairs of Black-crowned Night Heron, 2,000 pairs of Squacco Heron, 1,500 pairs of Glossy Ibis and 20,000 pairs of Whiskered Tern. And these are only a few of the 176 or so species that have been recorded breeding here. This truly is one of the great refuges for birds in the European continent.

The delta forms where the Danube, after flowing for 2,860 km, splits into three main branches – the Chilia, Sulina and Sfantu Gheorghe Rivers – while still 90 km from the Black Sea. Between these rivers lies a dense network of interconnecting channels with a remarkable combined length of 3,500 km. Within this maze the main habitat type is reed-swamp, but there are also areas of lake, sand dunes, scrub, woodland and meadow. One peculiar characteristic of the delta is the existence of 'plaurs', islets formed by decaying reed debris built up over the years to 1–1.6 m in height. The plaurs float and their position changes constantly according to the conditions of wind and current. Another local feature is the 'grind', a sand-dune system; many of the site's more substantial woodlands grow on these.

The only way to explore the delta is by boat, and on the

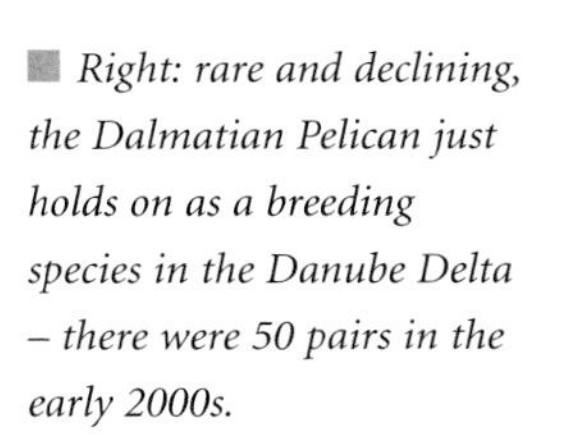

Right: rare and declining, the Dalmatian Pelican just holds on as a breeding species in the Danube Delta – there were 50 pairs in the early 2000s.

Above: dry country to the south of the delta proper holds large wintering flocks of Red-breasted Geese. Here the birds are mixed with the larger Greater White-fronted Geese.

Romanian side ecotourism is well advanced, with 'floating hotels' set up to cater for wildlife watchers (go in a group, though, because these are very expensive to charter). Some of these are luxurious, and the whole adventure can feel like a cruise. From the moment you leave port you are immediately plunged into a quite different world from the one you left, where reeds grow as far as the eye can see, the pace is slower and the wildlife holds sway. The abundance of birds in spring or summer is hard to take in at first; everywhere you look commuting flocks of Great White Pelicans, cormorants or Glossy Ibises will be flying in neat formations overhead, while Western Marsh Harriers quarter the reedbeds at every turn and Eurasian Hobbies vie with Whiskered Terns (both very common here) to chase dragonflies. In the waterside bushes you soon come upon mixed colonies of Pygmy Cormorants, Little Egrets and various herons going about their breeding business, and the sharper-eyed may spot Eurasian Penduline Tits at their remarkable hanging nests made from plant down. Throughout each day you will be besieged by a constant chorus of reedbed birds such as Savi's, Moustached and Great Reed Warblers and Common Reed Bunting, heard together with the sharper notes of Common Moorhen and Water Rail. It is a good place to learn the art of listening.

A little more effort is needed to see some of the special species. One of these is the endangered Dalmatian Pelican, which still holds on in the delta, but only in a few places, such as Lacul Sinoie in the south-east, where the main breeding island has recently been saved from destruction due to erosion; there are now probably fewer than 50 pairs in the delta. Other choice species include the White-tailed Eagle (about five breeding pairs) and the Red-footed Falcon (150 pairs), both of which range widely while hunting and should be seen, given a little time. Some of the more secretive marsh birds, such as Little and Spotted Crakes and Little Bittern, may also require patience to see, especially from a boat. Often the best strategy is to find some dry land and watch from there, where mud abuts the bottom of the reed stems.

Waders can be hard to find although, as with everything on the delta, there are huge numbers here at times, mainly on passage. If you find a suitable area of shallow water, look out for three typical 'eastern' migrants: Temminck's Stint, Broad-billed Sandpiper and Marsh Sandpiper.

In winter the Danube Delta is just as impressive for numbers as it is in summer, although the birding is less comfortable. Nonetheless, winter counts of 500 Lesser White-fronted and 45,000 Red-breasted Geese (mainly in the southern part of the delta) should make most birders come running. Other counts that would be almost unimaginable in other parts of Europe include 32,000 Red-crested Pochards, 13,000 Ferruginous Ducks (also common in summer), 40,000 Northern Shovelers and 970,000 Common Pochards. All these birds, not surprisingly, attract raptors. The number of White-tailed Eagles may shoot up to 50, there will be a few Rough-legged Buzzards, and even the odd Greater Spotted Eagle may show up.

The delta has recently gained some attention as a migration hot-spot. Its convenient position on the Black Sea Flyway makes it well situated to receive migrants from any points north or east. In the last few years Sahalin Island, at the mouth of the delta, has been acquiring a reputation as a good place for scarce birds such as Paddyfield Warbler and Red-throated Pipit.

On a discordant note, in the last couple of years the Danube Delta has suffered from a series of setbacks. In 2005 the delta was closed to visitors when bird flu was discovered nearby and not long afterwards devastating floods hit the area. Then, disastrously, the Ukrainian Government sanctioned the construction of a ship canal through the northern part, potentially destroying some of the wildest and most bird-rich sections of the whole Biosphere Reserve. To the coldest hearts, it appears, not even the title of 'best place for birds in Europe' cuts very much ice.

Below: the Glossy Ibis feeds in the mud and shallow water, but nests colonially in trees.

Asia

For the purposes of this book Asia has been embraced to its fullest possible extent, working from the edge of Europe eastwards to the Bering Sea and Japan, and from the deserts of the Middle East to the tropical savannas and rainforests of India and Malaysia east to the archipelago of the Philippines. Such a definition is merely for convenience, and bears little resemblance to the avifauna 'on the ground', so to speak. The Middle East and Borneo, for example, have little in common ornithologically. The Middle East, indeed, shows an affinity with Europe, as does northern Eurasia. The region south of the Himalayas, including the Indian subcontinent, South-East Asia and the islands west and north of Wallace's line, is known as the Oriental region, and has its own distinctive set of birds, which are completely different from those to the north.

There are many highlights in such a mixed area. The mighty tundra and taiga belts of Eurasia are equivalent to those in Europe and North America, with the tundra supporting great populations of ducks, waders and gulls, while the taiga forests hold a variety of grouse, Old World warblers and chats. The mighty Himalayan chain and the adjoining ranges in China are phenomenally rich in species, and also in the kind of birds dreams are made of, such as brilliantly coloured pheasants and babblers, and Ibisbill and Wallcreeper. The only family endemic to Eurasia, the accentors, also does well here. Meanwhile, the deserts and steppes of Central Asia hold their own special magic and birds, while the temperate forests of China and Japan seethe with flycatchers, thrushes, tits and woodpeckers.

In the Oriental region, monsoon forests and savannas in India give way to true tropical rainforest in South-East Asia. These forests, characterised by Dipetrocarps, have the highest diversity of tree species in the world, but not the greatest richness of birds. Further east, the small and large islands from the Sundas to the Philippines all have their own endemics. The Oriental region has several endemic families, including the leafbirds, ioras and Filipino rhabdornises. However, its main attraction is its superb array of babblers, hornbills, pittas, pheasants, broadbills and drongos, among many others.

Right: Red-crowned Cranes in Japan.

Eilat

HABITAT Coast, cultivated areas, saltpans, scrub, wetlands

KEY SPECIES Lichtenstein's Sandgrouse, Hume's Owl, Dead Sea Sparrow, Arabian Babbler, White-eyed Gull, migrants

TIME OF YEAR Good all year round, but most popular in spring (from March to May) when the greatest variety of species can be seen

Eilat is a modern seaside resort in the far south of Israel, popular with those seeking all-year-round sun and the nearest major coral reef to Europe. A quick glance at a map also underlines its potential for migratory birds; its position close to the eastern end of the Mediterranean and at the north-eastern tip of the Red Sea places it right in the middle of one of the major migration routes in the world, between Eurasia and Africa, and at a point where birds are bound to be funnelled between these two great bodies of water. No wonder, therefore, that Eilat is one of the most exciting migration hot-spots in the Western Palearctic, and host to a spring birding festival that attracts visitors from all over the world. An amazing 420 species have been recorded, including a number of extreme rarities. Although spring and autumn are the best times to visit, there is plenty of interest throughout the year.

Below: difficult to see anywhere else in Israel, Lichtenstein's Sandgrouse is one of Eilat's most sought-after species; flocks can usually be seen coming in to drink at a favoured site as night falls.

One of the most exciting features of Eilat birding is the sea-watching. Its location on the Red Sea enables the site to host species that are very rare elsewhere in the Western Palearctic, and southern winds even deliver tropical seabirds from the Indian Ocean. In recent years, for example, two new taxa for the Western Palearctic have been seen: the Streaked Shearwater and, in 1992, even more amazingly, the 'Mascarene Shearwater', a race of Audubon's Shearwater, which was so unexpected that the finders originally thought they had discovered a species new to science. More regular visitors include Brown Booby and Red-billed Tropicbird, while the North Beach, the best area for sea-watching, more or less guarantees sightings of such birds as White-eyed Gull and Western Reef Heron.

Another major draw is the superb passage of raptors, which easily exceeds the volume of traffic through the more famous Straits of Gibraltar or the Bosporus, and rivals almost every other raptor migration site in the world; up to one million birds may pass each spring. The top three travellers are Common (Steppe) Buzzard, European Honey Buzzard and Black Kite, but the Levant Sparrowhawk is also a feature – 50,000 have been recorded in one day in April. This bird is one of just a handful of *Accipiter* hawks that migrate in spiralling groups, or 'kettles', rather than singly. Other raptors recorded in numbers include Pallid Harrier, Steppe Eagle and Lesser Spotted Eagle.

Behind the beaches, Eilat is effectively a big oasis in the midst of a desert landscape and, as such, attracts large numbers of passerine migrants. Just about every migratory breeding species from Europe and western Asia can be and has been recorded here, and visitors can expect to have their identification skills tested to the limit. The fields may contain tricky larks and pipits, for example, while the scrub may contain a profusion of challenging warblers. Specialities include such gems as Oriental Skylark, Bimaculated Lark, Citrine Wagtail, Bluethroat and Cyprus Warbler (these last two also winter here).

Along with its scrub and farmland, Eilat is also blessed with saltpans and reservoirs and these, hardly surprisingly, add gulls, terns, herons and waders to the list. The saltpans are situated just north of the beach, and almost always host Greater Flamingos, as well as their regular fellow paddlers, Black-winged Stilt, Pied Avocet and Slender-billed Gull, plus Caspian and Gull-billed Terns on the water's edge. Indeed, connoisseurs of gulls will have plenty to keep them esoterically occupied, as they pick out Heuglin's, Armenian, Yellow-legged and Lesser Black-backed Gulls among the hordes.

Among the waterbirds, waders and crakes are also

very well represented. Greater Sand Plovers regularly visit the beach, fields or saltpans, while such species as Caspian Plover and both Spur-winged and White-tailed Lapwings are somewhat rarer. Spotted, Little and Baillon's Crakes can be easier to see here than on their breeding grounds. The drainage canal opening out into North Beach can be a good spot.

Even on days when the migration is slow, there is always plenty of interest at Eilat. The parks and gardens host exotic-looking Green Bee-eaters, White-throated Kingfishers, Namaqua Doves and Palestine Sunbirds, while the desert profile of the area means that a number of specialized arid-zone species can be seen relatively easily. One of the most famous is Lichtenstein's Sandgrouse, an intricately marked bird of the stony desert that is usually very elusive owing to its habit of visiting waterholes after dark. On the north-west side of town, however, is a pumping station close to the desert where, for many years, the birds have come in at dusk to quench their thirst, to the delight of generations of admirers. In recent years it has not always been possible to visit the site because of disturbance, but 50 or more sandgrouse can show up at times, accompanied by birds such as the supremely well camouflaged Sand Partridge, Trumpeter Finch and the highly localized Arabian Babbler and Blackstart. These days, special drinking trays have been provided for the birds.

In these same cultivated areas are a number of other arid country specials. The delightful Desert Finch, for example, doesn't really live up to its name, being more of

Right: Eilat's parks and gardens host resident Green Bee-eaters of the smart blue-headed subspecies cleopatra.

Right: Caspian Plover is a rare but regular visitor to Eilat en route to and from its breeding grounds in central Asia and wintering sites in southern and eastern Africa.

Opposite top: the subtly plumaged Hypocolius is a rare and declining winter visitor to Dubai; it breeds across the Straits of Hormuz, in Iran.

Opposite below: the Cream-coloured Courser is a classic desert bird, and was one of the few that occurred here before the city came into being.

a scratch cultivation species than a denizen of true desert. Similarly, Desert and Isabelline Wheatears, Dead Sea Sparrow and Desert Lark are not afraid of a little luxury in this man-made oasis, while Tristram's Starling adapts to buildings as an alternative to its usual cliffs and rocky outcrops. These birds are often easier to see here than in their true core habitat.

One of the many pleasures of birding in this area is that you are rarely alone in your hobby. It is easy to visit the main ringing research station on the east side of town, and in the commercial centre of town is the International Birdwatching Centre, open from 5pm to 7pm daily, the nerve centre of birding in the area and always up to date with all the latest sightings. The centre also organizes trips further afield to visit well-known stake-outs for species such as Hume's Owl and Nubian Nightjar.

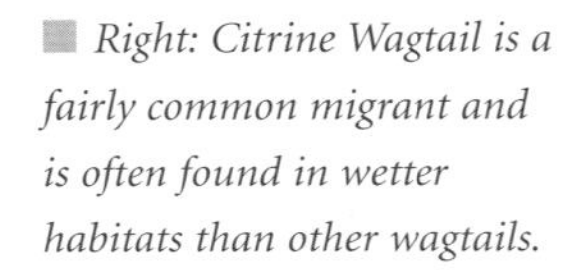

Right: Citrine Wagtail is a fairly common migrant and is often found in wetter habitats than other wagtails.

Dubai

HABITAT City parks and gardens, golf courses, mud-flats, desert, artificial wetlands

KEY SPECIES Migrant passerines, Hypocolius, Broad-billed Sandpiper, Red-wattled and White-tailed Lapwings, Pallid Scops Owl

TIME OF YEAR The migration seasons are best, from March to May and September to November

The confident, brash, gleaming city of Dubai is probably not a place one would immediately associate with birding. Yet this growing urban hub of commerce and construction, said to house 15 per cent of all the working cranes in the world, is a rare example of how people and birds can coexist as opportunists in the same setting. While people flock to the city to gaze at the remarkable new buildings, or to shop, or to make their fortune, the birds come here to utilize the completely new habitats that have been created out of the desert landscape. Both birds and people are largely transients: the city does not grant citizenship to foreign workers, so labourers come and go, while for their part a high proportion of the birds use the city only as a stopping-off point on their migration between the Palearctic and Afrotropical realms.

Dubai lies on the eastern end of the Arabian Gulf, not far from the northern tip of the Arabian Peninsula where the Straits of Hormuz separate Arabia from Iran. It has existed as a settlement since at least 1799, and became a port of some note before the 1930s. However, it is only in the last 20 or so years, after the economy was diversified away from dependence on oil, that Dubai has developed so rapidly as a centre for commerce and enterprise. The tallest hotel in the world, the famous Burj Al-Arab built in the shape of a sail, bears testament to the growing tourism industry, and there are plans afoot to build the largest hotel complex in the world, with 6,500 rooms. The upwardly mobile image is heightened by the many huge skyscraper developments, which include the tallest building in the world, the Burj Dubai, scheduled for completion in 2008, and massive new shopping malls and business parks.

However, as the urban area has expanded, so have the green spaces that are necessary to make life in the human melting-pot bearable. There are dozens of parks and gardens, golf courses and well-watered areas scattered throughout the city, and these artificial oases can provide shelter for an astonishing number and variety of visitors. Nearly 400 species have been recorded from the region in all.

One of the best-known green spots in Dubai city is

Right: the elegant, long-legged White-tailed Lapwing breeds in some of Dubai's newly created marshes.

Safa Park, a large area of grassland and woodland in the Jumeirah district, with a pond and boating lake. It is a classic migrant trap and can be alive with birds on any spring or autumn morning. Over 200 species have been recorded here, including such regulars as Ménétries's Warbler, Bluethroat, Semi-collared Flycatcher and Pied Wheatear. An extraordinary list of rarities includes White-breasted Waterhen, Black-throated Thrush and in April 2004 the first Eleonora's Falcon for the UAE, so it seems that almost anything can turn up here. The same applies to the Pivot Fields east of the city centre. There is permanent water on this site and, of course, plenty of grass. It is arguably now the best site in Dubai for birds, and ground species such as larks and pipits feature strongly. In a good spring or autumn you could expect to see Richard's, Blyth's, Tawny, Meadow, Red-throated, Tree, Water and Buff-bellied Pipits, together with Eurasian and Oriental Skylark, making it an excellent place to hone your identification skills. Water birds at the Pivot Fields include most of the common Eurasian herons, together with Glossy Ibis, Mallard, Garganey, Black-winged Stilt, Red-wattled and White-tailed Lapwings (both breed) and Moorhen. The list of rarities is huge, and includes such species as Bimaculated Lark, Asian Desert Warbler and Cinereous Bunting. Another excellent site for birds is the Emirates Golf Course, where the Dubai Desert Classic tournament is held each year; special features here are the desert species, including Chestnut-bellied Sandgrouse, Cream-coloured Courser and Greater Hoopoe-Lark.

There is also one 'official' nature reserve in Dubai, the Ra's al-Khor Wildlife Sanctuary, a 0.5 sq km patch of tidal estuary and saline lagoon in the centre of town, surrounded by the buzzing metropolis. There is a visitor centre here, and three hides that were opened in 2005, all with telescopes installed and with guards who offer visitors drinking water. The reserve is best known for its Greater Flamingos, which are fed each afternoon by the rangers, but for birders the sight of large numbers of Broad-billed Sandpipers (up to 4,000 in autumn), Lesser Sand Plovers (3,000) and Kentish Plovers (3,500, including some breeding birds), with smaller numbers of Pacific Golden Plovers, will be of more interest. In winter there may be 50,000 birds on the reserve, perhaps including Greater Spotted Eagles.

Slightly away from the main city are several excellent areas. The Mushrif National Park, 5 km east of Dubai Airport, actually preserves some indigenous woodland, and is a famous stake-out for the Pallid Scops Owl, as well as Arabian Babbler and, in summer, Yellow-throated Sparrow. Further up the coast, the mudflats at Khor al-Beidah often host coastal specialities such as Crab-plover (500 in winter), Great Knot and Greater Sand Plover, while on the coast road between Dubai and Abu Dhabi, the Seih ash Sha'ib lagoons hold such Gulf delights as Socotra Cormorant, Saunders's and White-cheeked Terns and Slender-billed Gull.

Not far from these lagoons is a village called Ghantoot. The acacia plantation at this spot is the most regular site in the United Arab Emirates to see one of the star birds of the

Right: Dubai's Ra's al-Khor Wildlife Sanctuary is a prime site for Broad-billed Sandpiper during the migration periods.

region, the Hypocolius. A hundred or more of these sleek, greyish, Waxwing-like birds used to winter in the UAE, but are now becoming more erratic in their appearance, and turning up in smaller numbers. Some winter roosts have disappeared. They arrive in the UAE after breeding in the riverine woodlands of Iraq, Iran and Afghanistan where, not surprisingly, their population is unknown.

To many birders the smart and charismatic Hypocolius, a bird of uncertain affinities and placed in its own family, offers sufficient reason on its own to visit Dubai and the other Gulf States for birding. However, there is much more to be seen in this region, and much yet to discover, so it could well be that the burgeoning city of Dubai will become more and more popular with birders.

Below: the bizarre Crab-plover, which specializes in feeding on fiddler-crabs, winters along the coast to the north of the city.

Korgalzhyn

HABITAT Steppe and wetlands

KEY SPECIES White-headed Duck, Pallid Harrier, Sociable Lapwing, Demoiselle Crane, Black and White-winged Larks, Black-winged Pratincole, Paddyfield and Booted Warblers

TIME OF YEAR Breeding season is best, particularly from May to July

Right: Korgalzhyn is a major breeding site for the globally endangered Sociable Lapwing.

The popular view of Central Asia is of a land of vast steppe stretching as far as the eye can see. These days the reality is quite different, much of the original vegetation in this part of the world having long since disappeared under the plough. However, in northern Kazakhstan, just 160 km to the northwest of the newly burgeoning capital city Astana, lies an area of steppe that has thankfully and wondrously survived, more or less intact, until the present day. At 2,589 sq km, the Korgalzhyn State Nature Reserve is the largest area of protected land in Kazakhstan.

This superb area is dominated by flat plains of feather grass. The lack of rainfall prevents the growth of any trees, although some have been planted along roads as shelter belts and are used by colonies of Red-footed Falcons; however, on the steppe itself many small bushes manage to hang on here and there, providing habitat for such birds as Booted Warbler. The plain is also dotted with a large number of lakes and marshes, some very large (Lake Tengiz has an area of 1,590 sq km) and quirks of geology have ensured that both freshwater and saline systems can be found virtually alongside one another. This allows for some curious juxtapositions: nowhere else do Black-throated Divers and Greater Flamingos occur together as breeding birds; the former is mainly a bird of tundra lakes, while the latter is found here at its most northerly outpost in the world.

It is the steppe, however, that holds the greatest interest for birders. The star species on the reserve – for all the wrong reasons – is the Sociable Lapwing. This species is considered to be critically endangered, having suffered a population crash of 95 per cent over the last 20 years, and it was considered to be down to perhaps 500 pairs (although the discovery of at least 3,000 wintering in Turkey and Syria offers hope that there are more breeding somewhere). The colonies here (the bird lives up to its name) have been intensely studied since 2004 in an effort to understand the problem; preliminary results suggest that trampling of eggs by cattle is a major cause of the decline, because the birds are drawn to the *Artemisia* stands that flourish in areas with a supply of cattle dung. In 2005, however, a highly encouraging 73 pairs attempted to nest at Korgalzhyn.

Sociable Lapwings are difficult to find on the steppe, but the same cannot be said of larks. Three species are abundant here: the widespread Eurasian Skylark, plus White-winged and Black Larks, two spectacular species that are largely endemic to this habitat. In summer the grassy plains ring with their high-pitched, trilling songs.

Right: Pallid Harriers hunt for rodents by quartering low and slowly over areas of tall grass.

Below: with its bold plumage and hurried, fizzing song, the Black Lark is hard to miss.

The Eurasian Skylark rises high in the sky in a drawn-out up-and-down display flight, an accompaniment to a sumptuous outpouring of song, while the White-winged Lark's party trick is to hover with such deep, slow beats that the wings, which are boldly patterned black-and-white, meet above and below. Yet, of the three, the Black Lark is the most impressive performer. It often sings on the ground in a curious posture, with tail cocked high and wings open and drooped, the wingtips trailing on the earth; every so often it will launch into the air and perform a few wing-claps, like a miniature, sooty-black pigeon.

Another special bird of the plains here is the rare Pallid Harrier. Its favoured foods read like a catalogue of the abundant small mammals, such as voles, hamsters, ground squirrels and steppe lemmings, which the harrier catches as it makes a slow, low flight over the grass. If mammals are in short supply it can always switch to birds, and its favoured method of hunting is to flush these high so that they may be caught in a mid-air chase. As a sideline this species also takes some of the many large insects that flourish on the steppe, including the ubiquitous grasshoppers and dragonflies.

Above: Demoiselle Cranes are dry-country birds that breed on the steppes at Korgalzhyn and shun the area's many lakes.

The steppe lakes are an important habitat for birds in their own right. Two of the most characteristic species are colonial birds which also chase insects on the wing. The White-winged Black Tern breeds on the highly productive lakes, where it can place its nest on floating vegetation, while the Black-winged Pratincole nests beside salt lakes on dry, lightly vegetated ground. The former usually catches aquatic food, such as dragonflies and damselflies, with nimble darting flights over the water, while the latter hawks over the grass to catch beetles and locusts, often hunting in noisy groups. In winter, the Black-winged Pratincole swaps the steppes here for the highland plains of southern Africa, where it can feed on much the same foodstuffs; it probably makes the journey in one or two long-distance legs at high altitude.

Apart from these specialities, the mixture of species on the wetlands could best be described as eclectic. There are truly wild Mute Swans here, shy migrants far removed from their tamed counterparts in Europe and North America. They breed side by side with the more conventionally wild Whooper Swans. All five European grebes breed, including important numbers of Red-necked, as do small numbers of the saline lake-inhabiting White-headed Duck, a species of high conservation concern. In the reedbeds adjoining the more productive lakes live Bearded Reedlings, Eurasian Bitterns and something of a steppe lake specialist, the Paddyfield Warbler. Also interesting are the two crane species here: the Common Crane breeds on the wetlands, while the dainty Demoiselle Crane is an inhabitant of the steppe itself and feeds mainly on grass seeds.

Not surprisingly, with so much wetland habitat, this area is also a good migrant stopover. Almost anything can be recorded here, and it is a noted site for waders and wildfowl, including such gems as Terek Sandpiper and Garganey. Indeed, on a spring or autumn visit one of the most prominent species you might see is a migrant: the Red-necked Phalarope. These peculiar, buoyantly swimming waders, which often spin around on the water to stir up food, migrate between their tundra breeding sites and the waters of the Arabian Sea. They may stop off here for a few weeks in their hundreds of thousands, outnumbering all the other birds.

Pontic Alps

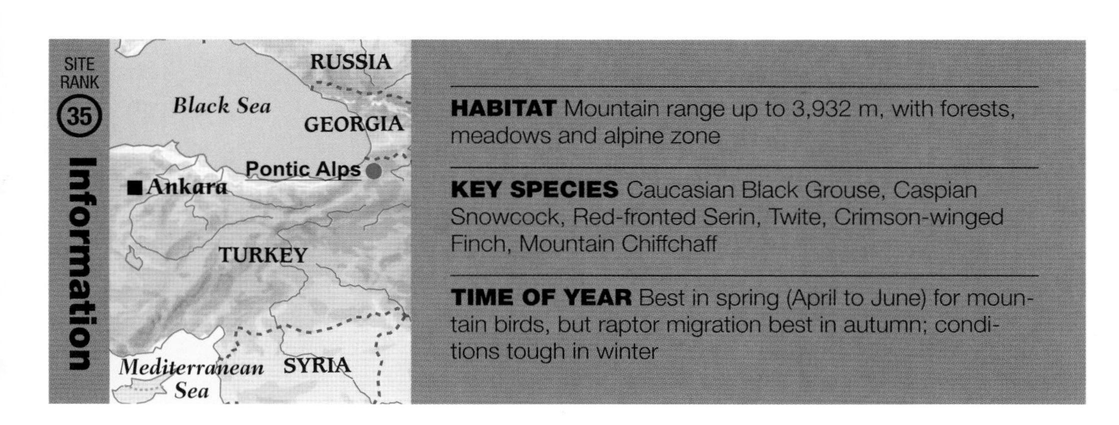

HABITAT Mountain range up to 3,932 m, with forests, meadows and alpine zone

KEY SPECIES Caucasian Black Grouse, Caspian Snowcock, Red-fronted Serin, Twite, Crimson-winged Finch, Mountain Chiffchaff

TIME OF YEAR Best in spring (April to June) for mountain birds, but raptor migration best in autumn; conditions tough in winter

Above right: Caucasian Black Grouse is such a celebrated resident of the mountains that it even has a festival in its honour. This is a female.

When Noah's ark came to rest on Mount Ararat after the Biblical flood, one of the first birds to find its way back home would undoubtedly have been the Caucasian Black Grouse. A short flight down the mountain and a little further west, and it would soon have returned to its western headquarters in the Pontic Alps, one of the key places to see this charismatic yet poorly known bird.

Nestling in the north-east corner of Turkey, close to the Georgian border and less than 300 km from Mount Ararat, the Pontic Alps rise spectacularly from the southern shores of the Black Sea to form a hinterland wilderness of high peaks, mountain streams, scree, rhododendron thicket and forest. This formidable range, which has formed an effective barrier to human movements for centuries and thus spawned a distinct culture among its people, provides a dash of the Himalayas on the very edge of Europe. As is usual in Turkey, the mixture of birds is eclectic, with familiar species from western Europe sharing space with some from much further

Right: patient and dedicated observers might get a glimpse of the shy Caspian Snowcock on the higher slopes of the Kackar Range.

Above: flocks of Red-fronted Serins are found in the mountains, where they can be quite confiding.

east. It is therefore a heady brew that greets the adventurous birder who finds a way to this isolated, and largely neglected, birding hot-spot.

It seems that everyone who visits inevitably makes a pilgrimage to the quaint village of Sivrikaya, the most famous stake-out for the Caucasian Black Grouse. On the western side of the Kackar Mountains (as the Pontic Alps are more formally known), and three hours from the Black Sea resort of Trabzon, this delightful spot, with its streams fringed with scrub that is full of Mountain Chiffchaffs, its patches of north-facing rhododendron scrub and its backdrop of snow-capped peaks, is the perfect scene for a classic birding adventure. Seeing the grouse is only really guaranteed in May and June, when you might still find yourself wading through thick snow to reach the site. Although the birds can be seen distantly from close to Sivikaya, for the best views you really need to be up among the tree-line meadows by sunrise, and that involves setting off from the village by 3 am, when it is dark and bitterly cold. The rough walk, best done with the help of a local guide, takes a couple of hours, anticipation and exhaustion competing for supremacy.

The black-and-white birds themselves, though, do have a certain star quality. They are neater than the field guides ever quite capture, and their celebrated 'flutter-jump' display on the lek, an advertisement to rivals and females alike, is a startling performance. The male suddenly takes off from the display arena with a rapid flutter of the wings, turns around 180 degrees in mid-air with tail spread and then glides down to where it started, whereupon it stands proud and upright for a few moments, like a gymnast holding position at the end of a routine. The manoeuvre is accompanied by a far-carrying swish of the wings, and creates an unmistakable strobe-like flash of the bird's white underwings. In the razor-sharp crisp air, and with the backdrop of mountain scenery occasionally peeping from behind the early morning mist, the experience of watching this most exclusive of leks is worth every bit of the hard climb to get up there.

Although Sivrikaya is the most famous lek, there are up to 2,675 Caucasian Black Grouse in Turkey in at least 42 locations (2006), and nowadays visitors are being encouraged elsewhere, too. In view of the bird's threatened status (an estimated population of 70,000 worldwide, but still classed as Data Deficient), an action plan for the grouse's welfare has been drawn up, which includes the development of ecotourism sites such as Posof or Ayder, within the Kackar Mountains National Park. These days there is even an annual Caucasian Black Grouse Festival to celebrate the bird and to increase awareness of its existence in the region.

There are plenty of other birds in the area besides the

Right: the famous raptor migration watchpoint of Borcka may yield sightings of such rarities as the Eastern Imperial Eagle. This is a juvenile.

grouse. Indeed, those waiting in the cold at the tree-line are bound to be rewarded with such birds as the charming Red-fronted Serin, a black-headed sprite that feeds mainly on the ground, the Mountain Chiffchaff (which sounds oddly like a Coal Tit), Common Rosefinch with its 'pleased to meet you' song, Green Warbler, Ring Ouzel and Twite, while in the alpine zone above the meadows there is a highly impressive list of montane species, including Blue and Rufous-tailed Rock Thrushes, Shore Lark, White-winged Snowfinch, Water Pipit, Red-billed and Alpine Choughs and Alpine Accentor. Such variety in this high-altitude zone is impossible to equal in Europe. A bit of harder work may also be rewarded by a sight of the decidedly scarce Crimson-winged Finch, while later in the day patient scanning may even reveal the red-and-white flash of a Wallcreeper. As if that list were not enticing enough, another very special gamebird also occurs here – the Caspian Snowcock. These birds ascend to the very highest rocky, snow-spattered slopes in the spring and, while their guttural calls can often be heard, seeing these pleasingly dappled grey-and-white birds is much more of a challenge.

Towards the eastern side of the range the climate gets damper and the north-facing forests are more sumptuous, attracting such species as White-backed Woodpecker and Red-breasted Flycatcher. However, there is still much to discover about this part of the range, and surprises are likely to be in store for anyone taking the trouble to explore.

No discovery, however, is likely to compare with the amazing events of 1976, when a group of birders decided to make a sustained raptor watch around the village of Borcka, on the eastern side of the Pontic Alps close to the Georgian border. Prompted by old reports of impressive raptor migrations, what they discovered was stunning: nearly 400,000 raptors of 28 species were seen to pass over in September and October, dwarfing the figures from the far more famous and celebrated Bosporus, on the western end of Turkey. Since then there have been other impressive counts and watches, culminating in the first record of Crested Honey Buzzard for the Western Palearctic in September 1979. Most migrating raptors are Common Buzzards of the eastern 'Steppe' race, European Honey Buzzards and Black Kites, but the many other species recorded include Pallid Harrier, Lesser Spotted Eagle, Eastern Imperial Eagle and Saker Falcon.

It seems that the raptors are funnelled past this part of the Pontics, avoiding the Black Sea to the west and the Greater Caucasus to the east. Yet oddly, despite the enormous potential of watching migration here, Borcka is still comparatively neglected. Surely there can be few areas of the world where such possibilities are so neglected.

Keoladeo Ghana

HABITAT Lowland marshes, scrub, woodland, grassland

KEY SPECIES Bar-headed Goose, Indian Spotted Eagle, Greater Spotted Eagle, herons, storks, ducks, Dusky Eagle Owl, nightjars, migrant passerines including Siberian Rubythroat and various thrushes, flycatchers and warblers

TIME OF YEAR Winter is best, October to March

■ *Above right: Bar-headed Goose is a winter visitor that flies over the Himalayas in order to reach Bharatpur.*

■ *Below: the wealth of waterbirds at one of Bharatpur's renowned jheels is a dream come true for birders.*

If you like birding to be ridiculously easy, there may be no better place in the world for you than the Keoladeo Ghana National Park in northern India. Usually known simply as Bharatpur, this site is small (only 29 sq km in area), flat, easy of access, and it simply teems with birds for most of the year. Over 400 species have been recorded here, and it is not at all unusual to see 150 in a single day (in the winter). Not much effort is required to observe the masses of herons, ducks, cormorants and storks on the flooded lakes (*jheels*); they are there in front of you, and often allow a close approach. Your only problem is to identify them all.

Keoladeo Ghana lies on the Gangetic Plain 180 km south of Delhi and 60 km west of Agra. It is an 'island' in a vast flat area of cultivation and its continued existence in such a populous region is a quirk of history. It seems that there were always marshes in the area, but in 1890 the Maharajah of Bharatpur, who was keen on duck shooting, extended and enclosed the local wetland by setting up a network of canals and earthen embankments, known as bunds. He henceforth used his newly created reserve for 'sport', inviting various visitors and dignitaries to join him on regular massacres of the wildfowl. But happily, despite the persecution, the birds kept coming, and by the 1960s they were afforded protection by the government of India. Then, largely thanks to

■ *Above: Painted Stork about to regurgitate food to its young in its treetop nest.*

pressure from the great Indian conservationist Dr Salim Ali, the national park was declared in 1982.

The most obvious inhabitants of Bharatpur are the waterbirds, which can be divided into residents and migrants. The monsoon season lasts from July to September, and this encourages many tree-nesting species to set up colonies in the branches of the acacias that grow on islands within the *jheels*. Throughout the reserve some 50,000 pairs of large waterbirds nest, including Little and Indian Cormorants, Darter, four species of egret, including Intermediate Egret, Black-crowned Night-Heron, Grey Heron, Painted and Black-necked Storks, Asian Openbill, Black-headed Ibis and Eurasian Spoonbill. Many colonies are mixed, allowing superb comparisons of all the species. They are so easy to see and so close that you have to pinch yourself to realize you are not in a safari park or zoo.

In the surrounding marshes there are dozens of other colourful or interesting breeding birds to see. These include the delightful long-toed Pheasant-tailed and Bronze-winged Jacanas, which trot energetically over the emergent vegetation, plus ubiquitous Indian Pond Herons, White-breasted Waterhens and Purple Swamphens. Sarus Cranes walk sedately over the marshes, towering over everything else, while less obvious 'birders' birds' well worth searching for include the weird Greater Painted Snipe and the highly secretive Black Bittern.

From October onwards these resident species are joined by many thousands of wintering Palearctic water birds. One of the commonest, the Bar-headed Goose, comes here after an epic flight that takes it over the peaks of the Himalayas; it has been recorded flying at 9,000–10,000 m altitude, and studies have shown that it has four types of haemoglobin in its blood, each working at different partial pressures of oxygen. Alongside it, many of the ducks sharing the *jheels* in winter will be familiar to visitors from Europe or North America, and include large numbers of Gadwalls, Northern Pintails, Eurasian Teals and Northern Shovelers. However, mixed in among them are good numbers of more typically Asian species, including the resident Indian Spot-billed Duck, Lesser Whistling Duck and Cotton Pygmy Goose.

Bharatpur is also famous for its birds of prey and, again, these can be divided into residents and winter visitors. One of the most important of the former is the sedentary Indian Spotted Eagle, which is now a very rare

■ *Above: Greater Spotted Eagles are probably easier to see here than anywhere else in the world – this juvenile is having a bad day.*

bird indeed; there is usually one pair on the reserve. The crisis among Indian vultures has hit here as everywhere; the White-rumped and Indian have disappeared, while the more solitary Red-headed Vulture still hangs on.

Winter is the best time for birds of prey, when dozens may come to Bharatpur from Eurasia to spend the season harassing the water birds. One of the most numerous, often numbering 30 or more, is the Greater Spotted Eagle, a bird that can be extremely difficult to find in its breeding haunts: Bharatpur is probably the best place in the world to see it. There can also be a few Eastern Imperial, Steppe and Bonelli's Eagles on site, all of which loaf about in the trees for hours on end, causing identification headaches for visiting birders. Less tricky are the numerous Western Marsh Harriers, and the snake-eating Short-toed Eagle, which is found around the drier parts of the reserve.

Once you have got to grips with the larger birds you begin to notice that there are plenty of small passerines around as well. Not content with being fantastic for waterbirds and raptors, Bharatpur in winter is a superb place for catching up with mouth-watering migrants from northern and central Asia. One of the best places is known as 'The Nursery', close to the barrier where you present your tickets, where there are scattered bushes and trees. Such gems as Bluethroat, Siberian Rubythroat and Red-breasted and Taiga Flycatchers can be found here, along with Tickell's and Orange-headed Thrushes. If all these birds are too easy for you, a fine range of difficult wintering warblers will tax your skills to the full: these include Sykes's, Dusky, Hume's Leaf, Blyth's Reed and Paddyfield Warblers, together with two specialities, Smoky Warbler and Brooks's Leaf Warbler.

In short, there is something for everyone here, whatever their birding ability, and the sheer numbers of birds can be almost overwhelming. You could easily spend a month at this marvellous reserve and still be seeing new species right to the end.

In recent years Bharatpur has suffered a major water shortage, mainly caused by its source of water being siphoned off for other uses, leading to fears that it would become degraded and lose its value as a reserve. Hopefully, this crisis can be remedied so that Bharatpur can continue to be flushed with water and birds.

Goa

HABITAT Beaches, mangroves, rice paddies, cultivations, lakes, scrub, lowland forest

KEY SPECIES Lesser Adjutant, Malabar Trogon, Malabar Grey Hornbill, Malabar Pied Hornbill, Blue-bearded Bee-eater, waders, terns, gulls, kingfishers, Indian Pitta, Sri Lanka Frogmouth, Malabar Whistling Thrush, Heart-spotted Woodpecker

TIME OF YEAR Good all year round, but the best time is November to March

A former Portuguese colony on the west coast of India, Goa is its country's smallest state, sandwiched between Maharashtra to the north and Karnakata to the south and east. Although blessed with a rich history and cultural mix, in which Catholic and Hindu influences ensure that opulent churches can sit beside Hindu temples, it is actually best known for its package holidays. Tourists from all over the world come to play on its long, sandy beaches and enjoy its seafood restaurants and family hotels, lending the place something of a hedonistic feel.

It was quite some years after the tourist industry was well established here that Goa's potential wildlife interest became apparent. No doubt bored birders on the beach looked up and began to notice healthy populations of gulls and terns mooching about at high tide, and holidaymakers with an interest in wildlife started noticing Coppersmith Barbets, Purple-rumped Sunbirds and Asian Paradise Flycatchers in their hotel grounds. Before long the keener visitors started spotting pipits on the rice paddies and ventured into the dry forests, and the burgeoning list of birds soon encouraged more and more people to visit. These days, Goa is a major ecotourism destination. Local agencies offer short birdwatching excursions, and tour companies arrange specific birding packages that can accumulate lists of 250 or more species in two dedicated weeks.

Goa's small size and rich variety of habitats make it perfect for birdwatching. Nowhere in the state is more than two hours' travel time from coastal resorts such as Baga, and most are actually concentrated near to the capital, Panaji, and thus easily accessible by a colourful

Right: it is easy to see why the beaches of Goa are an attraction to tourists.

choice of cheap public transport. In few other places is it as easy to combine a relaxing holiday with such rich birding opportunities.

The Indian Ocean beaches, especially Morjem Beach north of the Chapora River mouth, are often a good place to start. In the Palearctic winter the sandy shoreline can be excellent for shorebirds, gulls and terns. Birders can throw themselves into the complexities of distinguishing Greater from Lesser Sand Plovers and Kentish Plovers. Terns roosting on the sand-bars can include Sandwich, Gull-billed, Swift and Lesser Crested, while the gulls should include Brown-headed, Black-headed, Slender-billed, Heuglin's and, occasionally, Great Black-headed. Brahminy and Black Kites circle overhead while, out to sea, Ospreys and White-bellied Sea Eagles often appear.

Many of the river mouths in Goa support areas of mangrove where, at low tide, there are more waders to be seen, especially that arch crab-specialist, the Terek Sandpiper. Cruises up the Zuari River should also be good for kingfishers. Pied, White-throated, Stork-billed and Black-capped are common enough, while the Collared is rare and is a bonus. Green and Blue-tailed Bee-eaters will also be seen on these trips, on wires or treetops.

Away from saltwater, there are several excellent freshwater sites on the coastal plain that are full of birds, including Carambolim Lake near Old Goa and Maem Lake to the north-east of Panaji. The dense floating veg-

Opposite: *Sri Lanka Frogmouths roost in the understorey in dry forests, and pairs often have the endearing habit of huddling together.*

etation on the former quite literally supports Bronze-winged and Pheasant-tailed Jacanas, while ducks such as Lesser Whistling Duck, Garganey, Indian Spot-billed Duck, Northern Shoveler and Cotton Pygmy Goose hide away on the still water. Little Cormorants and, sometimes, Darters ply their trade under the water, while the mass of egrets should include Great White, Little and Intermediate. Storks are usually present, often just standing around or perched on the adjoining trees. Lesser Adjutant, Asian Openbill and Woolly-necked Stork are all recorded regularly. Another mainstay of these habitats is the delightful Little Pratincole, a common species close to the coast, often seen in wheeling flocks.

On the coastal floodplain, the many rice paddies and drier grasslands offer excellent opportunities for viewing terrestrial species such as larks and pipits. In the winter you could find yourself having to come to terms with Richard's, Blyth's, Red-throated, Tawny, Tree and Paddyfield Pipits, posing identification challenges which certainly make a change from being dazzled by Plum-headed Parakeets and Golden-fronted Leafbirds. A local speciality, the Malabar Lark, can also be found in this sort of habitat. The Dona Paula plateau, south of Panaji, is good for this species, and also offers the chance to see Oriental Skylark, Ashy-crowned Sparrow-Lark and, by way of light relief, Eurasian Hoopoe.

Despite the attractions of the coastal plain, no birder visiting Goa should neglect to venture inland to the woods and forests of the foothills of the Western Ghats. They introduce a completely new element into the birding, with plenty of regional specialities. The excellent Bondla reserve, an hour's drive from the coast, is a mere 8 sq km of dry deciduous and evergreen woodland, yet is often the easiest place in Goa to see such delights as Grey Junglefowl, Malabar Grey Hornbill, Bar-winged Flycatcher-shrike, White-bellied Blue Flycatcher, Scarlet Minivet, Nilgiri Wood Pigeon and Orange-headed Thrush. Meanwhile, the Bhagwan Mahaveer National Park (usually called Molem) on the south-eastern edge of the state, is much larger (250 sq km) and, although birding is often slower here, the diversity is somewhat higher. This is usually the most reliable place to find some truly world-class species such as Indian Pitta, Malabar Trogon, Sri Lanka Frogmouth, Malabar Pied Hornbill, Blue-bearded Bee-eater, Rufous-bellied Eagle, Malabar Whistling Thrush and Heart-spotted Woodpecker.

Right: *the Malabar Trogon often follows mixed species flocks of small birds through the forest.*

Opposite: *a careful search of the forest floor at Bhagwan Mahaveer may reveal an Indian Pitta. The species also winters at sites along the coast of Goa.*

Taman Negara

HABITAT Primary tropical rainforest from lowlands to 2,000 m

KEY SPECIES Pheasants including Great Argus, Crested Argus and Malayan Peacock-Pheasant, pittas including Garnet, Banded, Blue-winged and Giant, hornbills, Malaysian Rail-babbler, woodpeckers, babblers

TIME OF YEAR Good all year round

Too famous to be called by any other name (Taman Negara simply means national park in Malay) this treasure-trove 300 km north-east of Malaysia's capital, Kuala Lumpur, encloses the largest area of preserved lowland rainforest in mainland South-East Asia (4,343 sq km). Not surprisingly, much of it is quite a wilderness and, away from the main headquarters on its south-eastern fringe, there are no settlements and no roads at all within its borders. Instead there is just fabulous pristine lowland rainforest brimming with spectacular and remarkable wildlife.

The forests of South-East Asia are the most ancient rainforests of all – those within Taman Negara are estimated to be 130 million years old – and, at least in terms of their flora, they are also the richest on earth. There are more tree species per unit area than in Amazonia, providing habitat for a bewilderingly rich avifauna. More than 360 species of birds have been recorded at Taman Negara, and one of the real features of birding here is that you keep finding new species, even when you check over the same areas again and again.

Visitors stay at the plush new resort by the park centre, Kuala Tahan, with all comforts, including air-conditioning. The park was once accessible only by boat, a pleasant 60-km ride down the Sungai Tembeling, but a road has now been built to cope with the park's burgeoning tourist traffic. Another recent addition is a very impressive canopy walkway that winds through the forest for 500 m or so, 30 m above the forest floor near the centre, giving a unique view of this inaccessible part of the ecosystem. Of course, the forest proper is all around; an excellent network of trails leading from Kuala Tahan takes you to all parts. You can wander the Swamp Loop in 20 minutes or you can take a guided hike to the park's highest point, Gunung Tahan (2,179 m), which takes nine days.

One of the park's major attractions for the birder is its charismatic and colourful pheasants. Visitors cannot fail to hear the far-carrying, impressed-sounding *Oh, wow!* exclamation of the Great Argus, but seeing one of these peacock-sized but incredibly shy creatures is quite another matter; usually they are just scuttling rapidly

Opposite top: the Sungai Tembeling river is surrounded by pristine forest.

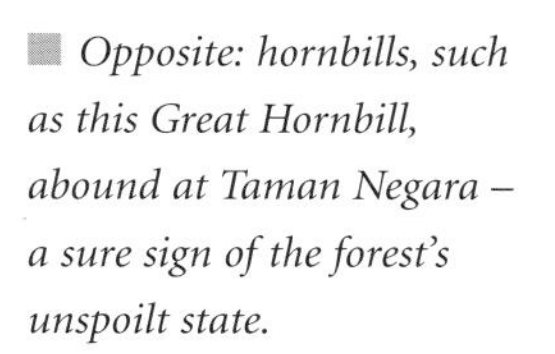

Opposite: hornbills, such as this Great Hornbill, abound at Taman Negara – a sure sign of the forest's unspoilt state.

Right: Garnet Pitta inhabits the forest floor, where it is very difficult to get a good view of its glittering colours.

across the trail, pulling 1.5 m of tail behind them, or slipping silently through the undergrowth. These pheasants are solitary, and the males invest much time in clearing display grounds of bare earth on the forest floor, blowing leaves away with rapid bursts of wing-beats. When females arrive the males give an extraordinary display, facing the females and lifting up their tail and wings to present rows of hitherto concealed orange-yellow ocelli, their equivalent of the peacock's 'eye'. Mating takes place straight away and the male takes no further part in the breeding cycle. At higher elevations, above 600 m, the rare Crested Argus can be found at its only outpost in Malaysia. It has an even longer tail than the Great Argus, but a less spectacular display in which just the head feathers and crest are used. Still another species, the endemic Malayan Peacock-Pheasant, completes an astonishing trio of polygamous pheasants in which the males are decorated by ocelli – in this case iridescent green.

Another very different group of forest-floor birds awaits the discerning birder at Taman Negara – the pittas. These gorgeous birds debunk the trend that ground-living forest dwellers are dull in colour, and they do it in style by being outrageously gaudy and brightly patterned. Many peoples' favourite is the Garnet Pitta, a delicious combination of crimson (on the belly and crown) and iridescent mauve (on the upperparts), garnished with a blue wing-bar for good measure. The Banded Pitta is another stunner, the male adorned with a brilliant cobalt breast and orange supercilium, but they are all special. Pittas make long hops over the forest floor and toss leaf-litter aside with their powerful bills. Several species, including the Giant Pitta and the migrant Blue-winged Pitta, eat a lot of snails, which they smash against a hard 'anvil', such as a rock, before eating.

Even the tourists spot the hornbills. These are perhaps the most obvious of all the birds at Taman Negara, and there may be more of the really big, impressive species of this family here than anywhere else in Asia, catching the eye as they fly through the air with regal, slow, loudly swishing wing-beats. The Great Hornbill, with its black, white and yellow plumage and large casque, is a typical member of the family, usually seen visiting ripening fig trees in the lofty forest canopy, where it deals expertly with even the smallest pieces of fruit despite the size of its capacious bill. Different hornbills occupy different niches: the Black Hornbill feeds mainly in the middle and lower strata so, although common, it is harder to see than the Great. The improbable-looking White-crowned Hornbill, with its snow-white tail and utterly absurd bushy white crest, which makes it look as if it is wearing a furry white, Russian style hat, feeds in small groups in dense lower growth, primarily on animal food. The hornbill from hell is the amazing Helmeted Hornbill, which appears to be part hornbill and part vulture, with its ugly fleshy red face and throat. This species consumes a high proportion of animal food, and not just invertebrates, either. It is adept at catching squirrels and snakes, and has been even known to take smaller hornbills.

These headline birds are but a small part of the abundant birdlife here. Other important groups that are well represented include the babblers, woodpeckers (including the huge, ant-loving Great Slaty Woodpecker), trogons and broadbills. Many of these can be seen as they join mixed-species feeding flocks moving through the rainforest.

One other highly sought-after species at Taman Negara is the Malaysian Rail-babbler. A taxonomic riddle that wanders over the forest floor, moving its head in time with its step like a small, long-billed chicken, this oddity is currently nestled in the same family as the whipbirds, all the others of which are Australasian. It is not colourful, nor is it particularly rare; most birders like to see it because it is just plain weird.

Below: to birders, the elusive Malaysian Rail-babbler is a major find. It is very shy and is best located by call, which is very similar to the whistle of the Garnet Pitta.

Gomantong Caves

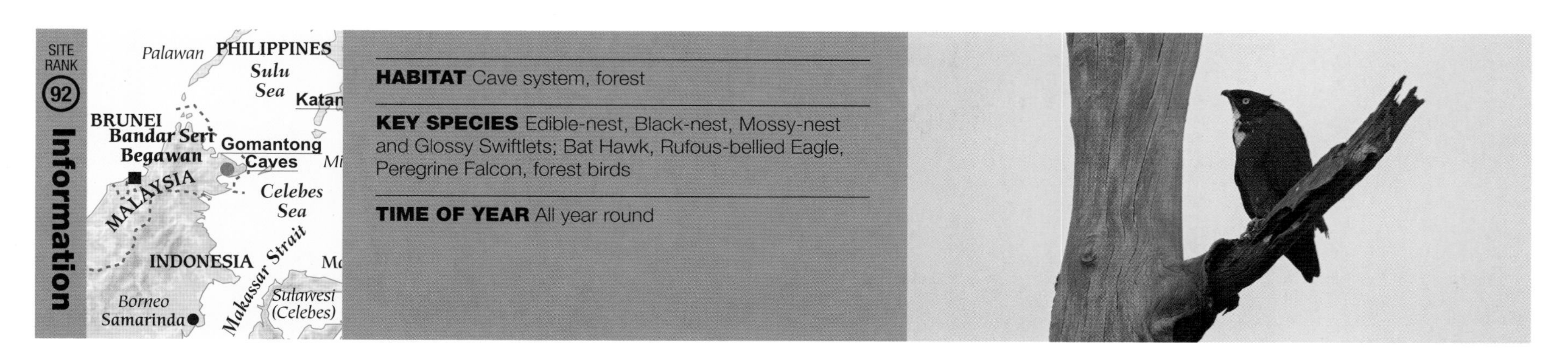

HABITAT Cave system, forest

KEY SPECIES Edible-nest, Black-nest, Mossy-nest and Glossy Swiftlets; Bat Hawk, Rufous-bellied Eagle, Peregrine Falcon, forest birds

TIME OF YEAR All year round

Above right: the caves have a population of one million bats, ensuring an abundant food supply for the attendant Bat Hawks.

There can be few more insalubrious surroundings for great birding than within the Gomantong Caves, on Sabah on the island of Borneo. It is dark, except for 20 minutes around noon when shafts of sunlight pierce the main chambers from above. It smells appalling, the result of countless years of guano accumulation on the cave floor. It is unwise to look up, for fear of being pelted with new supplies of that same guano. And should you stray off the paths, you could find yourself knee-deep in the deposits, with thousands of cockroaches, long-legged centipedes and other leggy invertebrates for intimate company.

What a place, though! Within these huge caves of cathedral-like proportions and all-enveloping hush, there is a special ecosystem. The cave ceilings provide habitat for about one million Sheath-tailed Bats, and for approximately the same total number of swiftlets, of which four species breed. The caves also form the basis of a major human industry, that of collecting the nests of the swiftlets to make bird's-nest soup.

The swiftlets are not much to look at, being essentially featureless dark brown, but they are extraordinary. For one thing, they possess the ability to find their way about by echolocation, one of only two groups of birds to have evolved this ability (the other is the Oilbird). This is the technique that bats use, but the swiftlets are less sophisticated, using it mainly for manoeuvring about the cave system, and much less, if at all, for catching mobile food. The clicks they produce are perfectly audible, sounding rather like a high-pitched trill or rattle, as if one was running a finger along the teeth of a comb, and they punctuate the silence of the caves when the birds enter or leave the caves at dusk and dawn.

Below: harvesting nests is a dangerous job – here the collectors are making their preparations.

The other remarkable aspect of these birds' lives concerns their celebrated nests. Each species builds its own particular type, each out of slightly different materials. The Glossy Swiftlet is rather the odd one out, making its nest out of plant material merely fixed to the substrate by saliva; these swiftlets cannot echolocate and build only on well-lit sites outside the main caves. The Mossy-nest Swiftlet makes a nest out of small pieces of vegetable matter, the entire structure glued together by saliva; the nest is not rigid so, unlike the other swiftlet nests at Gomantong, cannot be fixed to a vertical surface. The Black-nest Swiftlet's nest is a bracket-shaped structure made out of translucent saliva mixed with feathers, and the famous Edible-nest Swiftlet's nest is made of nothing at all other than saliva, and tends to look whitish in the torchlight.

The making of bird's-nest soup is an enormous industry, probably the single most commercially valuable business worldwide that depends completely upon a wild bird. It is worth well over US$1 billion a year and is increasing. At the same time, hardly surprisingly, the swiftlets are declining in those areas where their nests are harvested unsustainably (often, appallingly, when there are eggs or young in the nests). At Gomantong Caves, however, the collection is regulated; the caves are managed by the Wildlife Department of the government. Only two collections a year are permitted, one between February and April, just after the birds have built their nests but not laid any eggs, and then again between July and September when breeding, with the newly-built nests, has been completed.

It is a gripping experience to watch these collections tak-

Right: the entrance to the caves is quiet during daylight hours but becomes a busy throughfare for swarms of swiftlets and bats at dusk.

ing place in the lofty heights of the cave. Small groups of men reach the cave tops using rickety rattan ladders, ropes and poles that look as though they might snap or snag at any moment. The nests are gathered in minute rattan baskets by the assigned collector. The nests may be gathered anywhere between 30–90 m above ground, in semi-darkness, at great personal risk. Watching it is heart-in-the-mouth stuff. The rewards are high for the owners, however, as 1 kg of Edible-nest Swiftlet nests – about 100 nests – can be worth US$1,000 or more.

If you visit these caves, make sure that you remain until the evening. This is when there is a vast mêlée around the caves, as the whole community swaps shifts. The bats stream out of the caves for the night's feeding in seemingly never-ending waves, and a little later the swiftlets return from their foraging in the twilight skies above the forest. This rush-hour makes for an impressive spectacle, especially in the surroundings of the remarkable limestone Gomantong Mountain.

Not surprisingly, the hordes of small flying bodies attract predators. These include some quite unlikely hunters, such as the Rufous-bellied Eagle, which makes a clumsy grab as it flies awkwardly through the swarm, as well as the more professional bird-catchers such as the Peregrine Falcon, which is swift and decisive. However, the most impressive and expert hunter of these masses is the Bat Hawk, with its long, pointed, falcon-like wings. This highly specialized species, whose main daily foraging period lies in the half-hour around dusk, flies with leisurely, powerful wing-beats as it approaches the flocks, but then, having selected its prey (bat or swiftlet), suddenly accelerates and, if necessary, twists and turns so that it can grasp its quarry from behind and above. Prey is quickly transferred to the bill and swallowed whole, so that the Bat Hawk can carry on feeding while flying; it may be only a minute before it kills again. One Bat Hawk was seen to catch 17 bats in a single session, a pretty useful evening's work. At any one time there may be three or four individuals working the caves and sharing the airspace.

It should be mentioned that, exciting though the caves are, the surrounding forests are by no means lacking in wildlife, and the whole area deserves a few days' exploration. The Gomantong Forest Reserve is a good place for Orang-utans, and Proboscis Monkeys can be seen along the nearby Kinabatangan River.

Right: looking down on the whitish nests of the Edible-nest Swiftlet, which are made entirely out of the birds' saliva.

Limithang Road

HABITAT High mountain pass at 3,700 m, forest at all levels down to 650 m

KEY SPECIES Satyr Tragopan, Blood Pheasant, Snow Pigeon, Ward's Trogon, Fire-tailed Myzornis, babblers including Bar-winged Wren-Babbler and Coral-billed Scimitar Babbler, parrotbills

TIME OF YEAR Best in spring (April to June)

Experienced tour guides have been coming out of recently-open Bhutan muttering about "pristine forests stretching to the horizon" and "birds dripping from the trees". They recount the eerie tameness of the birds here, with shy creatures like pheasants keeping long in sight when, everywhere else in the world, they would disappear in an instant; and they marvel at the peaceful experience of visiting this Himalayan kingdom. Many have declared that Bhutan is simply the best birding and travelling experience of their careers.

So what is it about Bhutan that has made so many gnarled veterans so very enthusiastic? Well, there are plenty of bird species – nearly 700 have been recorded – but then, many other places in the world have similar diversity. No, it is Bhutan's almost uniquely pristine state and gentle atmosphere that make a visit here so inspiring. This is no ordinary country. It has kept traditional ideas of development and prosperity at arm's length, and instead has remained faithful to the tenets of Tibetan Buddhism, whereby hunting is forbidden and vast areas of wilderness are kept just as they are. Up until the 1960s Bhutan kept itself almost completely isolated, until the Chinese invasion of neighbouring Tibet woke it up to the realization that it would have to engage with the outside world. Nonetheless, television was introduced only in 1999, and tourism has only recently been encouraged – even now, you have to visit in a tour party, there is a high daily tariff, and numbers are controlled. There is only one major road in the whole country, and that was only begun in 1962. The government, moreover, remains resistant to outside influences; in 1987 the king declared, in response to criticism of Bhutan's slow economic growth, that "Gross National Happiness is more

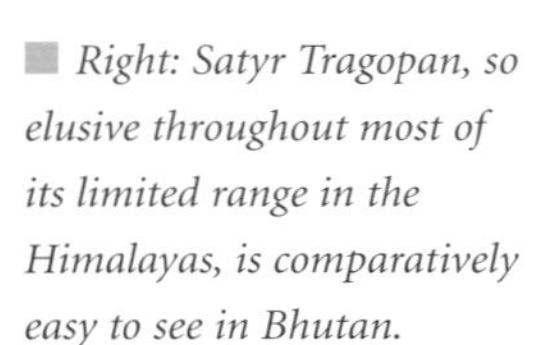

Right: Satyr Tragopan, so elusive throughout most of its limited range in the Himalayas, is comparatively easy to see in Bhutan.

Right: there are hardly any roads in Bhutan, and only one major one. This waterfall near the village of Namling is known as the 'death drop'.

important than Gross National Product", and Bhutan still officially produces happiness indices. This country has recognized that its own land is a priceless treasure, especially in its preserved state.

In this outstanding landscape, the mountains reach the dizzy heights of 7,700 m (the loftiest mountain, Gangkhar Puensum, is the highest unclimbed mountain in the world and the 20th highest in all) near the northern border, yet the land plummets down to less than 100 m above sea level on the Indian border. In between, the country is dominated by high passes, steep slopes and valleys, rushing rivers and tall forests. Owing to the high

■ *Above: a search of the rhododendron understorey in forest above 2,800 m may reveal the exquisite Fire-tailed Myzornis.*

rainfall, resulting from the monsoons that come up from the Bay of Bengal, the tree line is at an exceptionally high elevation – 4,000 m, one of the highest in the world – and this enables a high density of species even at this extreme altitude.

On the whole, forests in the Himalayas become more diverse as you go from west to east, and thus Bhutan's top birding site is indeed located in the eastern half of the country. The Limithang Road, only 'discovered' by birders in the early 1990s and now one of the most celebrated birding sites in Asia, begins at Thrumsing La, a pass at 3,720 m elevation, east of the small town of Jakar. Working its way east the road plunges down to 650 m in the space of less than 100 km, allowing birders to get a vivid picture of how the birdlife changes as altitude drops. Throughout its length, this road passes through rich, pristine forest.

So what of the bird highlights on this breathtaking route? Well, near the pass itself the cool forests, mainly made up of hemlock with an understorey of colourful rhododendrons, host some high-altitude species. Normally a bird like the superb Blood Pheasant would make the headlines, the male mainly a delicate blue-grey streaked with white, but with blood-red crimson leaking from the breast and undertail coverts, but in this enchanted place it is trumped completely by the remarkable Satyr Tragopan. The plumage of the male is simply extraordinary: it is mainly brilliant crimson red, dotted all over with tiny dark-rimmed white spots, with some blue skin on the face and chin, and is quite a large bird, as large as a Common Pheasant. Tragopans are famously hard to find anywhere, but the Satyr Tragopan is actually common here, and more likely to be seen than not. Incidentally the Blyth's Tragopan, never yet recorded by a commercial birding tour company, also occurs in eastern Bhutan and might turn up at this site.

Another very special bird found at this altitude, often feeding from the flowers of the rhododendrons, is the exquisite and much sought-after Fire-tailed Myzornis, like a cross between a babbler and a sunbird. Other birds found here at the top include the decidedly neat Snow Pigeon (a whited-out version of a Rock Dove) and both Maroon-backed and Rufous-breasted Accentors – classic high-altitude birds of the forest edge.

Following the steep slopes further down, the vegetation changes to warm broad-leaved forest, and the diversity of birds increases dramatically. This is reflected best among the babblers, one of Asia's most characteristic bird families. Simply dozens of species occur at these levels, including rare or choice species such as the marvellous orange-and-grey Golden-breasted Fulvetta (which looks rather like a tit), the tiny green Yellow-throated Fulvetta, the incredibly smart Cutia (with a clean white breast boldly barred black) which behaves in nuthatch-like fashion, the sweet-tongued Red-billed Leiothrix and the fetching Rusty-fronted Barwing. There is also a superb range of colourful laughing-thrushes, all boldly patterned and distinctive on the one hand, but skulking on the other: species include the White-crested, Rufous-necked, Striated, Scaly, Grey-sided and the elusive Blue-winged – all as different from one another as any set of North American warblers or Australian honeyeaters. Birders will also be seeking out the various head-bobbing wren-babblers, including the very rare Bar-winged Wren-Babbler, and several scimitar babblers – especially the very local Coral-billed Schimitar Babbler– with their unusual curved bills.

The higher forests hold other gems besides babblers. Prime among these is the splendid plum-and-raspberry-coloured Ward's Trogon, a species endemic to this part of the Himalayas. The fact that this was first recorded on a commercial birding tour in 1994 simply shows how recently the treasures of this forest have been discovered.

Further down, through the endless switchbacks, there is a subtle change to subtropical forest, where another set of birds awaits you. For example, these lower altitudes can be good for the splendid Rufous-necked Hornbill, in which the male is actually entirely rusty-coloured below and black above; this species is very rare and declining everywhere, it seems, except in Bhutan. In this warmer section, where the trees hang with orchids, several more typically lowland families appear, including drongos such as the Lesser Racket-tailed, and a range of flycatchers such as Little Pied, Pygmy and Ultramarine, but these are only a small fraction of the birds on offer. Yet another set of babblers occurs here, including a number of parrotbills in the stands of bamboo (Greater Rufous-headed and Black-throated Parrotbills, for example), plus various thrushes, woodpeckers, cuckoos, tits, owls, sunbirds and tree-haunting nuthatches and treecreepers. By then the diversity is such that, even in such a peaceful place, with the sweet smell of wildflowers and the luxuriant green of the forest, birders can begin to show signs of overstimulation. It's the only way to get stressed in Bhutan.

Bogani Nani Wartabone

HABITAT Lowland rainforest, rivers, ponds

KEY SPECIES Maleo, Sulawesi Masked Owl, Purple-winged Roller, Purple-bearded Bee-eater, kingfishers including Great-billed and Sulawesi Dwarf Kingfishers, Knobbed and Sulawesi Hornbills, Ivory-backed Woodswallow

TIME OF YEAR Good all year round

Right: endemic to Sulawesi, the Purple-bearded Bee-eater is a bird of forest clearings.

The island of Sulawesi is the sort of place a birder might care to dream up, and then dismiss with a sigh as too improbable to exist. Imagine a place with 70 birds that are found nowhere else in the world, and then paint them all in exotic colours and give them outrageous and outlandish forms – even extraordinary vocabularies. You might as well add in a few of the most bizarre mammals on earth, plus lush lowland and highland rainforests rich in palm trees and orchids to make the fantasy complete. Then you wake up – this place really exists.

Sulawesi (formerly known as Celebes) is a peculiar, almost starfish-shaped island sandwiched between Borneo to the west and New Guinea to the east. It is thought that, about 40 million years ago, a tectonic plate from the north of Australia floated away and crashed into an Asian plate to form eastern Sulawesi. Then, 25 million years later, the fused plate drifted north and hit another Asian plate, adding the western side. Thus, you might say that the island of Sulawesi results from the colliding of two realms, one Asian and the other Australasian, and this is indeed reflected in the fauna and flora of the island. To the birder the pleasing mix means that, for example, flowerpeckers, a family which is primarily from Asia, feed in the same flocks as honeyeaters, a family which is primarily from Australasia. The intermediate position of Sulawesi is further exemplified by the fact that Wallace's Line, and its deep oceanic trench, runs just to the west of the island, officially separating the Oriental and Australasian sides.

Much as the mix of species in Sulawesi is intriguing, however, the real ornithological highlights actually result not so much from the island's origin as from its isolation. Before fusing, each of Sulawesi's constituent plates would have spent a considerable amount of time well separated from other landmasses, with the result that natural selection ran riot and threw up plenty of original forms. It is these exciting endemic one-offs that really make Sulawesi special, and the sheer number of them is remarkable.

If you are going to look for birds in Sulawesi, a good place to start is the largest national park on the island, the 2,871 sq km Bogani Nani Wartabone National Park (formerly Dumoga Bone National Park) in the far north-east of the starfish. It is a tropical forest site with a huge bird-list of 195 species, including 30 endemics, and covers the

Right: one of a quartet of stunning endemic kingfishers at Bogani Nani Wartabone, the rare Lilac-cheeked Kingfisher preys on large forest insects such as mantids.

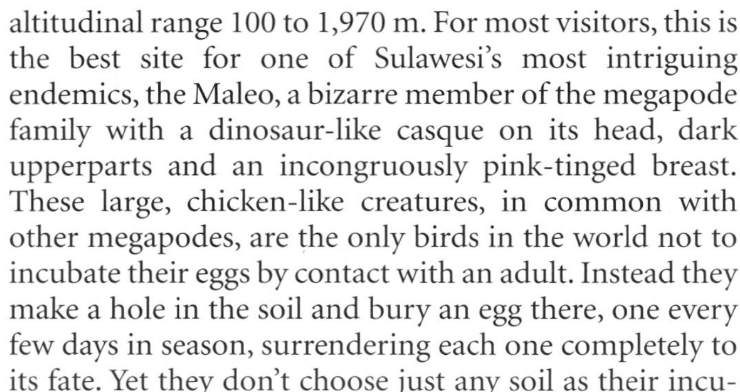

altitudinal range 100 to 1,970 m. For most visitors, this is the best site for one of Sulawesi's most intriguing endemics, the Maleo, a bizarre member of the megapode family with a dinosaur-like casque on its head, dark upperparts and an incongruously pink-tinged breast. These large, chicken-like creatures, in common with other megapodes, are the only birds in the world not to incubate their eggs by contact with an adult. Instead they make a hole in the soil and bury an egg there, one every few days in season, surrendering each one completely to its fate. Yet they don't choose just any soil as their incubator – it must be heated by geothermal activity to a temperature between 32 and 39 degrees, or alternatively warmed consistently by the sun. In the latter case, birds selected black soil for their excavations, which absorbs heat more effectively. In either case, after about 80 days the well-formed chick miraculously digs its way out of the soil.

Below: an external heat source – either geothermal or solar radiation – is used to incubate the Maleo's eggs.

Not surprisingly, suitable sites are at a premium, and Maleos, rather like turtles, come from far and wide (from up to 20 km away) to lay their eggs in the same shared patch of warm earth. This makes the eggs exceedingly vulnerable to predators, including pigs, monitor lizards, crocodiles and, worst of all, people. All seven Maleo sites in the national park are illegally farmed by the local community.

Not all the birds at Bogani are quite as weird in their behaviour as the Maleo, but the rather common Blue-backed Parrot (not endemic) is admittedly highly unusual among its family for being semi-nocturnal (it sometimes raids crops at night). However, the bird population tends to be notably striking, if not stunning. Take another set of species high on everyone's wish-list – the quartet of endemic kingfishers, boldly and unusually patterned. The tiny Sulawesi Dwarf Kingfisher is pleasingly pinkish, and hunts small invertebrates in undergrowth close to the forest floor; the Green-backed Kingfisher is tastefully clad in grass-green on the back and tail, rufous on the breast and brilliant iridescent blue on the head – it uses its large red bill to catch fearsome centipedes and beetles; the bold Great-billed Kingfisher uses its curiously sooty-black bill to catch crabs and crayfish in streams; and the incomparable Lilac Kingfisher, a hunter of insect megafauna such as mantids and cicadas, has the best colour mixture of all. It has a red bill, cobalt blue wings, neat brown cap and tail and

■ *Above: the large Ashy Woodpecker is another endemic and is thought to subsist mainly on termites.*

a flashy blue eye-stripe that is bordered above and below by the most remarkable, delicate, streaky lilac colour.

Sulawesi is, indeed, particularly well blessed with kingfishers and their allies, the Coraciiformes. Two other treasured species at Bogani are found mainly in clearings: the Purple-winged Roller (Sulawesi and neighbouring Helmahera are the only islands in the world with endemic rollers) and the fabulous Purple-bearded Bee-eater, a predominantly dark green, long-tailed species with deep purplish-blue underparts, the feathers of which are long, giving the bird its 'bearded' appearance. Meanwhile, in the forests, there are two endemic hornbills, the small, yellow-cheeked Sulawesi Hornbill and the very large Knobbed Hornbill. Nearby Tangkoko-Dua Suadara National Park has the highest breeding density of hornbills on the island.

Honourable mention should also go to some other exceptional species: the Yellow-breasted Racket-tail, a perky small parrot with a lot to say; the majestic Purple Needletail, the only really colourful swift in the world; the huge endemic Ashy Woodpecker, not far behind the Great Slaty in sheer size; the brilliant Red-bellied Pitta, commoner here than anywhere else; and the supremely smart Ivory-backed Woodswallow, a species of unimpeachable Australasian pedigree. The forests also have a superb selection of neatly patterned pigeons and fruit-doves, an amazing collection of starlings (including the bizarre Grosbeak Starling, which lives in colonies in dead trees and excavates its own nest-holes with its unusually swollen bill), and a hat-trick of owls, including the highly sought-after Sulawesi Masked Owl, which is almost impossible to see anywhere else.

This treasure-trove of birds, together with its collection of unusual mammals (including a mini-buffalo, the Anoa, and the Babirusa, a hairless pig with backward-pointing tusks) is under threat. Unbelievably Bogani Nani Wartabone, one of the finest and most important national parks in the whole of Indonesia, is little more than paper-protected. Illegal rattan production, logging, hunting and egg harvesting are all rampant, and even the park's boundaries are under dispute. Sulawesi has one of the fastest deforestation rates in the world, and it is quite possible that places like Bogani will not only suffer terrible depletion, but could even be wiped off the map, along with their dependent fauna. That, indeed, would be a sad end to any birder's dream.

Katanglad Mountains

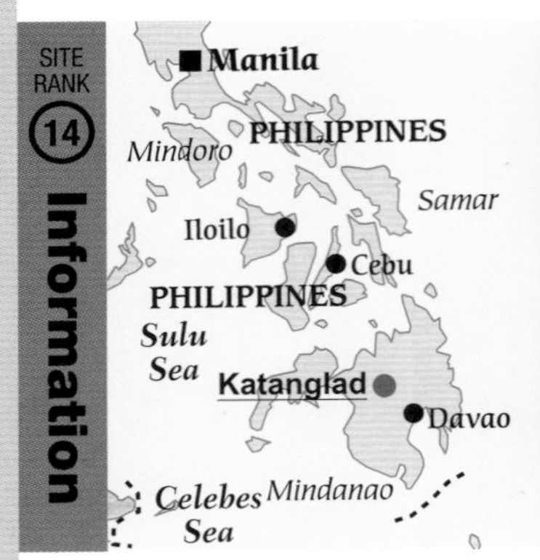

HABITAT Montane forest, forest edge

KEY SPECIES Philippine Eagle, Bukidnon Woodcock, Giant Scops Owl, Mountain Racket-tail, Apo Sunbird, Apo Myna

TIME OF YEAR All year round, although this site would be hard work in the rainy season, July to December

Of all the sites included in this book, this could be the one that you should hurry to first. Protected as a national park only on paper and threatened on all sides, it holds some of the last remnants of primary forest on the southern Philippine island of Mindanao, the second-largest island of the archipelago. Given the Philippines' appalling record of undervaluing and abusing their natural heritage, it is possible that, as a birding site of substance, this one could disappear pretty soon. The slash-and-burners are edging further uphill with every passing year, and the appearance of new settlements suggests that hunters and trappers will soon penetrate the most pristine parts of the range. It is hard to be optimistic about the birds of the Katanglad Mountains, or those of the Philippines in general.

The great sadness is that the Philippines hold an absolute treasure trove of marvellous birds. On the archipelago as a whole there are about 200 endemic species, many of them with restricted ranges. As this book is being written, 66 species are considered threatened, with 12 in critical danger of extinction and 12 ranked as 'endangered', the two latter figures being among the highest for any country in the world. Three of the 'critical' species are bleeding-heart pigeons, which just about says it all.

It is not easy to get to Mount Katanglad. First you need to get to the village of Dalwangan in the central highlands, and then you must pay your respects to the tribal elder, usually by the sacrifice of a chicken (yes, really!), before being allowed up the track. Then your luggage is loaded on to horses and you begin the long, arduous and often very wet hike up the mountain, depressed at the sight of the grassy, deforested countryside. You finally reach a 'lodge', with very basic facilities, in a grassland plateau at 1,368 m, from where each morning you leave for the two-hour hike to the best parts of the remaining forest.

Right: cultivation is progressively eating away at the priceless forests of Katanglad.

Opposite: the huge bill of the Philippine Eagle enables it to tackle medium-sized mammals, including the occasional monkey.

Above: Bukidnon Woodcock photographed by Simon Harrap, one of the tour leaders who first recognized it as a species new to science during the 1990s.

Above right: the gorgeous Steere's Pitta may be glimpsed as it forages in the forest litter.

The Katanglad area is considered to be one of the richest in the whole of the Philippines, with at least 30 of the endemics present in the two types of upland forest, tall Dipterocarp-dominated forest on the middle slopes and mossy forests at higher elevations above about 1,600 m. Birding here is typically stop-start, with most of what you are seeking appearing as part of mixed-species flocks. These gatherings are often dominated by Black-and-cinnamon Fantails, Mountain White-eyes and Cinnamon Ibons, the three most characteristic species, but many other species join in, too, such as Elegant Tit (like a very yellow Coal Tit with blobs of white paint on the back), Philippine Bulbul, McGregor's Cuckooshrike, Sulphur-billed Nuthatch and Stripe-breasted Rhabdornis, the last-named a member of the only family endemic to the Philippines. These birds are possibly most closely related to the babblers, and they hop and jump about in the canopy, gleaning or fly-catching, snatching anything from insects to tiny tree-frogs. Highly sociable, they move through the forest in small groups, but will gather into larger parties to roost. Other notable birds of the lower forests include the splendid, rare Blue-capped Woodland Kingfisher, the Philippine Hanging Parrot, the Montane Racket-tail and the Bagobo Babbler, a bird that gives a whole new meaning to the word 'skulking'.

Further up, the trees become smaller and more laden with moss, and the character of the avifauna also changes. Once you have reached these dizzy heights some very special range-restricted birds appear, such as the White-cheeked Bullfinch and two birds named after nearby Mount Apo, the Apo Sunbird and the Apo Myna. The latter, in particular, is a real treat to see, with its yellow patch of skin surrounding the eye, mainly blackish bronze plumage and ridiculous wispy crest with feathers that it seems to have borrowed from another bird.

On the lower slopes of Mount Katanglad in 1993, something happened to a bird tour leader that must only have featured in his wildest dreams: he discovered a bird species new to science while leading his group. Actually, the bird had been collected before, but wrongly labelled at the time (the 1960s) as a Eurasian Woodcock. This bird, however, made a sound quite different from the widespread species, not a low croak but more of a rattle, described later like the sound of "an apparition jangling ethereal keys as it zooms down a ghostly passage". It was duly described as new to science in 2001, under the name Bukidnon Woodcock, and then found to be quite widespread in the Philippines above 1,000 m. However this incident certainly shows how scantily the fast-disappearing Philippine avifauna is known.

One that is most certainly well known is the world-famous Philippine Eagle, the islands' national bird. Mount Katanglad is currently the most accessible site to see it, which is saying something, and the world population may not exceed 300 birds, spread over four islands. The great birds are quite enormous, the second-largest eagle in the world, and their old name of Monkey-eating Eagle used to convey a frisson of danger which their enormous bill and blazing yellow eyes magnify with a vengeance. In fact these massive birds prey primarily on Colugos (which are also known as 'flying lemurs' but are no relation to the lemurs of Madagascar) and Palm Civets, with monkeys taken less often. However, a pair has been observed catching monkeys by co-operatively hunting, one individual distracting, the other grabbing.

It is anyone's guess as to how long the eagles can hang on here. They are slow breeders, laying one egg every two years, a reproduction rate that evolved before loggers entered the Philippine scene. There is already evidence that they are spending more time higher up the mountain than previously, and finding food scarcer. There are only a handful of pairs here, and it would take very little for them to disappear. Sadly, if this famous and iconic bird cannot be preserved, what hope have the Philippine birds that don't make headlines?

Sinharaja Forest

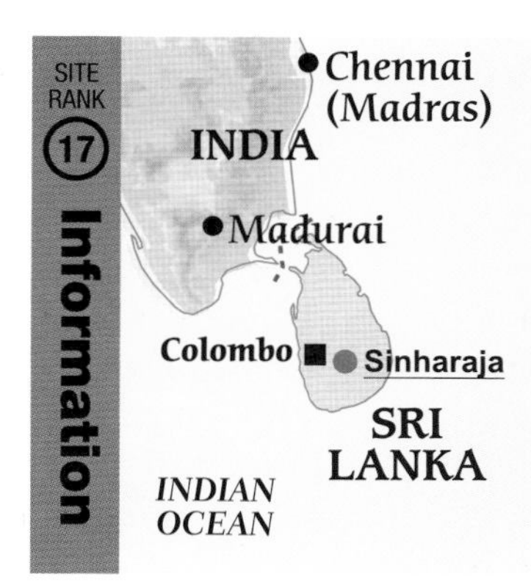

HABITAT Lowland rainforest, mainly secondary growth

KEY SPECIES White-faced Starling, Red-faced Malkoha, Sri Lanka Blue Magpie, Sri Lankan Spurfowl, Sri Lanka Junglefowl, Green-billed Coucal, Serendib Scops Owl, Spot-winged Thrush, Sri Lanka Crested Drongo

TIME OF YEAR Good all year round

Right: following up a series of unfamiliar calls led to the discovery of the Serendib Scops Owl in 2001.

Protecting some 50 per cent of the country's remaining wet zone lowland rainforest, the 190 sq km of Sinharaja Forest Reserve is simply the best place in Sri Lanka for birding. Almost every endemic on the island is found here, along with some restricted species shared with southern India (for example, Malabar Trogon and Sri Lanka Frogmouth) and several more widespread species that could be considered specialities, because they are especially easy to find (for example, Indian Pitta and Velvet-fronted Nuthatch).

Not only is this a great place for ticking off the various Sri Lankan endemics, it also has an exceptional quirk. As in many other tropical forests in the world, a feature of the

Right: Sinharaja protects 50 per cent of Sri Lanka's remaining lowland rainforest.

bird biology is that many inhabitants go around in mixed-species foraging flocks. It so happens, though, that at Sinharaja the flocks are the most variable in the world; the average of 42 species per flock is unparalleled, and some flocks contain as many as 48 (more than 10 per cent of all the birds ever recorded on the island!). Thus, the key to enjoying this site is to run into one of these bird gatherings. It can be very quiet here until you do; and then, once the birds are passing by, at all levels of the forest and often frustratingly high up, it will demand all your best birding skills to pick out all the various species.

The flocks form in the morning, usually at about 8 am, when the members are summoned by the whistles and *chack* calls of Sri Lanka Crested Drongos (recently split from the Greater Racket-tailed Drongo). These medium-sized glossy-black passerines with long outer tail feathers are soon accompanied by groups of Orange-billed Babblers, an entirely chestnut-brown, long-tailed babbler with a yellow eye and eponymous orange bill, and together these species form the key components of every Sinharaja flock. Within minutes, dozens of birds are attracted and, thus assembled, the flock moves off, apparently in no predictable direction, with each component species feeding in its own particular way and at its preferred altitude as the birds travel through the forest. Thus, low down in the bushes the Orange-billed Babblers are joined in their foraging by small numbers of the smart, pale-eyed Dark-fronted Babbler, while in the canopy groups of flame-coloured Scarlet Minivets play follow-my-leader from tree to tree, and Common Ioras and Golden-fronted Leafbirds keep maddeningly hidden in the leaves of the canopy. Meanwhile, woodpeckers such as the Lesser Goldenback join Velvet-fronted Nuthatches examining the surfaces of trunks and branches, while groups of the rare Red-faced Malkoha, each with bright yellow bill, red skin on the face, iridescent green upperparts and white belly, hop from branch to branch high in the trees searching for fruit and berries. A bit of diligent searching should also reveal one of the rarest of Sinharaja's birds, the White-faced Starling. This is not the most spectacular of the forest's birds, with its dark-grey back and streaky white underparts, but with a population of only 10,000 at the very most, and declining, it is classified as Vulnerable by BirdLife International.

Among the many other species at Sinharaja are several that don't join flocks but feed on the forest floor. Notable among these are two gamebirds; the splendidly plumaged, chicken-like Sri Lanka Junglefowl, which is easy to see, and often wanders around the park headquarters as if it were on duty collecting tickets, and the utterly slippery and tricky Sri Lanka Spurfowl, which is perhaps the easiest of the endemics to miss. Sharing the litter are two endemic thrushes, the Sri Lanka Scaly Thrush, which utters the occasional short whistling song, mainly at dawn, and the decidedly smart Spot-winged Thrush, which rather resembles the Eurasian Song Thrush or the North American Wood Thrush and has the same fabulous, melodic yet slightly melancholy song style that is also characteristic of both those species.

Other highlights at Sinharaja include what for many would be Sri Lanka's most attractive bird, the Sri Lanka Blue Magpie. This is a stunner, with a body of dazzling tropical sea-blue offset by neat warm brown head and wings; the tail is long and graduated, with a white trim. It is fairly easy to see at Sinharaja, especially by the guest-house on the edge of the reserve. Another is the rare Green-billed Coucal, which indeed does have a startling yellow-green bill in contrast to its black-and-chestnut body; this is a predatory species that feeds at all levels from the ground to the canopy. Other uncommon species recorded include the Sri Lanka Grey Hornbill and the smart Sri Lanka Wood Pigeon, a dark purplish pigeon with a dazzling black-and-white striped patch on the nape. This is traditionally considered to be a species of the highlands, but in June and July it makes seasonal movements down to Sinharaja (the forest reserve spans an altitudinal range of 300–800 m).

In 2001, a local professional bird guide, Deepal Warakagoda, managed to set eyes on a small owl that he had heard during the previous six years making puzzlingly unfamiliar calls in the wet forest zone of both Sinharaja and Kitulgala reserves. He realized immediately that he had found something special and, sure enough, the Serendib Scops Owl, a small, very rufous species that feeds in the undergrowth, was described to science for the very first time a year later. It was the first new species discovered in Sri Lanka since 1868. Highly endangered, this small insectivorous species has a range of only 230 sq km, and a population of a mere 45 birds was estimated in 2004. It is now very much a fixture on the birding ecotourists' wish-list.

This discovery, following hot on the heels of the scientific description of two new species of lizard in 2000, illustrates just how sketchy is the knowledge of the wildlife of Sri Lanka. Indeed, in recent years the list of Sri Lankan endemic birds has risen from 26 to 33 as a wide variety of forms have been assigned full species status as a result of new research. Visitors to Sinharaja, therefore, are not only treated to great bird flocks and a range of superb endemics, but also find themselves walking on the privileged ground of a scientific frontier.

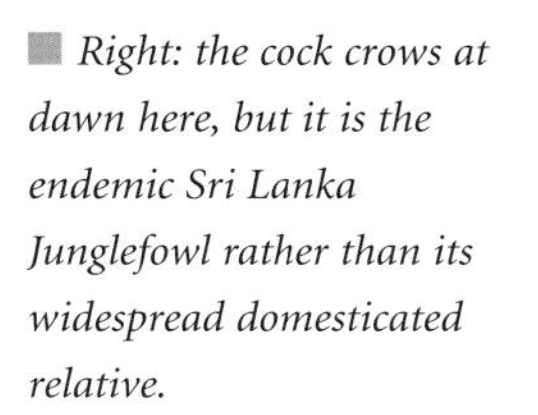

Right: the cock crows at dawn here, but it is the endemic Sri Lanka Junglefowl rather than its widespread domesticated relative.

Opposite: as a hot-spot for Sri Lankan endemics, Sinharaja is perhaps the best place on the island to catch up with the remarkable Sri Lanka Blue Magpie.

Mai Po

HABITAT Mudflats, fish and shrimp ponds, mangroves, reedbed

KEY SPECIES Waders, including regular rarities such as Spoon-billed Sandpiper, Nordmann's Greenshank and Asian Dowitcher; Black-faced Spoonbill, Saunders's Gull, Styan's Grasshopper Warbler

TIME OF YEAR April and May are peak months for waders; autumn and winter also excellent

Above right: one of Mai Po's shrimp ponds – a tranquil corner of the bustling city of Hong Kong.

At first sight it would seem that there is little room for wildlife in the sprawling conurbation of Hong Kong. In the north-west corner, however, close to the old border with China, is the world-famous nature reserve of Mai Po. Set beside a large tidal inlet known as Deep Bay, Mai Po provides a mixture of habitats, including mudflats, mangroves (which were left in place to keep illegal human immigrants out of the mainland), reedbeds and a network of old fishponds and *gei weis* (shrimp ponds). Surrounded by development and under constant pressure from Hong Kong's growing population, its high importance

Right: Mai Po is one of only three main wintering areas in the world for the endangered Black-faced Spoonbill.

for wintering and migratory birds was long recognized, but slow in being formally safeguarded. In 1985 the area was finally designated as a nature reserve and in 1995 as a Wetland of International Importance. Its 15 sq km are now run by the Worldwide Fund for Nature Hong Kong.

Mai Po's main claim to fame is its visiting waders, which occur mainly on the mudflats in Deep Bay, but also use the scrapes within the pond area at high tide. In spring and autumn a combined total of about 20,000 waders of up to 30 species pass through, mainly for short periods of resting and refuelling. The spring is best for numbers and variety, and at that season, given the right tides, one of Asia's greatest concentrations of shorebirds can be seen from hides around the site, often at very close quarters. Up until recently the best hides were along the boardwalk, but this area has now silted up and holds far fewer birds.

The commonest waders over the course of the year at Mai Po are Pied Avocet, Curlew Sandpiper, Common Redshank, Marsh Sandpiper, Red-necked Stint, Common Greenshank and Spotted Redshank, which together number more than 2,000 individuals. However, just about every migratory wader species in the Old World turns up during the season, some of which are highly sought after by birders. These include some great rarities: in spring 2006, for example, seasonal totals included 23 Nordmann's Greenshanks and 25 Asian Dowitchers. Both are threatened on a global scale, and can be difficult to see elsewhere. One of the pleasures of Mai Po is the process of sorting through the masses of waders to pick out the less common species.

The bird that has probably made Mai Po's reputation more than any other over the years is the Spoon-billed Sandpiper, the very mention of which will make a hardened shorebird enthusiast go weak at the knees. This peculiar sprite of a stint has an oddly swollen spatulate bill, and when feeding it sweeps it from side to side in spoonbill fashion, often walking around with the bill seemingly stuck to the surface of the mud. This bird breeds only in the far east of Siberia, and the world population probably numbers fewer than 2,000 individuals. Spring 2006 brought just one individual to Mai Po, but there were 16 in 2005, and most years bring a handful. Any observation of one of these birds inevitably brings an electric atmosphere to a hide full of ecstatic birders.

Although Mai Po is best known for its passage waders, this is by no means its only draw. About 60,000 waterbirds of various kinds use the site in the winter, includ-

Below right: the very rare Saunders's Gull breeds in north-east China, north-east Russia and Korea and is a regular passage and winter visitor to Mai Po in small numbers.

■ Right: a Nordmann's Greenshank (on the right) mingles with its commoner *Tringa relatives.*

ing various cormorants, herons, waders, gulls and wildfowl. Many of these are drawn to the on-site ponds, of which the *gei weis* are of particular interest, since they are also used for commercial shrimp and fish farming and are a striking example of sustainable agribusiness. In the spring and summer the *gei weis* house shrimps, but from November onwards they are drained one by one to allow an annual harvest of fish. After this, the ponds are opened to be replenished with shrimps and nutrients direct from Deep Bay, thus completing the cycle. During the draining phase more than 1,000 herons, egrets, cormorants and other fish-eating birds may congregate to feed on the non-commercial fish and shrimps left behind in the diminishing puddles. The nature reserve authorities typically set up bird-blinds opposite the particular *gei wei* at the draining stage, to allow the public to see the hordes of birds.

■ Right: no longer a guaranteed species at Mai Po, the Spoon-billed Sandpiper has become rarer in recent years and appears to be in severe danger of extinction, although a new wintering site was discovered in Korea in 2008.

Among the many species of waterbird present in winter are some great rarities. The most important of these is the endangered Black-faced Spoonbill; there were 350 at Mai Po in January 2006, constituting nearly a quarter of the world population of this diminutive member of its genus. This species, which breeds in Korea, winters regularly in only two other main sites worldwide. It feeds on the abundant crabs and shrimps at Mai Po, and often roosts by the fishponds – sometimes, it seems, hardly stirring all day long.

Another great rarity is the Saunders's Gull, which is usually found on the mudflats in Deep Bay. This small, elegant gull has a habit of making a low, sweeping flight down to a morsel of food, followed by the briefest of landings during which the item is snatched. Breeding only in east Asia, the world population of this gull is fewer than 10,000 individuals. In 2006 there were 50 at Mai Po, a slight increase after a few lean years.

In recent years mist-netting has revealed another great rarity, Styan's Grasshopper Warbler. This poorly known species breeds only on islands off Russia, Korea and Japan, and Mai Po is one of only two confirmed wintering sites. This recent discovery only serves to confirm how important this marvellous reserve is, especially for the birds of the highly pressurized east coast of Asia.

Liaoning

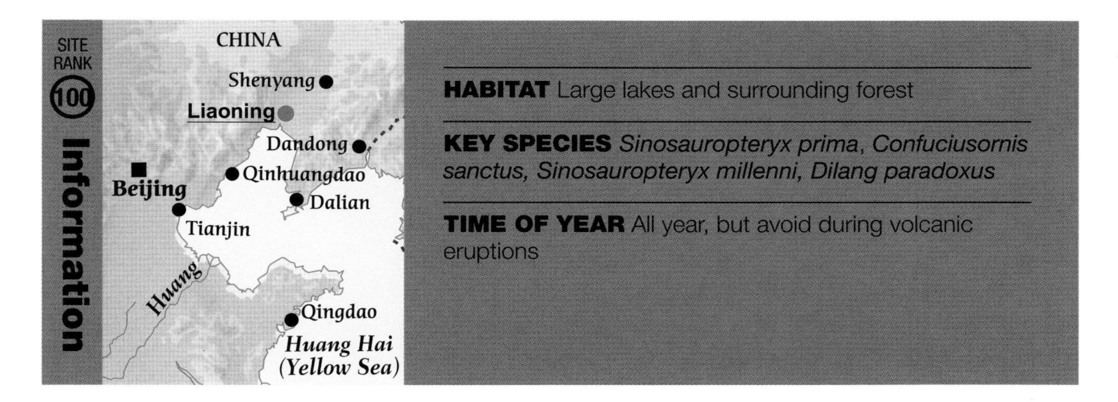

HABITAT Large lakes and surrounding forest

KEY SPECIES *Sinosauropteryx prima, Confuciusornis sanctus, Sinosauropteryx millenni, Dilang paradoxus*

TIME OF YEAR All year, but avoid during volcanic eruptions

There are not many bird species at Liaoning. Indeed, this province in north-east China is not good birding country at all. A day's drive from Beijing, it consists of low hills covered by farmland, together with cities and factories. The scenery is nothing special, and the environment is nothing special.

So why is it in this book? The answer lies in its rocks, and specifically in its shale deposits from 130 to 110 million years ago. Since the 1990s Liaoning has yielded some of the most important fossils ever recovered in terms of explaining the evolution of birds. The area has been described as a Mesozoic Pompeii, referring to the site in Italy where humans were famously entombed by a volcanic eruption and have been remarkably well preserved. In the few short years of its fame, with find after find being unearthed, Liaoning has cast more light on the early evolution of birds than anywhere else. Clearly, although it is a relative desert now, Liaoning was once one of the richest places on earth for birds.

The marvel at Liaoning is in the detail preserved in its fossils. There was once a shallow lake here where recently dead bodies would quickly be covered by mud and volcanic ash, with layers added from the repeated eruptions, shutting off the oxygen required for decomposition. The fine grains of sediment have allowed fossils to retain details that are missing from most other fossils elsewhere in the world, such as the soft parts of their internal organs. Some creatures have remains of their last meal still intact in their gut, while others give hints of colour patterns on the skin. And, most importantly for palaeontologists, a significant number of the fossils here have feathers.

In 1996 one discovery in particular caused a sensation. Named *Sinosauropteryx prima*, it was the fossil of a small bipedal dinosaur in the group known as therapods, among which is every child's favourite, *Tyrannosaurus rex*. What was special about this find was that it had feathers, or at least primitive structures that resembled feathers. At any rate, its body was covered with thin, hollow filaments, yet it had no wings and was clearly a reptile, not a bird.

It was not long before another discovery added evidence to what *Sinosauropteryx* had hinted at. The so-called 'Fuzzy Raptor' *Sinornithosaurus millenii*, discovered in 1999, was also clearly a reptile, but its feathers were far better formed. Some of its filaments were joined together into tufts, rather like modern-day bird down, and others were joined to a central shaft, or rachis, the same structure that most feathers have today. This, in the opinion of most palaeontologists, provided conclusive evidence that

■ *Below: a fossil of the earliest known bird with a toothless bill,* Confuciusornis sanctus, *which is thought to date from about 130 million years ago.*

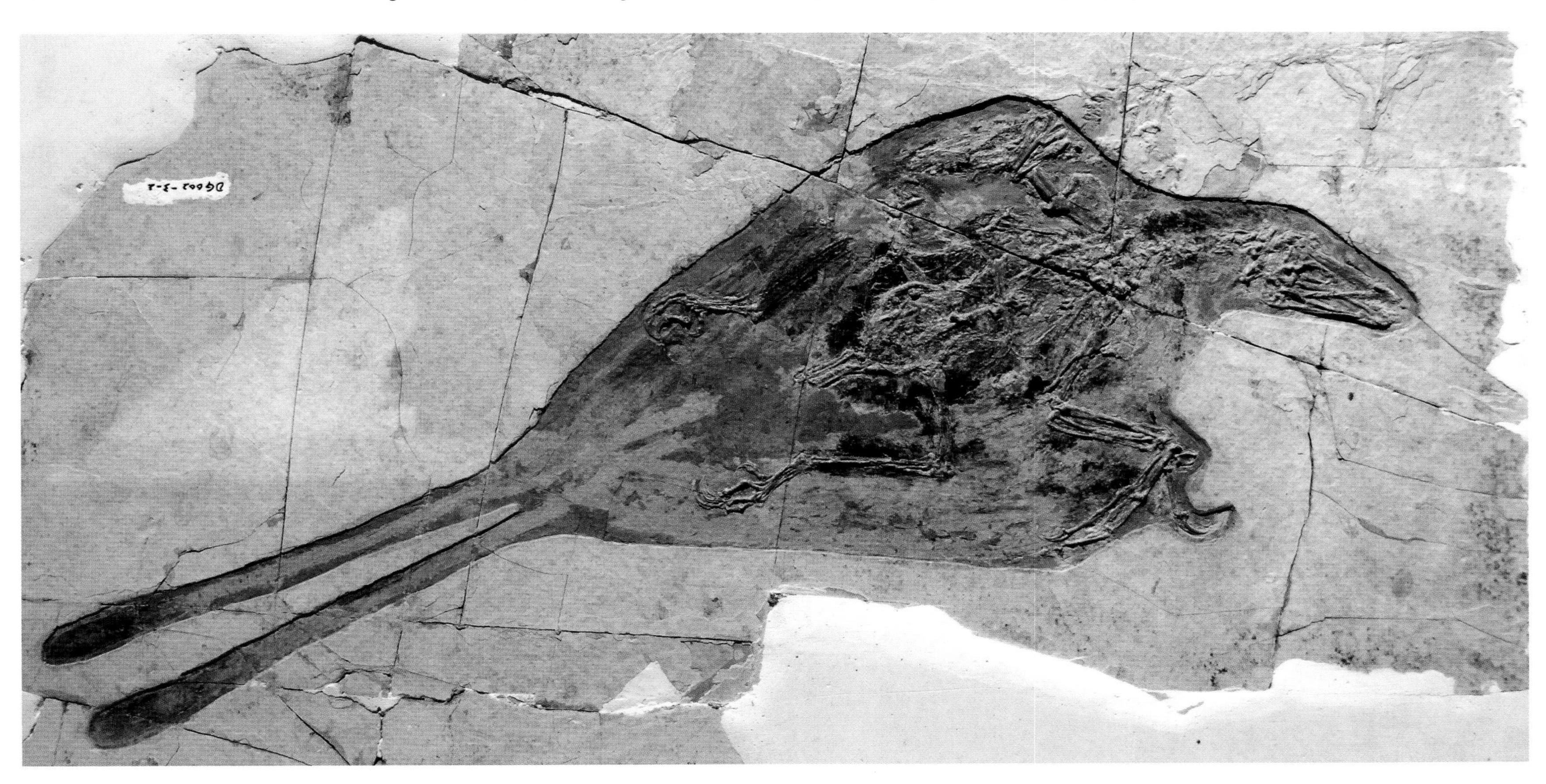

Above: *note the feather-like structures around the base of the tail of the fossil reptile* Sinornithsaurus millenii.

feathers are not, as has been long supposed, unique to birds, but are found on fossil reptiles, too.

The implications of the finds reverberated around the palaeontological world. If therapod dinosaurs had feathers, that was pretty strong evidence that birds and dinosaurs are very closely related and, indeed, that birds could actually be little more than modified reptiles. Secondly, if non-flying reptiles have feathers, that surely would refute the idea that feathers co-evolved with the need to fly or glide? Instead, it would lend weight to the theory that feathers were initially most helpful in keeping their owners warm, and became instruments of flight only as a secondary characteristic. And thirdly, the discovery at Liaoning of *Dilang paradoxus*, a tyrannosaur with primitive feathers, allowed some to suggest, somewhat playfully, that perhaps the mighty *Tyrannosaurus rex* itself had feathers, or that its young were actually fluffy chicks.

The finds at Liaoning are not just confined to reptiles; there are indisputable birds here, too. One of the commonest is *Confuciusornis sanctus*, first discovered in 1994. This is a species with feathers that differ little from those of present day birds, and it was the first fossil bird found with body plumage as well as feathers on the wings or tail. *Confuciusornis*, dated from about 130 million years ago (the precise age is uncertain), is also the earliest known bird with a toothless bill, a forerunner of all living birds. Details of the wings show that it could fly. There are now so many fossils of this species that both males and females have been discovered; one sex, although it is not known which, had long tail feathers, which would seem to have little function except in display.

Confuciusornis and the other bird fossils from Liaoning have muddied the water somewhat in respect of the oldest known of all bird fossils, the feathered but toothed *Archaeopteryx lithographica*, from Germany. The specimens of this bird are dated from around 150 million years ago, but their anatomy does not suggest any close links to the therapods. Thus, the origin of birds is a matter that is still not resolved, and it is to be hoped that, eventually, all the pieces of the jigsaw can be fitted together.

The discovery of these world-famous fossils has affected the formerly obscure province of Liaoning in a number of ways. It has put it on the world map, and has also given rise to a local industry. Bird fossils are very rare and, should a farmer unearth one, the prospect of great wealth beckons. In 1999 a single specimen, thought to be a link between reptiles and birds, was bought for US$80,000; it later turned out to be a fake. This, it hardly need be said, would transform the life of a worker used to earning a few dollars a month.

Unfortunately the illegal act of digging up fossils for trade is now rampant. At first sight it seems harmless enough, but the problem is that fossils out of context (for example, removed from soil that can be dated and linked with other fossils) are devalued scientifically. The treasures of the Laioning shale could be lost, together with their secrets. Thus, by a neat irony, birds that are long extinct are now threatened by human exploitation, just as so many of their present-day relatives are.

Beidaihe

HABITAT Coastal waters, gardens, woodland, sandy flats

KEY SPECIES Siberian Crane, Relict Gull, migrant passerines such as Siberian Rubythroat, Siberian Blue Robin, Thick-billed Warbler, Pallas's Reed Bunting and Daurian Jackdaw, Pied Harrier, Oriental Stork

TIME OF YEAR Best in spring and autumn

Above right: Elisa's Flycatcher is a recent split from Narcissus Flycatcher and is more subtly plumaged than its relative. It is a Beidaihe speciality and is one of several species of flycatchers that adorn the trees and bushes along this stretch of coast in the spring.

Right: the stunning male Siberian Rubythroat is a notorious skulker that is usually tricky to see well, even though it is a common migrant.

Beidaihe is an unprepossessing seaside town on the east coast of China, only 280 km from the explosively populous capital, Beijing. In the summer months it is frequented by hordes of holidaymakers, but in spring and autumn the only tourists present are birds – and there are lots of them. They stop here as they move between their breeding grounds in Russia or northern China and their wintering grounds in South-East Asia and Australasia. Beidaihe lies on a narrow strip of flat land bordered by high mountains to the west and the northern edge of the Yellow Sea to the east, and birds use this strip as a corridor for migration. Beidaihe's position, on a small peninsula jutting out into the Bay of Bohai, means that most travellers pass by this way. So, not only is it a superb place to witness migration, it is also a place to find large numbers of birds that, to visitors from Europe or North America, are only rare strays, or species imagined but never seen.

This is very much a migrant spot, and it has been estimated that only 15 of the 400-odd species recorded actu-

ally remain in the area all year round. Things begin in March, when the last of the winter visitors, such as Siberian Accentor, Eurasian Skylark and seaduck, depart the area. Migration begins and the first of the spring migrants arrive. The most conspicuous of the arrivals are the geese and cranes, mainly Bean Geese and Common Cranes, which come over in neatly organized V-shaped and W-shaped formations and announce their appearance with loud calls – deep honks and high bugle-blasts respectively. It is well worth checking every flock of cranes because several other species can join the flocks or make their own: the rare Red-crowned and Hooded Cranes pass by and Siberian Cranes, one of Beidaihe's specialities, still migrate over each season. These magnificent white birds come through when snow may still be lying on the ground on the Lotus Hills, the best watch point, south of the town.

In April and May the main northbound migration of smaller birds happens in waves, peaking in the middle of May. The attention of the birders then shifts to the small islands of vegetation in the town, such as gardens and groves of trees, where hundreds of small, tired sprites may be compressed into tiny clumps of bushes after rain. This is when some spectacular birds, all in fresh spring plumage, can be observed together. There may be dozens of Red-flanked Bluetails, Siberian Blue Robins, Siberian Rubythroats and Chestnut-flanked White-eyes, each offering their respective dash of colour, often found together with such gems as the smart but subtly plumaged Siberian, Dusky, Eyebrowed and Grey-backed Thrushes. The most intensely coloured birds include the Elisa's and Yellow-rumped Flycatchers, while the many warblers, with their relatively sober hues, will challenge visitors to the full. These include Pallas's Leaf, Yellow-browed, Eastern Crowned, Arctic, Pale-legged Leaf, Dusky, Radde's, Pallas's Grasshopper and Thick-billed. Beidaihe is a superb place to sort out the identification of these birds – or perhaps to become more confused.

The autumn migration begins in mid-August, when waders may be present on the sandflats to the north of the town, or at one of the estuaries to the south. These include such birds as Greater and Lesser Sand Plovers,

■ *Below: Siberian Thrush is a fairly common spring migrant at Beidaihe. The male (pictured) is black with a white eyebrow while the female is delicately scaled brown and buff.*

■ *Right: autumn is the best time to see corvids moving through, and it is worth checking flocks of Rooks for the striking black-and-white Daurian Jackdaw.*

Red-necked and Long-toed Stints and Grey-tailed Tattlers. Among them it is well worth checking the gulls: Heuglin's Gull is common, but if you are really lucky you might see a Relict Gull, a rare bird of the Mongolian Plains. From mid-September onwards this place can be good for pipits, including Richard's, Olive-backed, Red-throated and Pechora.

Autumn at Beidaihe is the best time to witness raptor migration and, as is the case for cranes, the best place for observing the birds in action tends to be the Lotus Hills. On suitable fresh sunny days with a north-westerly wind plenty of species can be seen in good numbers, including Eurasian and Japanese Sparrowhawks, Amur Falcon, Upland Buzzard (late in the season) and Crested Honey Buzzard. One of the star birds, though, is the superb Pied Harrier. Up to 14,000 of these range-restricted raptors have been counted passing in a single season (1986).

In common with what has been observed in other parts of Eurasia, the autumn at Beidaihe is better than spring for visible migration, allowing you to watch streams of smaller birds actually passing overhead during the morning rather than trying to make out shapes diving into bushes at dawn. These movements often involve pipits, rosefinches, larks and buntings. Indeed, Beidaihe is a great location for buntings during both main migratory periods. Ten or more species can be seen in a single day, although in autumn they are not as spectacular in plumage as they are in the spring. Black-faced and Pallas's Reed Buntings are the commonest at this time of year. Autumn is also a time to watch crows moving: among the Carrion Crows and Rooks there will be flocks of the grey-and-white Daurian Jackdaw. Numbers and variety of the smaller birds peak in October; the best time to see good falls of birds is a day or two before the arrival of a cold front from the north.

By late October and November the cranes reappear, just as the rest of the landbird migration is beginning to dwindle. They share the skies with some additional species that are strikingly apparent at this season, but less common in the spring. For example, now is the best time to observe the very rare Oriental Stork as it leaves its breeding areas in eastern Russia and northern China and makes the leisurely trip down to the plains not far to the south of Beidaihe, and it is also a good time to see small groups of the magnificent Great Bustard flying past. These regal birds, with their imperiously slow wing-beats, are among the highlights of days that begin to signal the advance of winter. The last migrants appear in mid-November, and then this seaside town goes quiet again for a few months, as only seaside towns can.

Wolong

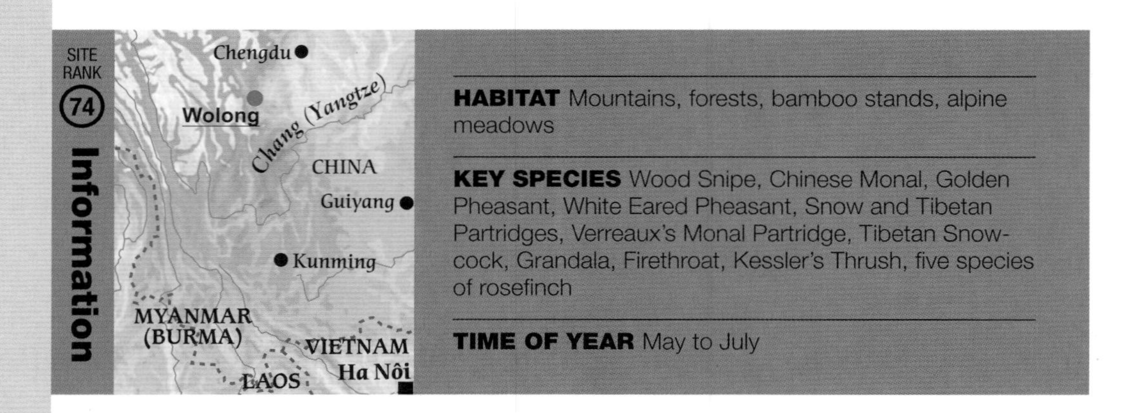

HABITAT Mountains, forests, bamboo stands, alpine meadows

KEY SPECIES Wood Snipe, Chinese Monal, Golden Pheasant, White Eared Pheasant, Snow and Tibetan Partridges, Verreaux's Monal Partridge, Tibetan Snowcock, Grandala, Firethroat, Kessler's Thrush, five species of rosefinch

TIME OF YEAR May to July

Above right: spectacular tragopans are among the most sought-after of Asia's pheasants for birdwatchers. Here a male Temminck's Tragopan shows the amazing colours of its facial skin during display.

Wolong National Park is famous as the site where the Worldwide Fund for Nature captive-breeds and studies the incomparable Giant Panda. These bamboo-guzzling black-and-white bears still occur in the wild here amidst the 1,700 sq km of mountain forest and meadow protected in this national park, but for the casual visitor they are almost impossible to see. Nevertheless, it is always worth a try and, by way of compensation, the park also provides refuge for some of China's most spectacular and exciting birds.

Indeed it is possible that birders, being somewhat single-minded people, might actually think of the word 'pheasants' before 'pandas' when they hear the name China. Certainly the mountains of China, especially those of Sichuan, hold more species of gamebirds than pretty much anywhere else, and Wolong is an excellent place to catch up with ten or more of them. Part of the attraction is that these secretive, shy birds are often quite difficult to see but, when you do, their amazing plumage can make even the most arduous search worthwhile. For most, the search involves moving around in some of the most beautiful scenery on earth.

Most tourists stay in the new hotel in Sawang, in the

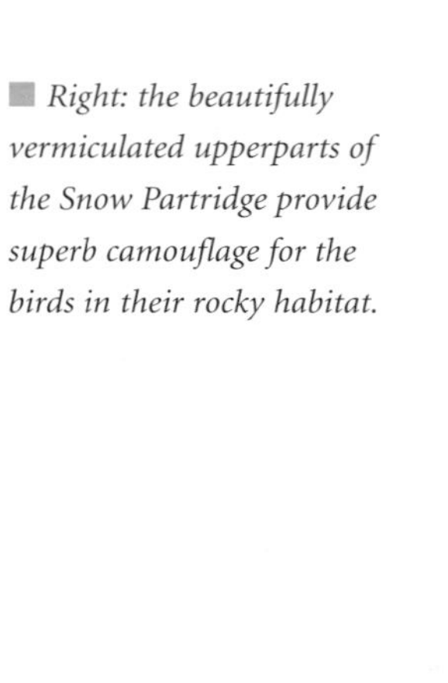

Right: the beautifully vermiculated upperparts of the Snow Partridge provide superb camouflage for the birds in their rocky habitat.

Above: Chinese Monal has a limited range in central China, mainly at altitudes between 3,000-4,500 m, and its population could be as low as 10,000 individuals.

lower parts of the national park. Here the forest and scrub on the steep slopes are home to the Golden Pheasant, a bird that is commonly kept in captivity and is familiar to many birders, but can be elusive here in its wild state. Rather a small, slim pheasant, it tends to skulk in the undergrowth, uttering its slightly strangled call. In common with several of the pheasant species at Wolong, and the pandas themselves, it tends to favour areas of bamboo thicket. The Common Pheasant can also been seen here.

From Sawang a trek to the Wuyipeng Panda Research Station brings birders into the truly sumptuous forests above 3,000 m, where the huge, moss-laden trees tower over rhododendron and bamboo thickets. This is the habitat for several other species of pheasant, the commonest being the superb Koklass Pheasant, the barking calls of which are a common sound at dawn. With its dark head, white neck patch and horizontal, spiky crest, the male rarely shows itself before darting away into the undergrowth. The Blood Pheasant is usually a little more generous, and it can be possible to admire its grey plumage suffused with narrow white streaks, like highlights. The main draw, though, is the remarkable Temminck's Tragopan, a blood-red pheasant (in the male) speckled with white spots, like sequins, which occasionally wanders across the forest tracks, usually at dawn. Several of these forest species include moss in their diet, a sensible move in such an environment.

To see the rest of the gamebirds, visitors have to take the road that winds in hairpins up to the Balang Shan Pass at 4,500 m, a lonely open area with scattered bushes, rocky outcrops and snowfields, overlooked by peaks of 5,500 m, and with the 6,000-m Siguniang peaks glinting in the distance. The forest thins out near to 4,000 m, and at this point it is possible to see two quite marvellous pheasants of high altitude. They are astonishing to look at, in different ways. The White Eared Pheasant is white like the snow, with a long, bushy, black-tipped tail and a small white ear-tuft below a red patch of bare skin on the face. The rare Chinese Monal, a local endemic, is a riot of iridescent colour, green on the face, with red, yellow and bronze on the neck and various shades ofblue and purple on the back, wings and tail. These two extraordinary species feed on plant material, often grubbing up bulbs of lilies and onions and other colourful alpine flowers. They have a habit of suddenly appearing, as if from nowhere.

These very high altitudes also play host to several

■ Right: flocks of Grandala feed on meadows and slopes at Wolong. This is the strikingly coloured male.

other gamebirds, including the splendid Tibetan Snowcock, a pleasing mixture of dirty white, grey and brown, resembling a slab of stone. The Snow Partridge is smaller and darker than the snowcock, with heavy brownish streaking on the breast; as its name suggests, it is rarely found away from snowfields, and feeds primarily on lichen, moss and grass – there isn't much else available on the rocky hillsides where it occurs. Among scattered bushes is the colourful Tibetan Partridge, with its chestnut neck and barred breast, while Verreaux's Monal Partridge is sometimes found on the rocky slopes. If you see all these gamebirds during your visit, you have either been remarkably blessed or you are lying!

Quite naturally, these verdant forests and pastures are host to plenty of other great birds. Wolong National Park is, for example, known as the very best place in the world to see the Firethroat (not quite a Giant Panda, but a cracking bird), a small chat with a bright orange throat, which frequents bamboo stands at lower altitudes. Up at the tree line, on the way up to Balang Shan Pass, early risers can find the increasingly rare Wood Snipe (which has a population now down to about 2,000 displaying birds worldwide, the males gathering together in leks), while up at these altitudes are such delights as Kessler's Thrush and a suite of splendid rosefinches, including Red-fronted, Beautiful, Dark-breasted, Dark-rumped and White-browed. At the very highest altitudes in the pass, flocks of the amazing Grandala feed on the ground near the snow, while other good birds here include Snow Pigeon, Brandt's and Plain Mountain Finches, Lammergeier and Himalayan Vulture.

■ Right: the often skulking White-tailed Rubythroat breeds in patches of scrub high in the mountains. It is an altitudinal migrant that winters in the lowlands.

Lena Delta

HABITAT Arctic tundra, coast

KEY SPECIES Ross's and Sabine's Gulls, Bewick's Swan, Brent Goose [Black Brant], King and Steller's Eiders, Long-tailed Duck, breeding waders including Grey Phalarope, Siberian Crane, Snowy Owl

TIME OF YEAR April to July

At 72° N, hidden by the vastness of the Siberian tundra, the Lena River spills out into the Laptev Sea to form the second-largest delta in the world. Rising just 15 km west of Lake Baikal, the Lena is the tenth-longest river in the world, flowing some 4,400 km in a north-eastward and then northward direction and widening to several km before finally reaching its mouth and fanning out over an area of 32,000 sq km consisting of tundra, numerous channels, 1,500 islands and at least 30,000 different lakes, large and small. The river brings 12 million tonnes of alluvia with it each year, and faithfully dumps this at journey's end to replenish the delta. Happily, a large proportion of this area is protected for future generations.

Tundra birds nest here in their hundreds of thousands, making this one of the richest places anywhere in the Arctic, both in terms of densities of birds (up to 641 birds per sq km) and numbers of species. In all, no fewer than 122 species have been recorded, of which 67 have been found breeding, an unprecedented number this far north. Some of the individual species occur in impressive numbers, too: there are at least 15,000 pairs of Greater White-fronted Geese, 10,000 pairs of [Tundra] Bean Geese, 6,000 pairs of Bewick's Swans, 5,000 pairs of Brent Geese, 7,000 pairs of Red-throated Divers and 25,000 pairs of Black-throated Divers. In addition, at least 200,000 pairs of ducks breed, especially Northern Pintails and Long-tailed Ducks. Not surprisingly, during the short breeding season the tundra hums to the rhythm of nesting birds.

The Lena Delta's position, east of the Tamyr Peninsula, allows the mix of birds to carry representatives from both west and east. Among the breeding waders, for example, there are Little and Temminck's Stints that are

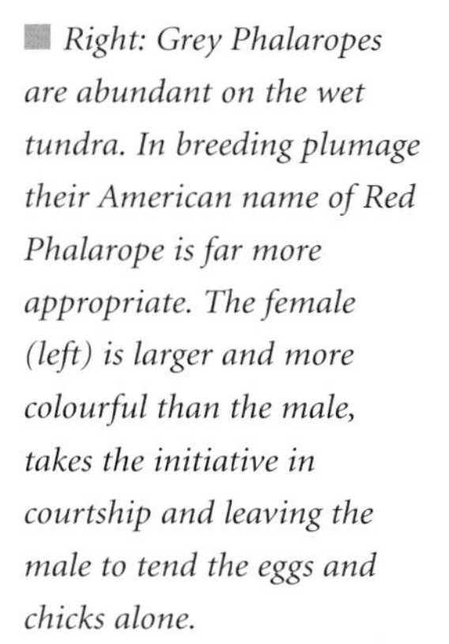

Right: Grey Phalaropes are abundant on the wet tundra. In breeding plumage their American name of Red Phalarope is far more appropriate. The female (left) is larger and more colourful than the male, takes the initiative in courtship and leaving the male to tend the eggs and chicks alone.

Right: better known for its pelagic lifestyle in the winter, the Sabine's Gull breeds at the Lena Delta, making a nest on the ground close to the water.

Right: King Eiders breed commonly on pools in the tundra, often far from the sea.

well known to birders in Europe, and Pectoral Sandpipers that are better known to North Americans. The Brent Goose is represented by two forms, of which the essentially North American subspecies, often known as the Black Brant, dominates. Recent breeding records of Spectacled Eider, Sharp-tailed Sandpiper and Long-billed Dowitcher suggest that these species are extending their ranges westward. Nevertheless, the core representation is Eurasian.

Within the delta ecologists recognize four principal habitats. The key community within the delta proper is wet polygonal tundra, so named because of the regular patterns formed by the frozen ice (the area is ice-free for only 120 days a year). In the Sokol district in the south-west of the delta, a drier tundra forms on the continental slopes rising up to 450 m, allowing the growth of such plants as Mountain Avens and Arctic Bell-heather, while an intermediate habitat, wet continental slopes, with hummocky rather than flat and polygonal tundra, occurs on some north-eastern shores. Finally, small sheltered valleys in the southern continental part of the delta hold some communities of alder and willow bushes, home to such species as Dusky Thrush and Willow Warbler.

The wet polygonal tundra is the key site for many of the waders, geese and ducks. One of its very commonest species is the Grey Phalarope, a truly glorious bird with spectacular orange-red breeding plumage, which is reflected in its American name, Red Phalarope. It reaches densities of nearly 50 breeding pairs per sq km in the central delta. Trailing some way behind in the same region are the Dunlin (11 pairs per sq km) and Pectoral Sandpiper (six pairs per sq km). Other common breeding waders in this zone include Grey Plover, Common Ringed Plover, Little and Temminck's Stints and Ruff, while characteristic non-wader species include the geese, Black-throated Diver, King Eider and Long-tailed Duck. However, dwarfing all of these in abundance is a passerine, the Lapland Bunting, which can show densities as high as 270 individuals per sq km. With the male's smart black-and-chestnut plumage and sweet but slightly strained song, it is a dominant bird of this windswept habitat.

A slightly different set of birds occurs on the continental slopes. While the Dunlin is even more abundant here than on the polygonal tundra, the Grey Phalarope is much scarcer and other species, such as the Ruddy Turnstone and Pacific Golden Plover, become dominant instead. In recent years the Curlew Sandpiper, usually much commoner on outlying islands off the coast, has been found here, and may be colonizing. Not surprisingly, with more dry ground, landbirds become more of a feature in these habitats, especially Willow Grouse, and Rock Ptarmigans, White Wagtail, Shore Lark and Snow Bunting. Another interesting breeding bird in this area is the Asian Rosy Finch.

While the above may be the key species in the delta area, it is some of the scarcer species that might fire the imagination of would-be birding visitors. These include the Snowy Owl and the Rough-legged Buzzard, whose numbers depend on the supply of Brown and Collared Lemmings, the Siberian Crane, which is mainly found further east, and two superb High Arctic gulls – Sabine's and Ross's. Both these species are difficult to see in both Europe and North America, but they are breeding species, if not especially common ones, on the Lena Delta, mainly in the northern section along the coast. Indeed, they often mix together in the same colonies, together with Arctic Terns.

The breeding season on the Lena Delta is very short. For seven months of the year the area is frozen (the permafrost layer here is 1,000 m thick), it can be dark throughout the day and the temperature regularly drops to -30°C and has been recorded at -53°C. The freeze-up begins in October and lasts until the end of April. Few birds breed before June, and chicks tend to hatch in July, when temperatures are most consistently warm and the vegetation and insect life are at their most abundant. This abundance, of course, ephemeral as it may be, is the key to bird life in the delta. Without it, the tundra lands would be strangely quiet.

Ussuriland

HABITAT Sea coast with offshore islands and mudflats; shallow freshwater lake with marshes and meadows; hills with rushing rivers; mixed forests

KEY SPECIES Oriental Stork, Red-crowned and White-naped Cranes, Scaly-sided Merganser, Mandarin Duck, Spectacled Guillemot, Styan's Grasshopper Warbler, Azure Tit, Pechora Pipit

TIME OF YEAR Best April to July

Opposite: the mighty Steller's Sea Eagle is a scarce winter visitor to the coast.

Ussuriland lies at the very south-eastern extremity of Russia, where the coast looks over the Sea of Japan, and where the hinterland borders the eastern edge of China. It is the richest area in the whole of Russia for birds, with over 250 species breeding and more than 400 recorded altogether, the diversity partly explained by the wide range of superb, often relatively untouched habitats, and partly by the mix of birds from northern and southern elements. Many common breeding birds from Siberia occur here, while a few species reach Ussuriland at the northern extremity of their world range. During the spring, when the trees are freshly in leaf and the meadows glow with wild flowers, there are few finer settings in which to watch birds.

Below: familiar from waterfowl collections the world over, the Mandarin Duck is found here in its natural home.

A visit almost invariably begins at the dreary, if historically interesting, city of Vladivostok. Happily it is possible to take a boat out from here into Peter the Great Bay and visit Russia's only marine reserve at Popov Island, where there are healthy populations of Spectacled Guillemot, Black-tailed Gull, Pelagic Shag and Japanese Cormorant, while Ancient Murrelet and Rhinoceros Auklet occur in much smaller numbers. Sailing around the area may also produce Streaked Shearwater and Swinhoe's Storm Petrel (here at its only breeding site in Russia), although these birds come to shore only at night, to visit their burrows under cover of darkness. Another great rarity found here is the peculiar Styan's Grasshopper Warbler, a bird that seems only to occur on offshore islands; the only way to see it is to land on one of the many small islets and wait for the bird to sing.

Going north from Vladivostok, flush to the Chinese border, the land flattens out into a broad, treeless plain, in which lies an immense, shallow freshwater lake, Lake Khanka, over 100 km wide. The sheltered waters here, together with the huge fringing marshes and vast acreages of meadows, support enormous numbers of birds, some of which are very rare. The reedbeds hold the bulk of the Russian population of two species of crane, the Red-crowned Crane – most famous as the national bird of Japan, with its mainly white plumage, black bushy tertials and neck and red crown – and the White-naped Crane, an altogether greyer bird with a white neck. Roadside trees provide nesting sites for the endangered Oriental Stork, while on the waters themselves are Falcated and Eastern Spot-billed Ducks, and occasionally the rare Baer's Pochard and Swan Goose. The marshes are quartered by the smartly patterned Pied Harrier, while Reed Parrotbill and Pallas's Grasshopper, Oriental Reed, Black-browed Reed and Thick-billed Warblers breed in the swamp or scrub. The denser patches of trees are good places to look for the angelic Azure Tit, while the meadows are good for Yellow-breasted and Pallas's Reed Buntings. The thin margin of trees around much of the lake can be a superb site for migrants on a spring day, when it offers the only cover for miles around and can attract thousands of northbound passerine migrants.

The open grassland should not be neglected; where trees grow, it is well worth a look for Daurian Starlings, Rooks, Daurian Jackdaws and the splendid Amur Falcon, which incredibly is a summer visitor here in eastern Asia, having come from its wintering grounds in tropical Africa. Another interesting species of this habitat is the Pechora Pipit; birds of the isolated population found around Lake Khanka have different calls and song to the birds further north, and will probably soon be split as a separate species, Menzbier's Pipit.

The central part of Ussuriland is dominated by the Sikhote-Alin Mountains, which run roughly north-east down to south-west, and spawn a number of large, fast-flowing rivers such as the Ussuri, Iman and Bikin. These rivers flow through a delightful landscape of forested hills at various heights; some of the woods are subtropical, while others are more reminiscent of the taiga belt. The former, with their growth of Manchuri-

Above: the swamp-loving, powder-blue and white Azure Tit is fairly common in Ussuriland.

Right: most of the Russian population of the highly endangered Red-crowned Crane breeds next to Lake Khanka.

an Oak mixed with cedar and pine hold such delights as Forest Wagtail, Chestnut-flanked White-eye, White-throated Rock Thrush, Pale Thrush and Oriental Dollarbird, while the latter hold such species as Radde's and Pale-legged Leaf Warblers, Mugimaki Flycatcher and Siberian Thrush.

Some of the rivers, such as the Bikin and Iman, have pristine gallery forest growing along their banks, with willow and alder next to elms, poplars and Manchurian welt-nuts, and these provide habitat for the tree-nesting Mandarin Duck as well as such conventional woodland species as Ashy Minivet, Eastern Crowned Warbler, Grey-backed Thrush, Yellow-rumped Flycatcher and Japanese Grosbeak. A few pairs of Blakiston's Fish Owls occur in these valleys, too, while the rivers themselves hold good numbers of the rare Scaly-sided Merganser and their shingle banks support Long-billed Plovers. It often requires a boat trip to see many of these fabulous birds, and it can be a bumpy ride.

Down to the south-west, in the tiny piece of Russia that lies south of Lake Khanka, is still another ecological zone, in which the foothills of low, oak-covered hills are skirted with birch and interspersed with flower-filled meadows. The Barachny Hills lie in this area, where several predominantly southern species have a foothold in Russia. These include Lesser Cuckoo, Manchurian Bush Warbler, Von Schrenck's Bittern, Yellow-legged Buttonquail and Vinous-throated Parrotbill. It is also a good spot for the rare Band-bellied Crake, although seeing both this and the buttonquail is no easy task. However, as you watch and listen to some of these subtropical Asian species, it is easy to appreciate that the Orient is not very far away.

Arasaki

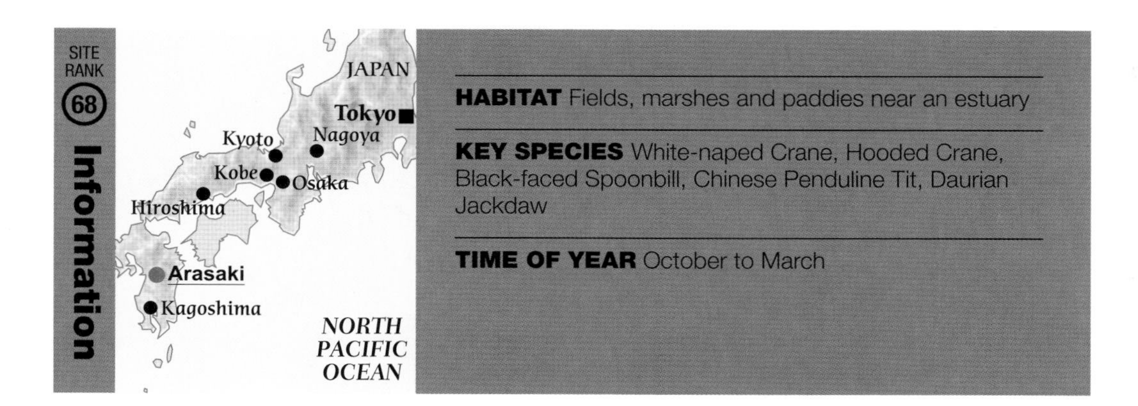

HABITAT Fields, marshes and paddies near an estuary

KEY SPECIES White-naped Crane, Hooded Crane, Black-faced Spoonbill, Chinese Penduline Tit, Daurian Jackdaw

TIME OF YEAR October to March

Below: the cranes of Arasaki are a major tourist attraction, and can be seen during the day from the visitor centre.

Cranes – the birds, as well as the machines – have an ability to transform landscapes. Even the dullest of damp fields can seem to come alive when these stately yet exuberant creatures set foot upon the inviting moist turf, and either stride serenely in search of food or leap unselfconsciously into the air to dance, lighting up the landscape with their patterns and movement. And that transformation is particularly strong in one small corner of Japan. The region of Arasaki, on the southernmost main island, Kyushu, is a flat agricultural area reclaimed from the sea in the 17th century. Close to the large town of Izumi, and next to the estuary of the Takano River, it is a patchwork of paddy-fields and marshes that, for some reason, are host to what is arguably the most impressive concentration of cranes in the world. From October to March about 10,000 of these birds settle at Arasaki, having travelled all the way from Siberia and northern China.

Cranes have been visiting the area since at least 1695, not long after the first sea defences were built. Since then, of course, much has changed, not least the wholesale loss of all the cranes' other refuges on the island, sacrificed to disturbance or development, which has left Arasaki as the only remaining site. There has also been an increasing appreciation of the birds. The area was designated as a National Natural Treasure in 1921, and then a Special Natural Treasure in 1952. A farmer began the regular feeding of birds not long afterwards, and

■ *Right: Arasaki's two main species: a smaller, darker Hooded Crane (centre) stoops beneath an arch made by two White-naped Cranes.*

now there is a special Crane Observation Centre (complete with two-storey telescope-bedecked observatory and restaurant) where the birds can be seen during the day, plus a Crane Museum and a Crane Park. Thousands of people come every year to appreciate in modern comfort what is, in effect, an ancient spectacle.

For birders, the delight is in the sheer number and variety of birds, as well as the ease of seeing them. The flocks are dominated by two species, the Hooded and White-naped Cranes. Both have quite dark plumage for cranes, especially the Hooded, with its slate-grey feathers covering the whole body up to the pure-white upper neck. The larger, longer-necked White-naped Crane has a similar dark grey colour on breast and back, but has soft white wings and upper tail coverts and a white patch from the nape down the neck to the upper back. While the Hooded Crane has a small red patch on the crown, the White-naped Crane has a large scarlet circular 'mask' surrounding the eye. It is thus easy enough to distinguish the two species. In the winter of 2005–2006 some 10,000 Hooded and 1,200 White-naped Cranes could be found in the region of Arasaki, which is about average. The two species mix freely together, although the White-naped Crane has more of a tendency to get its feet wet than its congener.

Each year brings some extra excitement for birders in the shape of some hangers-on. In 2005–2006, for example, there were two Sandhill Cranes and one Common Crane among the hordes, and indeed these two species turn up most years, the Sandhill Crane easily identified by its body plumage, a paler grey than the others, and the Common Crane by its striking black-and-white head pattern. Rarer visitors that come only every few years include the diminutive Demoiselle Crane, the Red-crowned Crane (which looks like a white-plumaged version of a Common Crane) and, as a special bonus, the mainly white Siberian Crane. Patient scanning through the flocks can also reveal an assortment of odd-looking individuals and hybrids. Thus, with luck, it is possible to see seven species of crane in a day at Arasaki, which is almost half the world total.

Besides cranes, the wide variety of habitats here, which includes reedbeds, coast and woodland, attracts what is an excellent variety of birds for Japan in the winter. Many ducks can be seen including, at times, unusual visitors such as Falcated Duck or American Wigeon mixed in with the many Northern Shovelers, Northern Pintails, Eurasian Wigeons and Eastern Spot-billed Ducks. In the creeks several species of heron occur, including Intermediate Egret and Black-crowned Night Heron; and sometimes a small flock of Black-faced Spoonbills also winters around the Crane Reserve. Look out too, for waders, which may include Kentish Plover and Northern Lapwing. The woodland areas hold Japanese Green Pheasant, the rivers have Brown Dipper and the splendid Crested Kingfisher, while on the fields and scrub every bunting should be checked: Meadow, Black-faced and Rustic Bunting are all regular, although their winter plumage is never quite so diverting as their spring finery.

Two special bonus birds also deserve a mention. A patch of reeds within the Crane Reserve regularly turns up Chinese Penduline Tit, the most migratory member of its small family. Recently split from Eurasian Penduline Tit, the males have a thinner black face-mask than their better known relatives. And on the fields visible from the observation tower, among the large flocks of Rooks, look out for the smaller Daurian Jackdaws feeding among them. These are also winter visitors from China; be aware that some, if not all, will be of the all-black morph.

The day's birding always begins and ends here,

Right: keen birders will look out for rarities among the multitudes. Up to seven crane species occur at Arasaki.

though, with the cranes. In common with flocks elsewhere in the world, the birds undertake regular daily movements at Arasaki. The roost site is opposite a guest house run by the crane warden and, in the pre-dawn blackness, you can be woken by the increasing excitement of the bugling calls. The birds then often rise as one to go off to the fields near the observatory, where fish and rice is provided for them daily. The sight and sound of the commuting flock is, without exaggeration, one of the finest experiences in the birding world.

Africa

As the second richest continental landmass for birds, with about 2,400 species, Africa is second only to South America and its diversity is hardly surprising since it also straddles the Equator and provides a vast range of habitats for birds.

Working from north to south, the belt of scrubby vegetation around the Mediterranean Sea is known as maquis, and has more affinities with southern Europe than with sub-Saharan Africa. South of this region lies the huge Sahara Desert, the largest on earth, fringed to the south by an arid grassland and savanna zone known as the Sahel, which is bisected by large rivers such as the Niger. To the east lies a zone of highlands, encompassing a number of ranges with their own unique set of birds. The Great Rift Valley, with its famous soda-lakes and grasslands teeming with mammals, runs from the Red Sea south to Tanzania, while to the west a huge belt of lowland tropical forest, the second largest on Earth behind Amazonia, runs down from Nigeria to the Democratic Repubic of Congo. Fringing the forest is a belt of rich savannah, which gives way in the west to desert-like landscape along the coast from Angola to South Africa. The Cape region of South Africa has a number of its own specialized birds, and the ocean thereabouts is exceptionally rich. Meanwhile, Africa has a number of interesting islands off its shores, none more spectacular than Madagascar, with over 100 endemic bird species and five endemic families.

Among the bird families unique to Africa are the mousebirds, turacos, woodhoopoes and ostriches, along with Shoebill and Secretarybird, while Madagascar adds in its celebrated vangas, ground-rollers, Cuckoo-roller, asities and mesites. Besides these, Africa is exceptionally rich in such families as sunbirds, weavers, shrikes, bustards and larks.

Right: Red-throated Bee-eater in The Gambia.

Merzouga

SITE RANK 72

Information

HABITAT Stony and sandy desert; salt lake (seasonal)

KEY SPECIES Desert Sparrow, African Desert Warbler, Egyptian Nightjar, Houbara Bustard, larks including Temminck's, Thick-billed and Bar-tailed, wheatears including Desert, Red-rumped, Mourning and White-crowned

TIME OF YEAR Spring (March to May) is best

Above right: one of Merzouga's star birds is the pastel-coloured Desert Sparrow which lives in small oases around the sand dunes.

This town, close to the Algerian border in south-east Morocco, offers rich birding in sublime desert scenery. It is a frontier area in every sense, bringing visitors daringly close to the Algerian border, and also giving a tangible sense of being on the edge of the deep, remote, inhospitable Sahara Desert to the south.

Many people's idea of desert is an empty, dead wilderness, but this area challenges that assumption. For a start, the landscape is varied and colourful, with a patchwork of stony plains, hammada (desert with tussocks), wadis, cliffs, settlements, plus the picture-postcard red-coloured sand-dunes and palm trees of the Erg Chebbi. It is also far from dead. Birds pop up almost everywhere

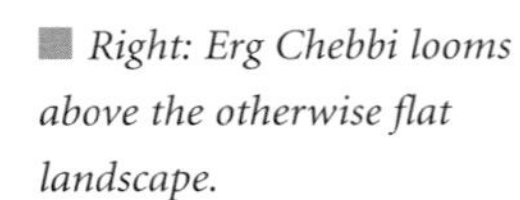

Right: Erg Chebbi looms above the otherwise flat landscape.

Morocco

can be found in sandy or gravelly areas on flat ground, while the closely related Desert Lark prefers rocky outcrops on slopes and avoids sandy areas. The Thekla Lark requires at least some scrubby vegetation to be present, while Temminck's Lark and the Thick-billed Lark are primarily birds of the stony desert. All these larks eke out a living by feeding on a combination of seeds and invertebrates, but the larks with short, thick bills, such as the Lesser Short-toed Lark, tend to specialize on seeds, while longer-billed species, such as the Greater Hoopoe-Lark, take in a higher proportion of invertebrates.

In this extreme landscape, birds often exhibit highly adaptive and fascinating physiology and behaviour. As a group, for example, the larks have an unusually low metabolic rate compared to other birds of the same size, allowing them to reduce internal water loss. Moreover, while the seed-eating larks need to drink, the insect-eating species obtain all their moisture from the body fluids of their prey, and may never drink at all throughout their lives. Larks are generally most active in the two hours after dawn and before dusk, taking a rest during the heat of the day. Two desert species, Temminck's Lark and Bar-tailed Lark, usually roost on mats of vegetation to keep cooler, while the Thekla Lark perches up on bushes to take advantage of any breeze in the air.

The local larks also have to be inventive while feeding: Crested and Thekla Larks are known to beat snails and other invertebrates against rocks in order to smash their shells or subdue them, while the Greater Hoopoe-Lark will take a snail high into the air and drop it to the ground, hoping to break the shell. This last species is also unusual for taking significant numbers of small vertebrates, mainly lizards.

Another group well represented in the Merzouga area is the wheatears, a group of chats. They are as common and conspicuous as the larks, if not more so, with their bold plumage and habit of sitting up on elevated perches. One of the commonest species is the smart White-crowned Wheatear, a large, bold wheatear that will tackle scorpions and will sometimes come to Berber camps to take hand-outs. Another is the Mourning Wheatear, an ant specialist.

To birders, what separates Merzouga from other parts of southern Morocco is the presence of rare specialists of the deepest desert. The most sought-after of these is the pastel-plumaged, somewhat sluggish Desert Sparrow, which hangs on in very small numbers here, having suffered something of a decline in recent years. It feeds mainly on the seeds of desert grasses, but is not shy of taking scraps where it can. For many years a favourite haunt has been the Café Yasmina on the edge of town.

The African Desert Warbler is much more difficult to find and erratic in its appearances. Although it inhabits sandy areas with scattered short bushes, it seems to be very fussy ecologically, leaving many suitable areas unoccupied. This skulker has a habit of scuttling across the ground from one patch of vegetation to another and then remaining hidden from view. It often follows wheatears and other birds as these perch on top of the same bushes in which the warblers are feeding, enabling

Above: excessive hunting has made the Houbara Bustard increasingly rare, but it just about hangs on in the area.

you stop, even in the most desolate places where it seems that nothing could survive.

Few bird groups typify these arid habitats more than the larks, of which there are nine species in the area, each with its own preferred niche. The Bar-tailed Lark, for example, the warblers to use them as sentinels.

Although this is the edge of the Sahara, it does rain from time to time in the area (average 200 mm annually), and one of the town's more curious attractions is a rain-fed lake, the Dayet Srij, 2 km outside town. This lake usually contains water between November and May; when it does there are often flocks of Greater Flamingos present, which are a great tourist attraction. During migration there can also be many waders on the salty waters, and wildfowl including Marbled Duck and Ruddy Shelduck. Looking for water birds floating incongruously on this improbable desert lake is something of a surreal experience.

Of course, being on the edge of the desert, this whole area can attract large numbers of migrant birds in the spring and autumn. In favourable conditions the greener areas near wadis can be alive with birds, including various warblers and flycatchers as well as a local speciality, the Blue-cheeked Bee-eater. The gardens of the hotels and *auberges* often provide great conditions for tired migrants, and there is always the chance of making a discovery here. Nearly 150 species have been recorded.

In years gone by, the desert beside the Erg Chebbi was a reliable place to find the rare and endangered Houbara Bustard. The birds still occur in the area, but at least some of these have been introduced deliberately to supply quarry for visiting falconers from Saudi Arabia. Birders can hire jeeps to take them close to the Algerian border to look for the birds, a trip that can turn out to be a real adventure. Anything that can persuade the locals that it is worth keeping these shy birds alive and unmolested, rather than being the victim of rich men's 'sport', may eventually help to save this globally threatened species from extinction.

Right: the Greater Hoopoe-Lark is common, and performs a superb, acrobatic flight-song in the breeding season.

Banc D'Arguin

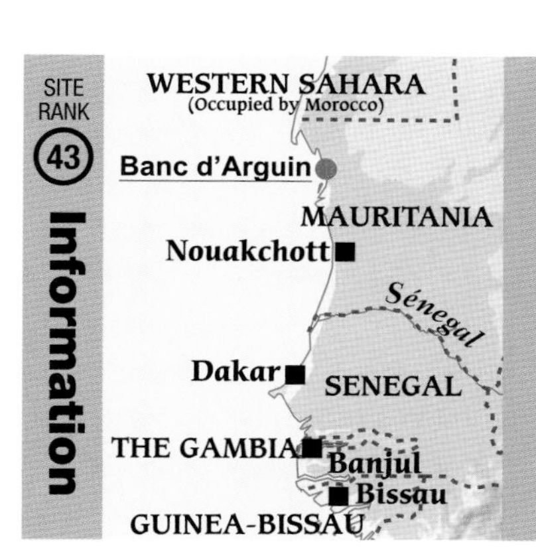

HABITAT Shallow sea, sea-grass beds, intertidal zone, islands, desert

KEY SPECIES Waders in huge numbers, White-breasted and Reed Cormorants, Mauritanian races of Grey Heron and Eurasian Spoonbill, gulls, terns, Sudan Golden Sparrow

TIME OF YEAR Superb all year, but most spectacular in winter (October to March)

If the birders of Europe, Greenland and Siberia ever wondered what happened to all their waders, having waved them goodbye from their breeding sites or staging areas in the autumn, they would find the answer here, among the vast mudflats and islets of the rarely visited wildlife wilderness that is the Banc d'Arguin. At least two million waders spend their non-breeding season on the coast, protected by this national park, the largest concentration of shorebirds anywhere in the world. Incredibly, it is the destination for one in three of all birds using the East Atlantic coastal flyway.

It is an overused word, but the sheer number of birds that can be seen on the Banc d'Arguin is breathtaking. The beaches and sand-bars throughout the park's 290 km of coastline are regularly crowded with birds of all kinds, all mixed in together – not just waders but also gulls, terns, cormorants, pelicans and flamingos. Wintering numbers for many familiar species are simply mind-boggling: 900,000 Dunlins, 540,000 Bar-tailed Godwits, 133,000 Common Ringed Plovers, 226,000 Curlew Sandpipers, 360,000 Red Knots, 102,000 Common Redshanks and 43,000 Little Stints, for example. Every few kilometres there are vast concentrations and flocks that would make the birding headlines in Europe, but are routine here. And the daily flights to and from roosting sites can be spectacular – especially if you take a sailboat out to one of the islands and simply drink in the sight and sound of the drama. This truly is an extraordinary place.

Not all the birds are visitors. About 40,000 pairs of fish-eating birds breed and are present all year round, representing what are probably the largest waterbird colonies in West Africa. These include 25,000 pairs of White-breasted Cormorants, nearly 7,000 pairs of Reed

Below: Bar-tailed Godwits probe into the deep mud, feeding by touch; more than half a million winter on the Banc d'Arguin.

Above: there are enough *fish to support a population of 1,600 breeding pairs of Great White Pelicans.*

Cormorants, 1,600 pairs of Great White Pelicans and 3,300 pairs of Grey Herons. The Grey Heron is of particular interest because it is of a highly distinctive race, endemic to the Banc d'Arguin, sometimes called the Mauritanian Heron, which is much paler grey than other races and is sometimes considered to be a separate species. It occurs alongside another intriguing speciality, the local endemic race of the Eurasian Spoonbill (7,000 pairs). This race lacks the yellow tip to the bill shown by its northern relative and also lacks any apricot-coloured marking at the base of the neck. These are the most southerly breeders of their kind, and they mix in winter with significant numbers of visitors from the north.

The Banc d'Arguin stands at the crossroads between the Palearctic and Afrotropical regions, and this is reflected in its wildlife. For example, the salt marsh grass *Spartina maritima* is present here at its most southerly point on the African coast, while white mangroves *Avicenia africana* are present at their most northerly. Several widespread Afrotropical bird species have their only Western Palearctic populations here, including the Grey-hooded Gull, Reed Cormorant and Royal Tern. The nomadic Sudan Golden Sparrow reaches its regular northern limit at Cap Blanc, a north-western extension of the park near the large town of Nouadhibou, and other Afrotropical species recorded include Yellow-billed Stork, Kelp Gull, Lesser Flamingo, Hooded Vulture and Speckled Pigeon.

The Banc d'Arguin National Park also straddles the transition zone between the Sahara Desert and the Atlantic Ocean. The actual park reaches 60 km out to sea and 35 km inland. On its eastern limits live a range of desert birds, which are known to include such treasures as Bar-tailed and Dunn's Larks, Greater Hoopoe-Lark, Brown-necked Raven, Cream-coloured Courser and the delightful Desert Sparrow. There have been some fairly recent records of Arabian Bustard, too, but the current status of this species in the area is not known. Much, indeed, has still to be learnt about the desert birds of Mauritania.

The offshore zone is fantastically productive. The Banc d'Arguin lies in an area where the warm Guinea Current meets the colder Canaries Current flowing from the north, resulting in strong upwellings bringing organic material from the sea bed. This supports large areas of sea-grass *Zostera*, which provide food for fish and invertebrates. The sea is remarkably shallow, only 5 m deep at low tide, even 60 km offshore, making the whole area ideal for the fishing birds. It supports people, too; the offshore fisheries attract an international fleet of trawlers, and within the park boundary the local people, the Imraguen, still subsist by fishing, either by wading in the water with nets (mostly for mullet), or using small sailing boats for catching sharks and rays. If more birders visited, they could probably make a handy income from ecotourism, too.

Notable among the birds in these shallow waters are the gulls and terns. Besides those mentioned above, there are significant numbers of Lesser Black-backed, Black-headed and Slender-billed Gulls, the latter both as breeding birds and winter visitors, plus a handful of Audouin's Gulls. Even more impressive are the terns. At least 1,000 pairs of Gull-billed, 2,500 pairs of Caspian and 5,000 pairs of Royal Terns breed here. There are even a handful of Common Tern nests, while many thousands of Black Terns winter offshore, completing their remarkable transformation from freshwater to saltwater specialist. Scarcer visitors include Lesser Crested and Bridled Terns. The latter used to breed in small numbers, but now it barely holds on.

The upwellings mentioned earlier can also attract pelagic birds, including skuas, Sabine's Gulls and shearwaters. However, the most interesting recent records concern storm petrels. In May 2003 a series of pelagic trips recorded good numbers of European and Madeiran Storm Petrels; then, in July 2005, a research cruise reported seeing about 10,000 Wilson's Storm Petrels, including some flocks of 700 or more. Clearly there are significant seabird numbers to be found here, and who knows what might turn up if regular pelagic trips were undertaken at all times of the year?

It is clear that the Banc d'Arguin, already a spectacular birding site, could still have more to offer.

Below: some of the 360,000 Red Knots that winter in the area.

Djoudj

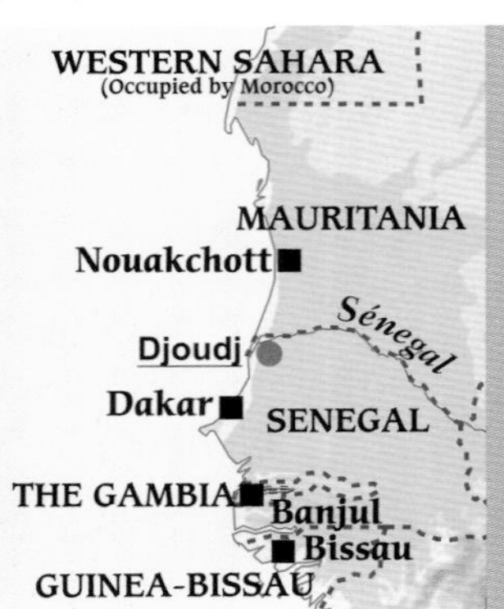

HABITAT Seasonally inundated lakes and ponds in river flood-plain; savanna, scrub

KEY SPECIES Great White Pelican, Aquatic Warbler, Arabian Bustard, Ruff, Lesser Kestrel, Golden Nightjar, Sennar Penduline Tit

TIME OF YEAR Flooded from July to February, with dry season between March and June. Receding waters make viewing good from December to January

The sight of the Senegal River must be a relief for millions of migratory birds making transcontinental journeys south from Europe to Africa, for this large waterway effectively marks the end-point of their very long crossing of the Sahara Desert. It forms an ecological boundary between the Sahara to the north and the semi-arid Sahel region to the south. Barn Swallows, which have been migrating by night to avoid overheating, can now switch back to daytime flying, while other migrants find succour in the relatively well-watered countryside adjoining the river. For some, the region even marks the end of their journey and they will winter here.

The Senegal River, the second-largest in West Africa, spews into a large delta on the border between Senegal and Mauritania. Some 60 km north-east of St Louis – Senegal's version of the river city – the land is seasonally inundated between July and February, and this is

Right: Great White Pelicans often hunt co-operatively for fish, swimming in masses and herding their prey into the shallows.

Opposite: a multitude of waterbirds, including egrets, hunt in the sheltered waters of the delta.

Below: the tiny Cricket Longtail is a speciality of the sub-Saharan Sahel biome. It is named after its shrill song, which resembles a stridulating insect.

where the 160-sq-km Djoudj National Bird Sanctuary is situated. Up to three million waterbirds may use this area as a stopover or wintering site, making it one of the 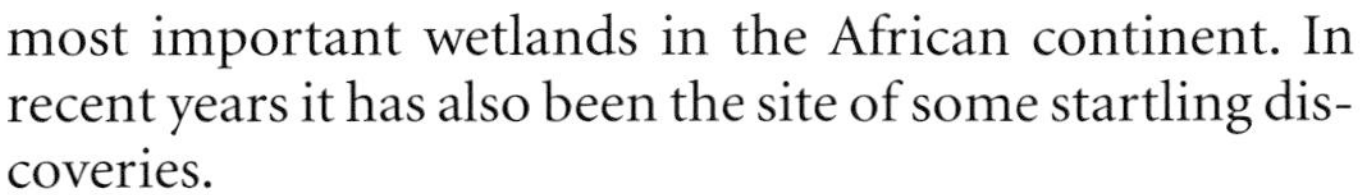most important wetlands in the African continent. In recent years it has also been the site of some startling discoveries.

Besides the lakes and marshes, the area contains much typical Sahelian vegetation, dominated by thorn bush savanna of *Acacia* and *Tamarix*, with areas of desert, grassland and low, salt-loving vegetation. A number of typical Sahel species breed, such as the rare Golden Nightjar, a deliciously plumaged member of the family which lives up to its name. Other dry-country specialities include the Kordofan Lark, a rare yellowish-brown species that breeds during the summer rains, the tiny, green-backed Sennar Penduline Tit, with its pepper-flecked forehead, and the splendid Cricket Longtail, a very small, effervescent bird with a long tail that is constantly spread and flirted. More widespread arid-zone species include the very smart Black-crowned and Chestnut-backed Sparrow-Larks and the Arabian Bustard, which is becoming harder and harder to find elsewhere. Birders will be equally delighted to see sub-Saharan staples such as Blue-naped Mousebird, Abyssinian Roller, Long-tailed Glossy Starling and African Silverbill in this typically West African landscape.

However, Djoudj is undoubtedly most famous for its wetland birds, which teem into the park during the wet season. For most casual visitors, the highlight will probably be a trip by pirogue to the island where an immense colony of at least 10,000 Great White Pelicans breeds, one of the largest gatherings of this widespread species anywhere in the world. Other abundant breeding birds include Darter, both Great and White-breasted Cor-

Right: large numbers of Palearctic wildfowl such as Garganey spend the winter at Djoudj.

morant, Black-crowned Night-Heron, Squacco Heron, Yellow-billed Stork and both White-faced and Fulvous Whistling Ducks. The River Prinia, an unspectacular but rare Sahel endemic, also occurs in the waterside vegetation, although most people would be hard-pressed to notice it.

Despite the abundance and density of the breeding birds on the lakes, marshes and flats of the Djoudj, even these numbers are dwarfed by the number of birds that simply come for the winter. Counts of half a million wildfowl are regular, and these have included no fewer than 180,000 Garganey, 240,000 Northern Pintail and 33,000 Northern Shoveler. Wader numbers can also be impressive, with a quarter of a million birds counted in the past, most of these Ruff, but with significant numbers of Black-tailed Godwit and Pied Avocet. When you mix this lot in with the regular wintering flock of about 20,000 Greater Flamingos and an increasing number of Lesser Flamingos (up to 12,000), you can see that, for a park of only 160 sq km, there are an awful lot of birds to be seen. That, of course, is just the larger birds. The counts for passerines have included two million Sand Martins passing through on migration, and 250,000 Western Yellow Wagtails.

Indeed, despite the eyebrow-raising counts of large waterbirds, and the attention that this has brought to the Djoudj over the years, it is actually a much less spectacular species that has underlined the great conservation importance of the park in recent years. In 2006 the Djoudj was revealed to be the single most important wintering site in the world of the highly threatened Aquatic Warbler, its lush marshes providing refuge for some 25 per cent of the world population. This fast-declining bird breeds in eastern Europe and Russia, and until the discovery in the Djoudj its wintering grounds were something of a mystery. Now that it has been found in an already protected area, the future for the Aquatic Warbler seems to be less bleak than it was just a few years ago.

Another remarkable discovery was made in this approximate area in 2007, although the precise whereabouts are being kept quiet. In January of that year, ornithologists from BirdLife International's local division, the Ligue pour la Protection des Oiseaux (LPO) in Senegal, discovered an enormous roost of raptors somewhere in the arid zone, one of the largest gatherings of the component species ever recorded. It comprised some 16,000 Scissor-tailed Kites and an amazing 28,600 Lesser Kestrels, the latter another of Europe's globally threatened species. What an astonishing sight it must have made. If ever proof was required about the exceptional importance of this site to Europe's wintering birds, this surely would be it.

Gambia River

SITE RANK 50 **Information**

HABITAT Mangroves, freshwater marshes, thorn scrub, forest and cultivations along a large river

KEY SPECIES Egyptian Plover, Red-throated and Northern Carmine Bee-eaters, African Finfoot

TIME OF YEAR All year round, although most birders visit from November to March, when Palearctic wintering birds are present. The rainy season, July to September, is very hot and humid

The Gambia, a tiny country whose boundaries cling to either side of the large Gambia River and is completely surrounded by neighbouring Senegal, has become a major birding destination. The beaches, wetlands, gardens and patches of forest on the coastal Banjul Peninsula have introduced many to the delights of sub-Saharan Africa, with colourful rollers, weavers, glossy starlings and sunbirds in the hotel gardens interrupting breakfast, lunch and dinner. Indeed, many go home satisfied with plenty of great birding memories having never left the coast. However, for the keener birder, there is something about inland Gambia that is even more special; perhaps it is the reality of leaving luxury behind, or of seeing the 'real' Africa, or of contemplating a truly different set of sightings. Whatever the reason, for many it is actually a trip up the wide Gambia River that marks the beginning of the truly adventurous part of their holiday.

At one time the best way to travel upriver was in one of the many ferries, but the Trans-Gambia highway now reaches inland and has cut down travelling times. It begins on the south bank of the river, crosses to the north at Farafenni some 100 km inland, and then crosses back by way of Janjangbureh (formerly Georgetown), a large town on an island in the Gambia River, before reaching Basse, 374 km from the capital, Banjul. There are many good birding sites along this highway, and the trip is enlivened by the two ferry crossings, where visitors can observe some of the comings and goings of West African life.

A good first stop could be at Bintang Bolon, about 50 km from Banjul. Here the Gambia River is still mainly saltwater and tidal, encouraging mangroves to grow along its banks. A short stop at this creek will often reveal birds

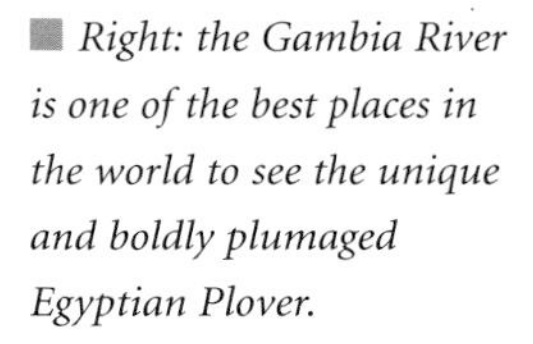

Right: the Gambia River is one of the best places in the world to see the unique and boldly plumaged Egyptian Plover.

Right: there is a large colony of Northern Carmine Bee-eaters at Basse. Besides bees, this species has a preference for locusts.

Right: the astonishing Exclamatory Paradise Whydah is a brood parasite; various species of pytilia are its chosen hosts.

similar to those that visitors may have enjoyed on the coast, such as various waders, Caspian and Lesser Crested Terns, Greater Flamingos and both Goliath and even Western Reef Herons. However, another 50 km further on and the whole area has a different feel. At Tendaba Camp, on the south side, the river is narrow enough for small boats to carry birders across to the other side, where two superb mangrove-lined creeks, Tunku Bolon and Kisi Bolon, hold specialities such as the exquisite African Blue Flycatcher and the far-from-exquisite Mangrove Sunbird, a distinctly underwhelming member of its dazzling family. This is one of the best places along the river to see the shy White-backed Night Heron, while you have to be very fortunate indeed to see African Finfoot or Pel's Fishing Owl, both of which are occasionally reported. Striated Heron, Spur-winged Goose, Yellow-billed and Woolly-necked Storks, Blue-breasted Kingfisher and Darter are much more reliable.

Away from the river, the Tendaba Camp area holds some dry-country birds, especially around the airstrip. It is possible to see the localized Abyssinian Ground Hornbill here, as well as the smart White-fronted Black Chat and White-crested Helmetshrike. Just to the west of here is one of Gambia's largest protected areas, the 110-sq-km Kiang West National Park, which is mainly dry woodland savanna. No fewer than 300 bird species have been recorded here, over half the total for the whole country. However, it isn't an easy place to work and such choice species as African Yellow White-eye, White-shouldered Black Tit, Brown Snake Eagle and Brown-necked Parrot can be hard to track down.

Winding Cisticola occurs by the ferry crossing at Farafenni and, indeed, it is the birds of freshwater wetlands like this that most birders will seek as they journey on the Trans-Gambian towards Janjangbureh. Just beyond the village of Panchang is a small swamp which retains its water in the dry season, and attracts such species as African Pygmy Goose, Lesser Moorhen, Orange-breasted Waxbill and the wonderfully named Exclamatory Paradise Whydah. Not far along is the famous Kau-ur Wetland, which is larger than Panchang and has the knack of attracting some exciting birds. Greater Painted Snipe often occurs here, along with such delights as White-faced Whistling Duck, Kittlitz's Plover and, in the winter, many Collared Pratincoles. However, the headline draw at Kau-ur and other wetlands in this area is the glorious Egyptian Plover. This boldly coloured species, a delicious combination of black, white, ash-grey and, on the underparts, ochre, is not a plover, but actually a courser, and has the unusual habit of covering its eggs, or even young, with sand in order to conceal them from predators.

Below: quiet backwaters might yield sightings of the shy African Finfoot.

Many birders spend at least a night at Janjangbureh. The main reason for this is to take a pirogue trip a couple of hours downriver in search of one of Gambia's most sought-after birds, the African Finfoot. In the quiet backwaters these birds are fairly easy to see, and bonuses may come in the shape of Shining-blue Kingfisher and Black Coucal.

Every birder makes the stop at Bansang Quarry on the way to Basse. One of the most attractive of a stunning family, the Red-throated Bee-eater, breeds at this site with a population of about 200 pairs. With their intense red throats and vivid purplish-blue undertail coverts, it seems impossible that they could be upstaged by any other bird. Yet, once the traveller has finally arrived in Basse, such trumping is perfectly possible. This is the best site in Gambia for the astonishing Northern Carmine Bee-eater, a longer-bodied, more elegant bird than its counterparts, with a pointed tail-streamer. An all-over deep, rich, pinkish-red, with an iridescent green head, these stunning birds can be seen along the banks of the rivers. Watching these birds, with their exotic appearance, in this rural setting deep into Africa, the holiday resorts on the coast of Gambia seem a long way away indeed.

Ivindo

SITE RANK 19 **Information**

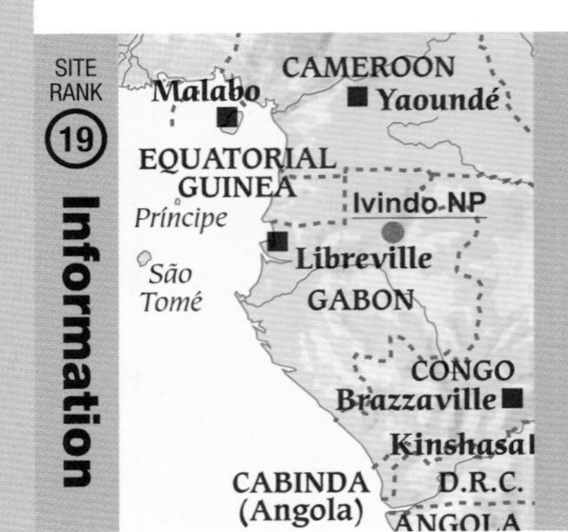

HABITAT Humid tropical rainforest; river

KEY BIRDS: Large species list (more than 450 species) includes African River Martin, Spot-breasted Ibis, Rufous-sided Broadbill, Ja River Scrub Warbler, Grey-necked Picathartes

TIME OF YEAR All year round

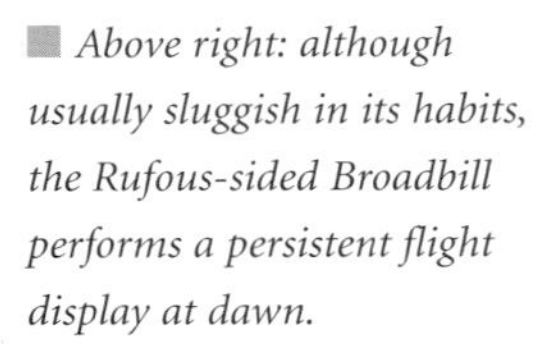

Above right: although usually sluggish in its habits, the Rufous-sided Broadbill performs a persistent flight display at dawn.

Few parts of Africa can compete with the extravagant species-richness of South America, but this corner of the mighty Guinea-Congo Forest comes pretty close. More than 430 species have been recorded in just the northern end of this newly created national park, prompting some admiring biologists to dub it "Africa's Amazonia". Many of the species present are spectacular, rare and restricted in range.

Before delving into the details of this site, however, it is worth reflecting on the modern miracle that is Gabon. At the beginning of the 21st century this West African country was bucking a worldwide trend by retaining a remarkable 80 per cent of its natural vegetation cover (mainly primary forest). Not surprisingly, the vultures of the logging industry were gleefully gathering over what was largely unprotected wilderness. Then, in August 2002, the president of Gabon, Mr El Hadj Omar Bongo, changed all that. With a bold and inspired stroke of the presidential pen, he lifted Gabon's tally of national parks from zero to thirteen. Over 28,500 sq km of land, mainly virgin forest, was thereby safeguarded for the future, covering no less than 11 per cent of the entire land surface of the country. Ivindo is one of the jewels in this sparkling new crown.

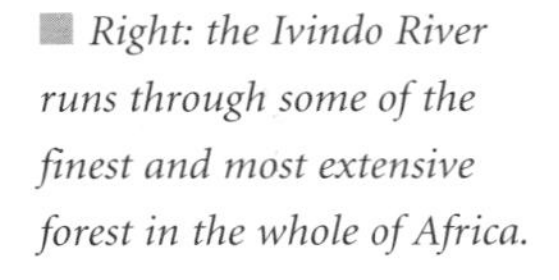

Right: the Ivindo River runs through some of the finest and most extensive forest in the whole of Africa.

These magnificent forests are similar in structure to those in other parts of the tropics. Mighty rainforest giants compete for space and create a dense, dark understorey below the well-lit canopy. Vegetation is thick and almost continuous from ground level to a height of 20 m or more, creating a wealth of micro-habitats for the birds. At Ivindo, some researchers have noticed that a slightly higher proportion of species than is typical of rainforests use the very lowest forest strata, close to the ground, and this means that the jungle birding is even more difficult than usual, with a host of skulkers hiding away in the undergrowth. Birders will have a tough task working through the many species of greenbuls (20 occur here) and warblers, but the rewards are great: such jewels as Green-breasted Pitta, Gabon Coucal and Black-eared and Grey Ground Thrushes adorn these lower levels, too, and if you are very lucky you may catch a glimpse of the two rare forest guineafowl, the Black and Plumed Guineafowl, as these shy birds scuttle around in the gloom.

There are many other treats in this forest. One of the memorable sounds at Ivindo is the odd persistent clockwork whirring made by the wings of the Rufous-sided Broadbill as it flies in a tight circle at a constant height before returning to its perch. This is a territorial signal performed by both sexes year-round; birds about to fly often give a couple of little vertical hops before take-off. The Grey-headed Broadbill occurs here, too, both species perching motionless in the greenery before making rapid fly-snatching sallies after passing insects.

At the Ipassa-Makokou Research Station in the north of the national park, close to the town of Makokou, birders can find some help in exploring the forest. This area has a distinguished history of ornithological explo-

Below: the peculiar Grey-necked Picathartes makes a nest of mud stuck to the side of forest cave; its choice of such habitat is the reason for its alternative name of Red-headed Rockfowl.

■ *Right: African River Martins appear at Ivindo in erratic numbers – they are thought to breed at as yet unknown locations upriver.*

ration; for 20 years up to 1984 it was the most intensely studied forest in the whole of Africa. Researchers cut a network of trails through the trees and many of these are still in existence today, so that visitors can use telescopes to watch the feeding flocks of birds pass by in the canopy, and hope to get glimpses of such gems as Rachel's Malimbe. Fruiting trees attract a wide range of species such as Afep Pigeon and Grey Parrot, as well as many primates. This is also a good area to look for the elusive Congo Serpent Eagle, a large raptor that specializes in snakes and lizards, but keeps itself well hidden. This raptor sometimes strikes at prey with its feet on the ground, in the manner of a Secretarybird.

The research station has been newly refurbished and the birdwatching begins on-site. Besides being next to the forest, the station also overlooks the Ivindo River and on most days Spot-breasted and Olive Ibises fly over, commuting between their roosting and feeding sites. Those with time and patience can check riparian thickets for White-crested Tiger Heron, and at night the strange booming call of the Nkulengu Rail can be heard. At certain times of the year you may even catch a glimpse of the exquisitely odd African River Martin. These blue-black sprites with waxy red bills and crimson staring eyes breed in large colonies in coastal Gabon. In February, after breeding, they migrate upriver and disappear for the off-season to unknown locations upstream, deep in the vast Guinea rainforest. They can appear along the Ivindo in large numbers; for instance, a flock of 15,000 was seen at Makokou in 1997.

Rivers and wetlands provide the only break from forest cover over much of the park's 3,000-sq-km area. Towards the south, however, there are also several 'bais', natural open areas where mineral deposits and soggy ground prevent the growth of forest trees. The best known of these is Langoue Bai, a couple of hours' drive (and then a long hike) from the town of Ivindo. Only discovered in 2000, this bai is frequented by large mammals, including Forest Elephant, Western Lowland Gorilla and African Buffalo, which are the main attraction and are likely to become an ecotourism trade in the near future. Nevertheless, the birding is great here too. Besides wetland species such as Hartlaub's Duck, the bai offers the chance to see large forest birds, such as Black-casqued Wattled, Piping and African Pied Hornbills and Great Blue Turaco flapping across the clearing. For those with more esoteric interests, the barely known Ja River Scrub Warbler is common here, with 30 to 40 pairs breeding. This is one of West Africa's various 'disappearing' species; it was first discovered in 1914 in Cameroon and then went missing again until 1997. It occurs only in forest clearings with dense growths of sedge.

One other highly sought-after species occurs in Ivindo National Park: the Grey-necked Picathartes. This is not merely one of the strangest birds in Africa, it is also one of the oddest anywhere, and nobody can quite decide what a picarthartes is; it looks like a cross between a rail and a thrush, and goes around with languid, otherworldly movements. It occurs only among rocky outcrops under the forest canopy, usually in rugged, hilly areas. Seeing this bird, like many of the species of this fabulous area, is a tough but rewarding challenge.

Bale Mountains

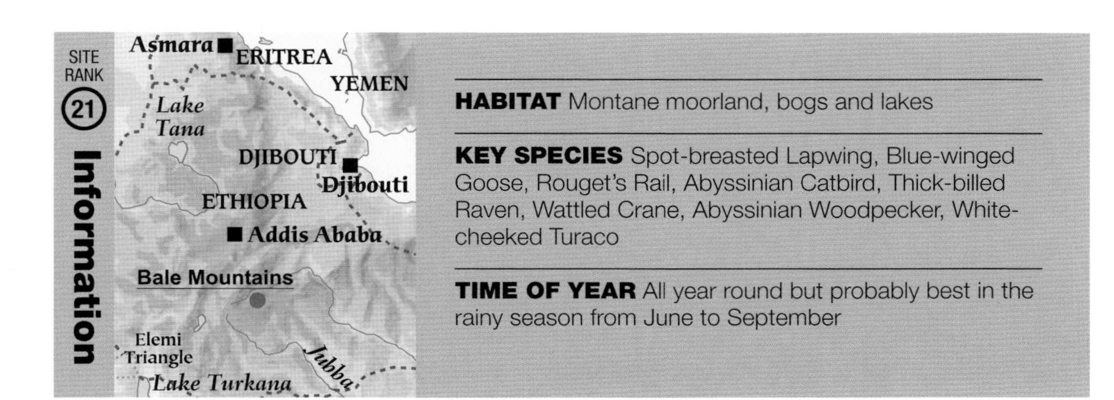

HABITAT Montane moorland, bogs and lakes

KEY SPECIES Spot-breasted Lapwing, Blue-winged Goose, Rouget's Rail, Abyssinian Catbird, Thick-billed Raven, Wattled Crane, Abyssinian Woodpecker, White-cheeked Turaco

TIME OF YEAR All year round but probably best in the rainy season from June to September

Below: the astonishing swollen bill of the Thick-billed Raven is a sight to terrorize the abundant rodents of the Bale Mountains.

The Bale Mountains National Park encloses the largest continuous area above 3,000 m in the whole of Africa. It is a dramatic world of mist-enshrouded open woodland, high altitude lakes and expanses of moorland punctuated by the tall, statuesque stems of Giant Lobelias. The range soars above its surrounding plateau as an altitudinal island, and holds a fine range of montane specialities, including no fewer than 14 of Ethiopia's 30 endemic birds. Indeed the Spot-breasted Lapwing is entirely endemic to this one area.

Most visitors will arrive from the north, and the national park headquarters is near the small settlement of Dinsho, on the northern border 400 km south of the Ethiopian capital Addis Ababa. This is one of the few parts of the park with open grassland, making it a good place to spot the Abyssinian Longclaw as it walks meadowlark-like on the ground, probing for food. Meanwhile its co-endemic, the Abyssinian Catbird, occurs in the open forest nearby. As its name suggests, it looks eerily like the unrelated Grey Catbird of North America – blue-grey, with chestnut undertail coverts. The Abyssinian Catbird is as much a thicket-skulker as its namesake, but can often be found in small family parties and has a much more powerful, melodious song. It appears to subsist largely on berries.

These open woodlands on the northern side of the park are dominated by juniper and tall *Hagenia* trees, with patches of dense scrub in between. One of the

Above: the endemic Spot-breasted Lapwing can often be located by searching close to herds of domestic cattle.

inhabitants here, the localized Chestnut-naped Francolin, is unusual among its relatives for being tame and easy to see. It also shirks the typical dawn and dusk feeding pattern of other francolins, possibly to avoid getting its plumage wet from morning dew in the sodden vegetation.

On the southern side of the park the forests have a different character. There are many more species of constituent plants, including large Podocarps, and the branches are festooned with creepers, moss and lichen. This, the Harenna Forest, is the best place to find the small Abyssinian Woodpecker and the delightful White-backed Black Tit, a bird which is indeed sooty black, with a pure white saddle.

The open woodland gives way to quite different vegetation above the tree line at 3,300 m. This is moorland, with scrubby patches of heathers reaching 0.5–1 m in height, with bare ground in between. A different francolin occurs here, the aptly-named Moorland Francolin, which restores its group's reputation by being shy and hard to see. A characteristic plant here is the famous red-hot poker (*Kniphofia*), which provides nectar for the occasional Tacazze Sunbird. Birding in these areas is also likely to produce flocks of the smart Black-headed Siskin, which can often be seen in yellow-flowering bushes of St John's wort, and also uses the tops of Giant Lobelias as lookout posts.

Despite these moorland delights, to birders the most exciting section of the Bale Mountains National Park lies at higher altitude still. Leaving the regional capital of Goba, the road continues southwards and climbs up to the Sanetti Plateau (3,800–4,200 m), and as it does so it becomes the highest all-weather road in the whole of Africa. The plateau itself is an area of heathland, stark volcanic mountaintops and high-altitude tarns, and it has its own highly distinctive birdlife. The star species is probably the Spot-breasted Lapwing, a somewhat irascible wader that, despite being very rare and doubtless highly specialized to this habitat, seems to like hanging around domestic cattle. The Rouget's Rail is another bird that confounds expectations; this splendid bird, a large red-brown species, is remarkably confiding and has been known to run around birders' feet, quite unconcerned.

The Wattled Crane, as stately as any species in its family, with a white neck, red hanging wattle behind the eye and a long 'morning suit' of grey that trails down almost to the ground, at least looks suitably wild; it is a seasonal visitor here, coming up during the wet season from June to September. It feeds almost entirely upon parts of sedges.

Sharing the Wattled Crane's predilection for sedges is the Blue-winged Goose, a portly goose that hardly ever swims. It can be hard to see, a grey-brown blob grazing in a grey-brown bog, but when it flies it shows off a huge patch of powder-blue on its forewing. Alongside it is a population of the primarily Palearctic Ruddy Shelduck, here completely isolated from the rest of its kind and at its southernmost outpost.

Curiously, two other Palearctic species also occur here in isolated populations. The nearest Red-billed Choughs in Africa are found in Morocco and Algeria, and the small population of Golden Eagles, only discovered in the early 1990s, is the only one in sub-Saharan Africa. Intriguingly, some of the plants here also have Palearctic affinities; the only wild rose in Africa, *Rosa abyssinica*,

Right: Harenna Forest is a good place to find the diminutive Abyssinian Woodpecker.

occurs on the heathland in the Bale Mountains.

The Golden Eagles, as well as the Eastern Imperial, Steppe and Greater Spotted Eagles which winter here, are well fed by the astonishing density and diversity of rodents on the plateau. An endemic, the Giant Mole-rat, may be found here at a density of 2,600 individuals per sq km, and there are also various other species of exotically named mice (for example the Giant Climbing Mouse) and rats (Ethiopian Narrow-headed Rat). These rodents play an important role in the ecosystem, churning up the soil and making numerous burrows. They also support the animal for which the Bale Mountains National Park was originally created, the magnificent Ethiopian Wolf. It is often possible to watch these carnivores hunting; they simply lie in wait at the entrance to a burrow and strike when one of the diurnal rodents pokes its head above ground.

Some mole-rats also fall prey to another real oddity, the Thick-billed Raven. Common in the highlands of Ethiopia, this is a huge raven with a massive bill and luxuriant throat-hackles. The bill is laterally compressed but hugely swollen above and below, so both mandibles are curved. The sight of one of these peculiar monsters grabbing a bizarre mole-rat simply sums up the uniqueness of this remarkable part of Africa.

Below: at altitudes above 3,300 m, moorland dotted with lakes is home to most of the region's endemic birds.

Rift Valley Lakes

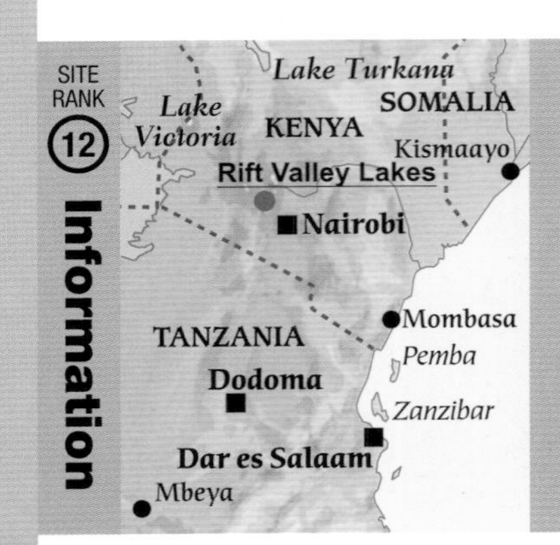

HABITAT Fresh and saltwater lakes, shoreline, scrub and woodland

KEY SPECIES Greater and Lesser Flamingos, Great White Pelican, African Spoonbill, waders, ducks, African Fish Eagle, hornbills including Northern Yellow-billed, Jackson's and Hemprich's

TIME OF YEAR Good all year round

Above right: the legion of colourful birds seen on safari has spawned many a keen birder – this is a Beautiful Sunbird.

Below: Africa's Rift Valley Lakes support exceptional numbers of African Fish Eagles. They feed on flamingos as well as fish.

Aside from lions and elephants, it is hard to imagine a scene more evocative of wild Africa than the sight of millions of flamingos, all spindly legs and craning necks, covering a shallow lake like some giant pink bloom. In Kenya, the scene of many of the famous photographs and feature films that first brought flamingo lakes to worldwide attention, it is still possible to enjoy this amazing sight, and listen to the flamingos' murmuring roar. However, despite the veneer of abundance that it presents, the fragile ecosystem that supports the flamingos is in danger of tipping over and disappearing into the memory. Kenya is changing, and as it does so, even its most iconic wildlife is being put at risk.

Large numbers of flamingos are still seen at two of Kenya's famous Rift Valley Lakes, Lake Nakuru and Lake Bogoria. Lake Nakuru is much the better known, having been a world-renowned bird sanctuary since 1960, but Lake Bogoria is nowadays a more reliable place to see them. Numbers at both lakes are subject to extreme fluctuations, but there are almost always several hundred thousand at Lake Bogoria, while Nakuru may at times hold just a few thousand. At their best both sites can hold a maximum of about two million birds yet, despite their look of permanence, the birds don't actually breed at either site. Instead there is a shifting Rift Valley population, many of which breed at Lake Magadi in the very south of the country and Lake Natron in Tanzania.

Nevertheless, the alkalinity and restricted depth of both lakes create ideal conditions for hungry flamingos. The potency of the water ensures that the foundation of the food chain is the blue-green alga *Spirulina platensis*, which grows abundantly, with little competition. This alga constitutes the main diet of the more abundant species, the Lesser Flamingo, while the Greater Flamingo, which often flocks with the Lesser and may number several thousand individuals at each lake, eats a much greater variety of slightly larger items, including chironomids, copepods and other aquatic invertebrates, which themselves depend on the algae. Flamingos are filter feeders, taking in water through the crack between the mandibles, using the tongue as a piston. The bills of the two species reflect their different diets; the bill of the Greater Flamingo filters items out by lamellae attached to the sides of the bill, allowing in particles up to 4–6 mm in size, whereas the upper mandible of the Lesser Flamingo has a deep 'keel' that is fitted with narrow plates and will only allow through items as small as 0.5 mm, preventing competition between the two species. Moreover, the items that Greater Flamingos eat are generally in the sludge at the lake bottom, so these birds usually submerge their heads to feed; the Lesser Flamingos are searching for floating prey, and submerge only their bills.

■ *Above: with many thousands of Lesser Flamingos on Lake Bogoria, things get a little cramped at times.*

Lake Bogoria is a spectacular lake, up to 42 sq km in extent, with hot springs and geysers and lava boulders on its shores, and surrounded by volcanic mountains. Besides the flamingos, there can be many other waterbirds present, such as Black-necked Grebe and Cape Teal, while the surrounding bush holds a good number of species typical of the Somali-Masai biome, including the wonderfully named White-bellied Go-away-bird, Northern Yellow-billed Hornbill and D'Arnaud's Barbet.

Lake Nakuru is a very different place, situated close to the important and burgeoning town of Nakuru, an agricultural and industrial centre. Over the years this 33-sq-km lake and its surrounds have played host to no fewer than 450 species of birds, and it still attracts such species as Yellow-billed Stork, African Spoonbill, Great White Pelican and Grey-hooded Gull, along with an impressive selection of Palearctic waders in season. Many of these species are fish-eaters – hardly what you might expect on an alkaline lake. But in 1960 a resilient fish known as *Oreochromis alcalicus* was introduced from Lake Magadi to curb mosquitoes; its effect on these was minimal, but it survived on algae and now provides food for the many large fish-eating birds.

It should be mentioned that these two lakes are not alone in attracting Rift Valley waterbirds. Not far from Lake Nakuru is the equally famous Lake Naivasha, which was formerly a freshwater paradise covered in colourful water-lilies and brimming with birds, but is now cursed by floating masses of introduced water hyacinth. It is still an excellent birding site, with 450 species recorded in and around it, and surveys regularly log 80 or more waterbird species. But the lilies have been savaged by introduced Coypu (now eradicated) and crayfish, and numbers of many species, such as Red-knobbed Coot, have declined. It still holds an exceptionally high density of the magnificent African Fish Eagle, the wild call of which is one of the great sounds of Africa. On Lake Nakuru these fine predators often specialize in catching flamingos instead of fish.

To the north of all the other lakes is another superb site, Lake Baringo, a freshwater wetland that manages to trump the other lakes by having 500 species on its list. About the same size as Naivasha, at 168 sq km, it is bounded on the western side by cliffs that hold several very rare species, including Jackson's and Hemprich's Hornbills and Bristle-crowned Starlings. The waters themselves are alone among the four lakes in still supporting a few White-backed Ducks and Darters, while an impressive colony of about 20 pairs of Goliath Herons dominates one of its islands.

It is a sad fact that every one of these lakes is suffering from one or more potentially serious ecological problems. Lake Baringo suffers from encroaching livestock denuding some parts of the catchment and allowing soil to enter the water. At Lake Bogoria there is a greatly increased presence of humans and their animals, while at Lake Naivasha, where the shores are mainly privately owned, the rich fringing vegetation and local rich soil is being given over to commercial agriculture and floriculture, the latter providing cut flowers for the European market and, unfortunately, contaminating the lake with agrochemicals.

Nowhere, though, is the future as bleak as at Lake Nakuru and, indeed, the site has gained international headlines recently for the mass deaths of flamingos. These are still unexplained, although it is thought that the die-offs may be caused by stress-related infectious diseases. Heavy metals from industrial waste, which are still dumped into the lake, could well be at the root of such problems. The human population has increased both at Nakuru town and in the surrounding countryside, and this has led to catastrophic deforestation in the lake's catchment. This process, if it is not halted, could well spell the end of the lake's viability as a place for birds. The flamingos could soon be no more.

Below: the flamingos of the Rift Valley have been attracting tourists for almost 50 years.

Serengeti

HABITAT Grassland, woodland savanna, acacia woodland, lakes, mountainsides

KEY SPECIES Common Ostrich, Secretarybird, vultures, Kori, Black-bellied and Buff-crested Bustards

TIME OF YEAR All year round

The Serengeti encapsulates most people's idea of Africa: the open plains dotted with acacia trees; herds of grazing animals running around and kicking up the dust from their feet; the great never-ending battle for survival between the carnivores and grazers, the hunters and the hunted; and the great icons of Africa wandering around, such as elephants, giraffes and rhinoceroses. Probably more footage of wildlife has been taken in this corner of Africa than anywhere else on earth.

The magic of the Serengeti is in no way exaggerated. The national park itself is huge (14,763 sq km) and is contiguous with the Ngorongoro Crater region to the east and the Masai Mara, in Kenya, to the north, so the populations of animals and birds are substantial, and it is perfectly possible to find corners of the national park that you can more or less have to yourself. Besides the plains and savannas, there are saline lakes and rivers, and the park is bordered by hills and mountains; in such areas you can indulge in specialist birding, in search of such unusual birds as the Grey-crested Helmetshrike, Red-throated Tit and Rufous-tailed Weaver.

It is impossible not to be impressed by the hordes of animals. The area is especially famous for its herds of White-bearded Wildebeests which, together with associated Burchell's Zebras and Thomson's Gazelles, undertake a remarkable seasonal migration from the Serengeti to the Masai Mara and back each year. During the rainy season (January to April) about two million wildebeests graze the short-grass plains of the south-east Serengeti and drop their calves; as the weather dries they then migrate towards the wetter margins of Lake Victoria to the west, and then spend the bulk of the dry season in the Mara. Oddly, when it rains again, and the grazing improves everywhere, they nonetheless up-sticks and return to the south. Nobody is quite sure why.

As these vast herds travel, some are inevitably picked off by the 3,000 Lions and 9,000 Spotted Hyaenas that

Below: big skies and big game – the Serengeti plays host to 2 million Wildebeest for much of the year.

■ *Above: the Common Ostrich, up to 2.2 m tall, has a remarkable turn of speed and is just as effective as any antelope in outrunning most predators.*

■ *Opposite: a White-backed Vulture gives away its identity in the midst of the scavenger scrum.*

occur in the ecosystem, not to mention Cheetahs and Leopards. In fact, with so much prey about, it would be a lazy predator indeed that could not fill its stomach. Indeed, death stalks so many grazing animals that opportunities also abound for the ecosystem's scavengers. This is where visitors can be distracted from mammals, as they watch the scrums of vultures that develop at carcasses – as much an iconic image of Africa as any other.

For the birder, the differences between the various species of vulture are especially interesting. Carcasses are spotted in one of two ways. White-backed and Rüppell's Vultures are colonial, and a couple of hours after daybreak each morning members of the colonies fly out to ride the thermals at heights of 200–500 m above the ground. They all spread out and, once a carcass has been found, the message is communicated by the very action of many birds wheeling down to the ground, and a crowd soon gathers. Other species, such as Lappet-faced and White-headed Vultures, are loners and essentially look out for their own needs, often getting by on scraps and road-kill, and even by killing smaller prey items of their own, which could be sick or injured.

However, it seems that neither White-backed nor Rüppell's Vultures are able to puncture the skin of large ungulates, so it is sometimes not until a Lappet-faced Vulture appears, with its fearsome bill, that the real free-for-all can begin. Once it does, the unseemly and gory scrum can begin: the colonial vultures swarm over the carcass, mainly gobbling up soft flesh and intestines, while the Lappet-faced, White-headed or Hooded Vultures tend to remain in the periphery. As far as the Lappet-faced and White-headed Vultures are concerned, this is because they prefer tougher body parts, such as skin, tendons and other coarse tissues and are content to wait their turn. The smaller Hooded Vulture, unable to compete in the mêlée, merely waits its turn.

Another group of birds that dedicated mammal-watchers cannot fail to notice are those that feed on or at the coat-tails of larger beasts. Good examples of this are the oxpeckers, two members of the starling family that

Above: a tender moment between two Southern Ground Hornbills. As they say, beauty is in the eye of the beholder.

specialize in removing ticks and other invertebrates from the hides of large mammals. The Red-billed Oxpecker, much the commoner of the two (the other is the Yellow-billed), is hardly ever seen away from its favourite hosts, namely buffalos, giraffes and rhinos; it will even roost on their backs. The only small ungulate on which it will graze for parasites is the Impala and, oddly, despite the abundant opportunities that must be present on elephant hides, this is one animal that is always avoided. Apparently, elephants have sensitive skin and will not tolerate the sharp claws of the oxpeckers.

Of course, the grazers are not the only animals that stride around on the grasslands. So does the world's largest bird, the Common Ostrich. This enormous bird, which has the largest eye of any bird (larger than a small hummingbird), is just big enough, tall enough and fast enough to cut it in the rough and tumble of the grassland without needing to fly. A fully grown ostrich stands up to 2.2 m tall, and it can run up to 70 km per hour when escaping a predator, weaving from side to side.

Not far behind in sheer size is the Kori Bustard, by most measurements the heaviest flying bird in the world. It stands 1 m tall and big males can weigh as much as 19 kg. Like the Common Ostrich it is largely omnivorous, and is famous for its ability to catch and eat snakes. It is one of several species of bustard that occur in these plains, the others including Black-bellied, Buff-crested and Hartlaub's.

There are two other quite fearsome, ground-dwelling, predatory birds that stalk alongside these. The Secretary-bird is a modified raptor with very long legs, a sharp bill and an equally sharp acceleration when chasing small mammals, reptiles and even birds on the ground. Like the Kori Bustard, it often despatches snakes, even deadly species such as mambas and puff adders. Southern Ground Hornbills, huge black-and-red monsters that only their mothers could truly love, tend to pick on the smaller stuff.

Thus, the celebrated megafauna of the Serengeti is not limited to animals with fur. There are big birds here to rival the big animals. And indeed, many people who have never taken an interest in birds at all finally notice them when on safari, and this can lead to a lifelong interest. After all, when they return home there aren't many life-or-death battles to witness, except on the bird table.

Western Caprivi Strip

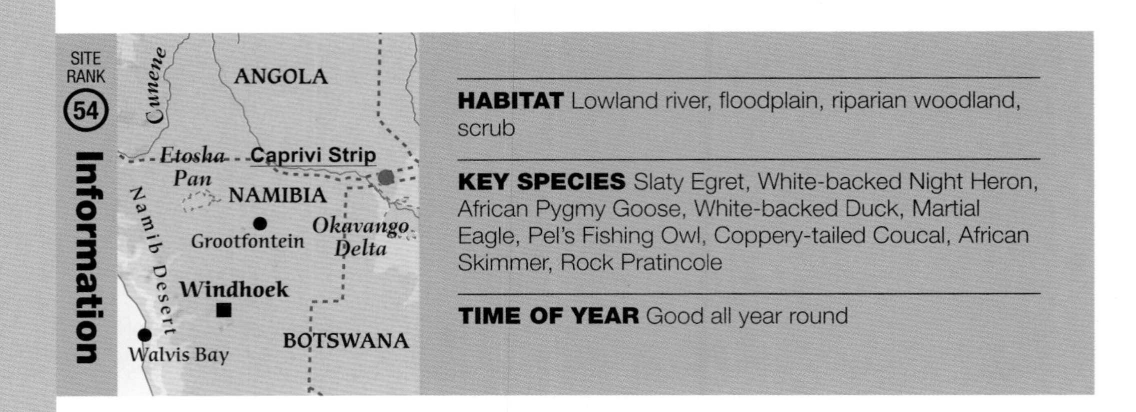

HABITAT Lowland river, floodplain, riparian woodland, scrub

KEY SPECIES Slaty Egret, White-backed Night Heron, African Pygmy Goose, White-backed Duck, Martial Eagle, Pel's Fishing Owl, Coppery-tailed Coucal, African Skimmer, Rock Pratincole

TIME OF YEAR Good all year round

The Caprivi Strip is a long, narrow extension of land connecting Namibia to the Zambezi River, ceded by the colonial British to the Germans in 1890. A completely flat, featureless landscape of floodplains, swamps, deciduous woodland and scrub, it is totally different in character from the rest of dry, parched Namibia; it is also far richer in birds. More than 450 species have been recorded, and it is not unusual for keen birders to see 300 in a single day. If you add the frisson of exploring a poorly known and wild part of Africa, where you can actually leave your vehicle and wander around outside among the game (admittedly risking your life if you are not careful), then it is easy to see why this area is fast becoming one of Africa's most popular areas for the discerning birder.

The western part of the Caprivi Strip is situated just to the north of the Okavango Delta, and provides access to every one of the special birds seen in that vast inland wilderness; it also offers a better opportunity to see extra woodland and dry-country birds. To begin with, one of the most popular stake-outs in the Western Caprivi is Popa Falls, a small tourist area near a section of the Okavango River that has rapids dropping down about 5 m. This is a very reliable site for the delightful Rock Pratincole, one of Africa's many birds that are widespread but difficult to see. The pratincoles spend most of the day roosting on rocks above the water, and tend only to hunt at dawn or dusk, when they hawk above the water for flies and other insects. Between March and July high water levels may force the birds to go elsewhere. A similar problem occurs with another special inhabitant of the river, the African Skimmer. This bird roosts on sandbanks instead of rocks and, like the pratincole, is a bit of a layabout during the day, feeding on fish in its

Below: Southern Carmine Bee-eaters breed colonially in holes in banks along the Okavango River.

Right: some lodges in the area have their own regular pair of Pel's Fishing Owls.

inimitable way at twilight and during a full moon. The birds cannot do without their roost sites, and go elsewhere when the water encroaches too high.

Just to the south of Popa Falls is the Mahango Game Reserve, or now, more correctly, the Mahango Section of the Okavango National Park. Four hundred species have been seen in the park's 250 sq km, with its combination of grassland flood-plain, general wetland, savanna and riverine woodland. In the eastern part of the park there is an excellent loop drive that covers all the habitats and allows convenient birding, even out of the car (take great care). On the lily-covered ponds and backwaters early in the loop look for the diminutive African Pygmy Goose, the secretive White-backed Duck and both African (very common) and Lesser (rare) Jacanas; with patience you can also pick out such cover-hugging birds as Allen's Gallinule and Dwarf Bittern. Meanwhile, in clumps of papyrus, check for both Greater and Lesser Swamp Warblers and both Chirping and Winding Cisticolas. A great deal of luck might land you a roosting White-backed Night Heron here, a nocturnal species that is widespread in the Caprivi Strip but nowhere easy to see. In contrast to the common Black-crowned Night Heron, these birds are not sociable and do not breed in colonies.

The eastern part of the loop passes a picnic site near a giant baobab, where you can view the Okavango flood-plain at your leisure. Here, on the open marshes, you might catch sight of one of the park's three pairs of Wattled Cranes, plus a superb variety of other long-legged wading birds. Alongside such common species as African Openbill Stork, Glossy Ibis and Goliath Heron, the Black Heron can often be seen folding its wings into the famous umbrella posture, providing shade below the bird that either attracts fish or makes them easier to see. A similarly plumaged species, the rare Slaty Egret, is also sometimes found here, often hunting in such long grass that it pops into view only occasionally. Other species occurring here include Long-toed and African Wattled Lapwings, and Collared Pratincole. Among the smaller birds found hereabouts are Luapula Cisticola, the splendid Rosy-throated Longclaw and several species of swallow, including Wire-tailed, the gorgeous Lesser Striped, and Grey-rumped.

The woodlands at Mahango can be excellent for birds of prey. Near the entrance gate, for example, the scrub and woodland may provide sightings of African Cuckoo-Hawk, Martial Eagle, Brown Snake Eagle,

■ Above: adult and juvenile Rock Pratincole; this delightful species roosts on rocks by day and hunts for insects at twilight.

Bateleur, Tawny Eagle, Common and Dickinson's Kestrels and African Hobby. Many of these species can be seen in the treetops in the early morning, while African Marsh Harriers and Ospreys hunt the marshes and ponds.

About 30 km to the south of Mahango, and actually within Botswana, is another excellent area of floodplain and wetland along the Okavango River, near the town of Shakawe. This is where you really need to go if you wish to see one of the area's most sought-after birds, Pel's Fishing Owl. Several lodges in the area have a resident pair, and these birds can be seen either roosting in dense foliage during the day, or fishing at night. They live up to their name by perching on a branch just above the water and intermittently dropping down, talons first, to grab a fish. These unusual owls, with their deliciously pinkish-brown plumage, sometimes hunt from a sandbank, making regular sallies into the water. Somehow they never seem to get their bellies wet.

There are other attractions beside this charismatic species. From September onwards, for example, you can enjoy the marvellous spectacle of breeding colonies of the Southern Carmine Bee-eater on the Okavango River, along with African Skimmers, while your chances of seeing Slaty Egret and White-backed Night Heron on the floodplain are probably better than at Mahango. You might also find the splendid Rufous-bellied Heron on the grassy marshes, usually not far from cover. For fans of smaller birds, the area is just as rich as further north, with the usual papyrus specialists and many species within the forests, too, including such delights as Brown Firefinch and the intriguing Retz's Helmetshrike. In most places such birds as these would form the highlight of the day, but in country blessed with this sort of abundance of birds, they are merely the sideshow of a very rich feast.

■ Right: Mahango combines several bird-rich habitats, including woodland and flood-plain.

Bwindi

HABITAT Forest from 1,160 m to 2,607 m, swamp, scrub

KEY SPECIES Grauer's Broadbill, White-headed Wood Hoopoe, Black Bee-eater, Neumann's Warbler, Grauer's Swamp Warbler, Regal and Purple-breasted Sunbird, Red-throated Alethe, Equatorial Akalat, White-starred Robin, Jameson's Antpecker

TIME OF YEAR All year round

The wonderfully evocative name of south-west Uganda's Bwindi Impenetrable Forest National Park gives the impression that you will be faced by a wall of vegetation too thick to pass. In fact, though, the name refers less to the nature of its forests, which are no more impenetrable than others in east Africa, than to its exceptionally tricky terrain. The area consists of alarmingly steep ridges and plunging valleys, all covered in gloriously untouched forest and made all the more awkward to negotiate by the humidity and by the slippery nature of some of the trails. It is one of the few places in east Africa where the transition from low-altitude primary forest to montane forest is fully encompassed within a single national park, untouched and seamless.

Although quite a small park, at 331 sq km, the Bwindi Impenetrable Forest is legendary among birders. This is not just because it is exceptionally rich in species, with 347 having been recorded so far, but it also plays host to a large number (23) of restricted-range birds that are endemic to the western slope of the Rift Valley, the so-called Albertine Rift. Since the rest of the range of these birds is within the politically troubled Democratic Republic of Congo (DRC), Bwindi is just about the only place where they can be seen and enjoyed.

This area is part of what used to be an enormous forest that once covered much of western Uganda, as well as Rwanda, Burundi and the nearby DRC. This great block is very ancient (25,000 years' old), and is thought to have been one of the Pleistocene Refugia that survived through the last ice ages. This, along with its altitudinal range, explains its diversity. In addition to its birds, the forest holds more than 200 species of butterfly (including the almost mythical Giant African Swallowtail) and an incredible 120 species of mammal. Of the latter, the Chimpanzees and Mountain Gorillas make Bwindi a major ecotourism destination. Several groups of gorillas in the forest have been habituated to people and can be

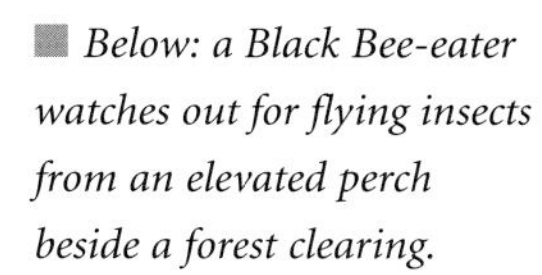

Below: a Black Bee-eater watches out for flying insects from an elevated perch beside a forest clearing.

Uganda

Right: the highly distinctive White-headed Wood Hoopoe is one of the more widespread forest birds at Bwindi.

visited for an hour a day by groups of up to eight people. Even the most obsessed birders inevitably find themselves on one of these incomparable gorilla-tracking adventures.

Nevertheless, once you've had your audience with the gorillas, you can concentrate on the birding. Most independent travellers and visiting groups allow themselves at least three or four days for this, beginning in the lowlands around Buhoma (near to the gorilla camp, where the treks begin) and finishing in the highlands around Ruhija. Most of the Albertine Rift endemics are found in the highlands, but the highest biodiversity is lower down.

The forests around Buhoma have an unusually visible canopy, so birds that would be very difficult to see in similar habitats elsewhere can sometimes be watched at leisure here. These include the Dusky and Olive Long-tailed Cuckoos, Western Bronze-naped Pigeon, Bar-tailed Trogon, various woodpeckers and the splendid White-headed Wood Hoopoe, which is usually encountered in noisy groups of ten or more. Such access also allows for scrutiny of that widespread forest phenomenon, the mixed feeding flock, and among the many small insectivores you can expect to find Shelley's and Red-tailed Greenbuls, Yellow-eyed Black Flycatcher and that delightful, brilliantly blushing *Phylloscopus*, the Red-faced Woodland Warbler. Meanwhile, a range of sunbirds, including Northern Double-collared, Green-throated and Blue-headed, are attracted to flowering trees, while six species of forest starling seek out fruit.

Above: the impenetrable nature of the forest refers more to its tricky, undulating terrain than the trees themselves.

These include the rare, glossy-blue Stuhlmann's Starling, Sharpe's Starling with its orange breast, and the Narrow-tailed and Purple-headed Starlings, here reaching the eastern limits of their ranges.

In the lower forest strata live a suite of altogether more skulking birds. Among these are a guild that have a habit of following swarms of driver ants or army ants, often spending much time perching motionless just above the ground. These include the unobtrusive Equatorial Akalat, with its truncated single-burst song, the White-starred Robin and two localized species, the endemic Red-throated Alethe, the dominant bird among the ant-followers, and the bicoloured Grey-chested Babbler. Other birds feed at ants' nests, including the Jameson's Antpecker, a peculiar estrildid finch of the forest undergrowth.

At Bwindi, you can tell how high up you are by the mix of birds. For example, if you are below 1,600 m, you can expect to see the European Robin-like White-bellied Robin-Chat gleaning the foliage. Above this height, however, a similar, related species takes over, the endemic Archer's Ground Robin, distinguished from its relative by a thin white supercilium. Similarly, in the misty highland forests, festooned with epiphytes and clothed in moss, flowering trees attract two sensational sunbirds, the brilliantly yellow-and-red-breasted Regal Sunbird, and the remarkable Purple-breasted Sunbird with its range of iridescent purple and bronze hues. Other endemics found at this height include the Stripe-breasted Tit, the smart Dusky Crimsonwing and the oddly-named Strange Weaver, which is undoubtedly rare but not particularly unusual to look at.

No trip to Ruhija, or indeed to Bwindi, is complete without the famously laborious three-hour trek down to the Mubwindi Swamp, in the east of the park. Many of the highland specialities can be seen along the way, while the swamp itself attracts a number of exceptionally rare birds. The most sought-after is probably the almost mythical Grauer's Broadbill, a small green-and-blue species whose main attraction is the fact that hardly anyone has ever seen it. In recent years, however, researchers have managed to find a few nests of this Albertine Rift endemic – quite a feat for such a small, quiet and annoyingly elusive bird. The other main attraction is Grauer's Swamp Warbler, a long-tailed brown skulker with a spotted throat and breast, which is found in marshy vegetation. Another skulker that can be seen here and elsewhere in the forests is the splendid Neumann's Warbler. It is a rare endemic which is very difficult to see – two irresistible attributes to the obsessive, and it also looks unusual; its tail is so short that it simply looks like a small bird that has had an accident at the back.

While you are searching for these difficult birds, Black Bee-eaters often provide some welcome glamour as they perch on dead tree stumps and sally for insects. At Bwindi you never know what might draw your eye next. That is one of the charms of this wonderful forest.

Dzalanyama Forest

HABITAT Miombo woodland; some grassland and rocky areas

KEY SPECIES Stierling's Woodpecker, Olive-headed Weaver, Spotted Creeper, Pennant-winged Nightjar, Boulder Chat, Anchieta's, Miombo Double-collared and Shelley's Sunbirds

TIME OF YEAR All year round

Miombo woodland is one of the major biomes of east and southern Africa, covering some 2.7 million sq km from Tanzania to Mozambique. A distinctly open type of woodland, with grass or sparse scrub for an under-layer, it is primarily made up of trees of the closely related genera *Brachystegia* and *Julbernadia*, which are broad-leaved and semi-deciduous, shedding some or all of their leaves for two to four weeks during the dry season which runs from April to October. Growing in nutrient-poor soils, miombo is not especially rich in biodiversity, but it does host a suite of birds that are found in this habitat and nowhere else.

There are few, if any, places in Africa better for birding in miombo than Dzalanyama Forest Reserve, just 58 km south-west of the Malawi capital, Lilongwe. Encompassing 100,000 ha of largely untouched woodland, among the largest in Africa, the reserve holds a superb range of miombo birds, including several that are very rare. The most sought-after is probably the highly restricted Stierling's Woodpecker, which is hard to find anywhere else. The peculiar Olive-headed Weaver is not far behind, closely associated with trees festooned with an 'old man's beard' type of lichen known as *Usnea*. Besides mainly feeding around clumps of *Usnea*, searching for insects, this inconspicuous weaver, yellow below and olive-green above, also builds an elaborate nest made entirely from this lichen and placed within a clump of it, usually below a supporting branch. Few nests in the world could possibly be better camouflaged.

The real delight in birding miombo, though, is in finding and following mixed bird parties, which are a major feature of this habitat and a spectacular aspect of this site. Strangely, unless you find a flock you may be tempted to think there are no birds here at all, for there can be long periods when nothing is happening. Then, quite suddenly, you will run into a bird party moving through the trees no faster than a person's walking pace, and then enjoy a marvellously exciting half-hour as you try to pick out all the different species. Some flocks here can hold 30 species, but usually you will find only about ten.

These flocks are not so different from the mixed feeding bird parties that can be found in tropical rainforest. They form at daybreak each day, and the component individuals are usually highly territorial, remaining in the flock within their own boundaries. Sometimes there can be a noisy changeover when the flock passes between territories, as is the case in the confirmed flock-maker, the Southern Hyliota, which is a small flycatcher-like warbler with a yellowish belly and dark upperparts relieved by a conspicuous white wing-bar.

Below: the diminutive Anchieta's Sunbird is a miombo specialist; it is endemic to central southern Africa.

One of the exciting aspects of these flocks is the wide range of different species that join them, each with its different niche. In the same gathering you will have confirmed canopy feeders, such as the almost tailless Red-capped Crombec, and ground-dwellers such as the Miombo Rock Thrush, which will pass its time searching through the leaf litter on the woodland floor (although, somewhat oddly for a rock thrush, it also regularly perches in the branches of trees). The differences in niche may be subtle: the Miombo Tit, for example, works the trunks and thicker branches at moderate heights, while the Rufous-bellied Tit prefers the small branches and twigs within the canopy. Meanwhile, the enigmatic Spotted Creeper creeps jerkily up and along

Right: the pennants of the Pennant-winged Nightjar are modified inner primary feathers, which trail behind the flying male.

branches, starting on each tree by flying to its base.

It is fascinating to notice the different techniques birds use to feed within the flocks. The Yellow-bellied and Southern Hyliotas, for example, are always on the move in the canopy, gleaning the leaves warbler-style for invertebrates, restless and quick. The uncommon Souza's Shrike, on the other hand, which looks like a half-adult, half-immature shrike, is a typical sit-and-wait predator, watching motionless in the sub-canopy for movements of food below. The delightful White-tailed Blue Flycatcher, a powder-blue species with a long, white-edged tail, behaves rather like one of the fantails of the Australasian region, forever bowing and curtseying when perched, and flicking open its long tail. Böhm's Flycatcher could hardly be more different; this is a sluggish, brown-coloured flycatcher with heavy spotting on the chest. It spends much of its time perched motionless, making occasional sallies at canopy height; its main food appears to be black tree ants. Yet all these disparate species choose to gather in roaming flocks; part of the reason, of course, is that there are more pairs of eyes to spot predators.

The social make-up of each species in the flock also varies. Thus Miombo Pied Barbets are usually found just in pairs or a family group, while Souza's Shrikes are often in threes, and Whyte's Barbets live in groups of pairs, all of which may join the flock as they search the branches for food.

Not all the birds in the miombo are confirmed flock-dwellers. The area is noted for its sunbirds, some of which appear to shun groups. The specialists include the Miombo Double-collared Sunbird and the highly localized Shelley's Sunbird, which seek out flowering creepers and other blooms. Neither of these sunbirds is particularly spectacular, being iridescent green above with a red band across the chest. This could not be said, however, of the gorgeous Anchieta's Sunbird, a stunning mite of a bird with brilliant crimson and yellow breast and iridescent blue throat. This localized species does join flocks from time to time.

Not all of the land protected by Dzalanyama is covered by miombo, and the very healthy total of 290 species is also drawn from riparian forest, *dambos* – patches of grassland in between the wooded sections – and rocky outcrops where, if you are fortunate, you may find the enigmatic Boulder Chat. If you walk in areas of short grass during the summer, you could even flush one of Africa's most spectacular birds, the Pennant-winged Nightjar. Yet, despite these wonders, your lasting memory from Dzalanyama will probably be from working the mixed feeding flocks in the miombo.

Bangweulu Swamps

HABITAT Large papyrus swamp, grassy floodplain, miombo woodland

KEY SPECIES Shoebill, Wattled Crane, Denham's Bustard, Swamp Flycatcher, Rosy-throated Longclaw, African Quail-Finch

TIME OF YEAR For Shoebills, the best season is May to July, but there is interest throughout the year

Sometimes the very nature of a site fits its star bird perfectly, and in that respect the Bangweulu Swamps of northern Zambia are the perfect location to find the Shoebill. It somehow seems right that, in these remote and inaccessible waterways, there should lurk a bird that is itself secretive and mercurial. Way out here, in this thinly populated part of the continent, one would expect something rare and peculiar to peer out from behind the clumps of papyrus.

Throughout the world, there are few more outlandish-looking birds than the Shoebill. It stands tall on the waterside vegetation and has rough grey plumage, but not even the briefest of views can overlook its remarkable bill, which dominates the bird's profile. Shaped rather like a Dutch clog, the bill is extraordinarily capacious, sharp-edged and fitted with a fearsome hook at the end, and it is clear from the start that it has a very specific task to perform: to catch and hold substantially-sized fish that form the core of the Shoebill's diet. The hook keeps them pinned in, while the sharp edges can disable the fish, and decapitate them where necessary. For its job description it is superbly effective, and provides enough large meals to make its sluggish owner's life tick slowly along.

The Shoebill is the quintessence of the patient hunter, and it can stand entirely motionless for more than 30 minutes, by which time one can hardly resist the desire to poke it to see if it still alive. But this is a strategy that works very well for catching the Shoebill's favourite food – lungfish. These large, primitive fish have to come to the surface periodically to gulp air, and when they do the Shoebill is ready for them, quite able to throw its whole body into an all-or-nothing lunge. If the bird misses it cannot recover in time to try again, so it moves right away and finds a new place to stand and resume its time-consuming vigil.

Being solitary and inconspicuous hunters, Shoebills can be hard to find, even where they are common. Fortunately, at Bangweulu, the allure of the Shoebill is understood – the naming of the lodge complex, Shoebill Island Camp, offers a clue – and in season it is easy to hire canoes and be guided to the target by expert guides who understand the birds and are up to date on their whereabouts. Thus, between May and July, while the seasonal waters are receding around the camp and lungfish are concentrated in ever shallowing pools, this becomes one of the most reliable places in the world to see them. After July, however, the birds retreat to the large, inaccessible, permanent parts of the swamp, and are much more difficult to track down.

The Bangweulu Swamp complex is one of the larger wetlands in Africa, covering some 10,000 sq km. About 4,000 sq km are designated by BirdLife International as an Important Bird Area, partly because of the population of Shoebills. An aerial survey undertaken in 2006 estimated the number of Shoebills at 470 individuals, which would undoubtedly rank as a significant proportion of the world population of this rare species.

Inevitably, however, there is much more to Bangweulu than just Shoebills. There are also significant populations of many other water birds, including Spur-winged Goose, Saddle-billed Stork, various herons and egrets (including Goliath Heron) and such ducks as Red-billed and Hottentot Teal. The rare Slaty Egret occurs irregularly, while the famous 'umbrellas' formed by feeding Black Herons (they fan their wings to form a canopy over themselves so that fish are easier to spot in the shade so created) are a reasonably common sight. In many parts of the permanent swamp there are abundant

Below: the highly localized Rosy-throated Longclaw feeds at the edge of wet areas.

Right: Bangweulu's star bird, the Shoebill, stands motionless on a riverbank waiting for a lungfish to come within reach.

growths of water-lily, where African and, for once, Lesser Jacana can be easily seen, along with parties of the exquisite Long-toed Lapwing. There are also some interesting passerines to be found here. Rosy-throated Longclaws wander around on the margins of wet areas, the Swamp Flycatcher, rather small and dull but for its white throat, makes sallies from the papyrus stems or other waterside vegetation, while the Greater Swamp Warbler and Chirping Cisticola remain hidden inside the forest of stems.

Some of the finest parts of the wetland are seasonally, rather than permanently, flooded, and these provide a rich grassland habitat for a different set of birds. The most obvious of these is the tall, stately Wattled Crane, which is hard to miss, and its population may number well into the hundreds. It runs a close second to the Shoebill for drama. Another less heralded speciality is Denham's Bustard, which has the conspicuous and endearing habit of leaping into the air and down again, with black-and-white wings rapidly beating as if making a vain attempt to fly. It does, however, rather drag its reputation down by habitually foraging through animal droppings for dung-beetles (there is plenty of game here, including a rare antelope, the Black Lechwe), and also will temporarily defend one of the abundant termitaria that dot this flat landscape. Other exciting species of the grasslands include the African Quail-Finch and the localized Grey-rumped Swallow, while in the wetter parts, between October and March, there are significant numbers of Palearctic waders, such as the globally threatened Great Snipe.

There is one further habitat at Bangweulu that deserves attention: its miombo woodlands, found on the edge of the swamp. Many specialities of the habitat occur here, such as Rufous-bellied and Miombo Tits, Yellow-bellied Hyliota and Red-capped Crombec. It is well worth giving this drier habitat some attention if you can drag yourself away from the attractions elsewhere.

Below: the uncommon and impressively large Denham's Bustard lives an idyllic life searching among piles of dung for beetles.

Sir Lowry's Pass

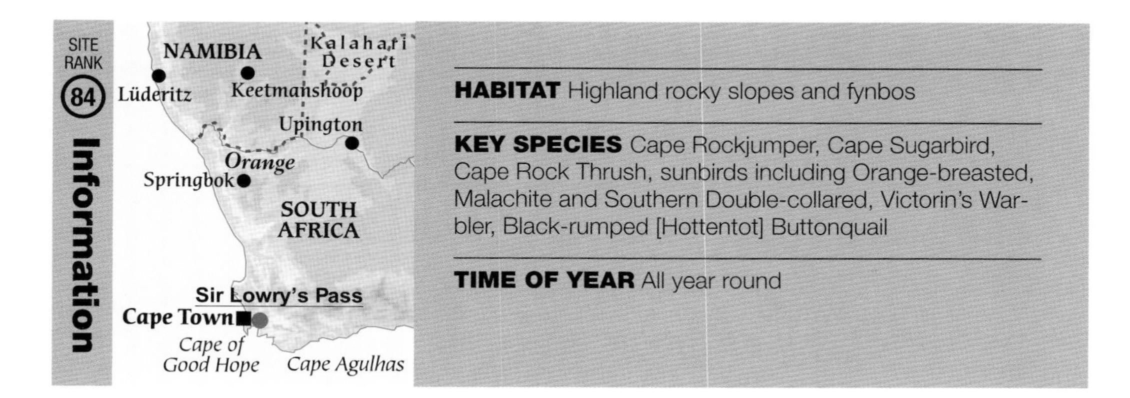

HABITAT Highland rocky slopes and fynbos

KEY SPECIES Cape Rockjumper, Cape Sugarbird, Cape Rock Thrush, sunbirds including Orange-breasted, Malachite and Southern Double-collared, Victorin's Warbler, Black-rumped [Hottentot] Buttonquail

TIME OF YEAR All year round

South Africa is famous as a great country for birding, but perhaps what is actually most exciting about the southern tip of the continent is its astounding floral diversity. The world is divided into an exclusive club of six 'floral kingdoms' – vast assemblages of plants linked by geographical region – and one of these, sitting at the same table as kingdoms covering whole continents, is the relatively tiny Cape Floral Kingdom, confined to the western part of South Africa. An area from the Olifants River, 240 km north of Cape Town, to Port Elizabeth, and extending only a few hundred km inland, hosts 8,700 species of plant in all, and some parts of the Cape, especially in the west, have more plant species than a comparable area of tropical rainforest.

Why should this be relevant to birders? The main reason is that the most characteristic habitat within the Cape Floral Kingdom is a type of low-growing, scrubby vegetation known as *fynbos*. As well as being floristically miraculous, *fynbos* hosts a suite of specialized and intriguing birds found nowhere else in the world.

One of the best and most convenient sites to go birding in the *fynbos* is Sir Lowry's Pass, in the Hottentots Holland Range, about 50 minutes to the north of Cape Town. Within minutes of arriving here you will notice

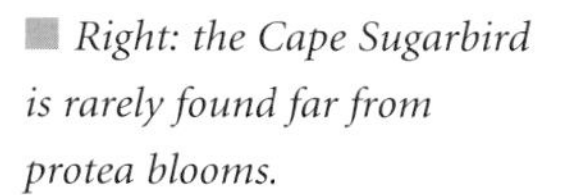
Right: the Cape Sugarbird is rarely found far from protea blooms.

the three main components of this unique habitat: the peculiar grass-like spikes known as restioids, the rubbery-leaved, heathy *Ericas* and, most famously, the spectacular, tall and sumptuous proteas. These plants are the very essence of the *fynbos*, and each has something to offer the various birds that live here.

The most celebrated of all the avian inhabitants of the *fynbos* is the magical Cape Sugarbird. It would look like your average nectarivorous bird, with its small size and long, decurved bill, were it not for its outrageously long, graduated trailing tail, which, in the male, is about three times as long as its owner. The tail transforms a fairly ordinary bird into something highly unusual, and indeed the sugarbirds, of which there are two species in South Africa, are something of a taxonomic riddle. They appear to have affinities with sunbirds, with the honeyeaters of Australia, and with babblers; as yet, no one is quite sure of their closest relatives.

What is in no doubt at all is the sugarbirds' utter dependence on proteas. Apart from feeding on their nectar and foraging for insects on their blooms and narrow leaves, the sugarbird also sites its nest in a protea, uses fragments of protea to build the structure, and also uses protea bushes for shelter. Not surprisingly, your average sugarbird is seldom found far from one of its favourite blooms, and the birds typically follow the sequential blooming of different species at different times of the year, leading to short migratory or nomadic movements.

■ Right: Orange-breasted Sunbird is more or less endemic to the Cape Floral Kingdom.

■ Opposite: a rare view of the normally skulking Cape Rockjumper.

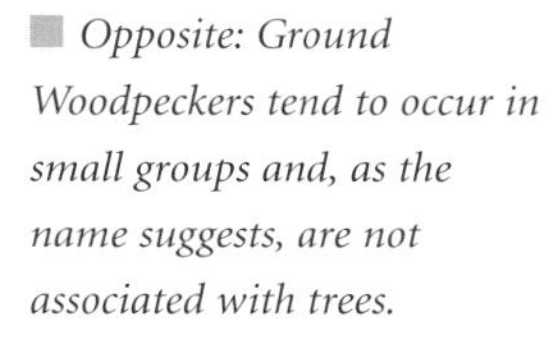

■ Opposite: Ground Woodpeckers tend to occur in small groups and, as the name suggests, are not associated with trees.

When feeding, a sugarbird deals with different plants in different ways, according to the physical characteristics of the inflorescences. When feeding on the platform-like flowerhead of *Protea multibracteata*, for instance, it sits atop the bloom and leans down to probe the outside; on the deep cup-like bloom of *Protea repens*, however, it almost disappears from view as it enters the enveloping arms of the flowerhead. Each of these manoeuvres ensures, of course, that the sugarbird is brushed with pollen, ready to pass on to the next plant.

With all these flowers about, it is hardly surprising that the *fynbos* hosts plenty of avid nectar-hunting sunbirds. One species here, the gaudy Orange-breasted Sunbird, is endemic to the region. Its particular preference is more for *Erica* species than for proteas, but this doesn't restrict it much, since there are 550 species of *Erica* endemic to the Cape. A small sunbird with an iridescent green head, olive back and tail and orange-red breast, it is often highly aggressive to nectar competitors, although it has to defer to sugarbirds, feeding lower down on the flowering spikes. It is often found in company with the spectacular, dark-green Malachite Sunbird, and the wide-ranging, common Southern Double-collared Sunbird, the latter with its iridescent blue and crimson breast-band.

The restioids are not great nectar plants, but they, too, have their avian admirers. One such is a seed-eater, the rather dull, yellowish and streaky Cape Siskin, which keeps a low profile, feeding in small parties amid shrubs and bushes and on the ground. The restioids also contribute to the dense, scrubby nature of the *fynbos*, a characteristic allowing this habitat to hold quite another set of specialists – those that skulk in thick vegetation. These include the Hottentot Buttonquail (recently split as a species from the widespread Black-rumped Buttonquail), which is usually seen only when it flushes away at your feet, and the ultra-secretive Striped Flufftail, which is so reluctant to show itself that there are records of it being stepped upon rather than flushing away. Both of these species will drive you mad as you look for them.

You will probably have better luck with another skulker, the Victorin's Warbler, because this species will occasionally deign to leave cover and perch on a bush-top when it needs to sing its pleasing, lilting song, after which it is liable to disappear again for hours. For a bush warbler it is a handsome species, rich brown with grey cheeks and distinctive yellow, staring eyes. It is another endemic of this part of South Africa.

Sir Lowry's Pass is a rugged, hilly area with plenty of rocky outcrops in addition to its *fynbos* cover. This provides excellent habitat for several other specialists. The unusual Ground Woodpecker, an ant-guzzler that hardly ever perches on trees, is common, often being seen in small groups as each bird digs and holds its bill underground, lapping up legions of the ants. Birders may also find two endemic rock thrushes, the ashy-grey Sentinel Rock Thrush and the rich cinnamon Cape Rock Thrush. However, the main target bird of the rocks here is the highly localized Cape Rockjumper, for which Sir Lowry's Pass is one of the best sites. This splendid, flame-eyed, chat-like bird, in which the male has a bright chestnut breast and rump, a white-tipped black tail and a long white submoustachial stripe on a black face, more or less lives up to its name, in that small parties can be seen hopping from rock to rock or flying low from one outcrop to another. When perched they often hold up their tails, and give loud whistling peeps. Rockjumpers are co-operative breeders, young birds helping at the nest of an adult pair. They are shy, cover-loving birds that feed on the ground, and may take an effort to see. However, as with the other unique inhabitants of this superb place, the effort of finding them is well worthwhile.

Mkhuze

HABITAT Sand forest, fig forest, savanna, rocky outcrops, wetlands

KEY SPECIES Southern Banded Snake Eagle, Crested Guineafowl, African Broadbill, Narina Trogon, Trumpeter Hornbill, Neergaard's Sunbird, Green and Pink-throated Twinspots

TIME OF YEAR Any time, although July to November is probably best

Above right: the delightful Pink-throated Twinspot is found in the sand forest.

Below: Green-winged Pytilia's preferred habitat is thorny scrub.

Although not very well known outside South Africa, this 400-sq-km game reserve in KwaZulu-Natal boasts the second-longest bird-list of any site in the country, not far behind the gigantic Kruger National Park. Its 450 species represent a quite astonishing total for such a comparatively small area, and the sheer abundance of birds makes this site a complete delight to visit.

The reason for this diversity is the great variety of habitats packed in together. Each, quite naturally, has its own set of birds and, if you add Palearctic migrants, intra-African migrants and local dispersing species to the mix, it is just possible to see how the number of species reaches such dizzy heights.

Much of the reserve consists of the acacia savanna that is such a familiar sight in Africa, depicted in endless safari brochures and wildlife books. If you visited this game reserve as your first-ever look at the African bushveld, your eyes would be popping at the sight of such birds as Lilac-breasted Rollers, Little Bee-eaters, Southern Yellow-billed Hornbills, Village Weavers, Grey-headed Bushshrikes and Blue Waxbills. However, such iconic birds are just the hors d'oeuvre of a visit to Mkhuze. A closer look will soon reveal the scarcer inhabitants: Broad-billed Rollers in riverine woodland, White-fronted Bee-eaters along the banks of the Mkhuze River, Trumpeter Hornbills in the forests, Lesser Masked and Thick-billed Weavers in the marshlands, the amazing Gorgeous Bushshrike – a brilliant red, orange, black, green and yellow miracle – in the woodlands, and the colourful Green-winged Pytilia in thorny thickets.

Everything here requires a closer look. With such variety on tap, a visit to the same place on consecutive days, and at different times of day, will bring new rewards. It is amazing how new species just seem to pop up from nowhere and, as a result, this game reserve is one of the most popular sites in South Africa for keen birders.

Mkhuze has some quite specialized habitats along with the more typical ones of the region. The most famous is known as the sand forest, which occupies a small area close to the northern edge of the reserve. Consisting of woodland growing upon old sand-dunes, it supports some highly localized species, including the range-restricted Neergaard's Sunbird, a diminutive member of its family with a short bill and mainly green and brown plumage save for a bright crimson breast-band. Occurring only in eastern South Africa and southern Mozambique, Neergaard's Sunbird is fussy about its dry habitat, and Mkhuze is one of the very best places to find this rarity. The same applies to the much more widespread African Broadbill, which is commoner here than almost anywhere else. Its frog-like call is the best giveaway, although it often requires a bit of effort to find.

Other good birds in the sand forest include Crested Guineafowl, Narina Trogon, Square-tailed Drongo and two glorious seed-eaters, the Green and Pink-throated Twinspots. Shy and elusive, the twinspots are sometimes seen from the hides in the area (Kubube or Kumasinga); both birds have black breasts studded with white spots, while they differ in the rest of their plumage, the Green Twinspot being primarily pale green, and the male of the Pink-throated having a face and breast as perfectly mauve as it is possible to imagine.

Another prize habitat at Mkhuze is another forest, this one consisting mainly of fig trees and fever trees. It lies at the southern end of the reserve, and is nowadays accessible only on guided walks arranged by the park authorities. Among the good birds found here are, once again, Narina Trogon and Square-tailed Drongo, plus the skulking Green Malkoha, the fabulous, flame-winged Purple-crested Turaco, the bright green African Green Pigeon and the pert Blue-mantled Crested Flycatcher. Another prize to be had is the scarce Southern

■ *Below: classic acacia savanna can be found at Mkhuze.*

■ *Above: the glorious Purple-crested Turaco occurs in the fig forest.*

Banded Snake-Eagle, a raptor that primarily catches small snakes by still-hunting from a perch, while very lucky birders may even catch sight of a roosting Pel's Fishing Owl.

The presence of a fishing owl indicates that there is plenty of water hereabouts. In fact, there are rivers throughout the park, and some of these drain into shallow pans that hold water throughout the year. In places these pans are fringed with reed-swamp and papyrus. The pans offer some great game viewing and, of course, attract large numbers of birds. The largest of them, Nsumu Pan in the south-east of the park, astride the Mkhuze River, is good for such fish-eaters as Great White and Pink-backed Pelicans, Yellow-billed Stork, African Openbill and the African Fish Eagle, all of them pretty common. In season, depending on water levels, a visit between November and March can produce a good list of Palearctic waders such as Marsh, Wood and Common Sandpipers, Ruff and Common Greenshank, while locals such as African Wattled Lapwing, Spur-winged Goose and White-faced Whistling Duck occur throughout the year. Meanwhile Inhlonhlela Pan, near the northern boundary of Mkhuze, has the right sort of lily-covered waters to attract African Jacana and the exquisite African Pygmy Goose.

There are yet more habitats to explore, including dense thorn-scrub, inhabited by the scarce Grey Penduline Tit; palm forest, home of the Lemon-breasted Canary; and rocky outcrops, favoured by such species as the Striped Pipit. The list goes on and on, and so do the birds.

Andasibe

SITE RANK 9 **Information**

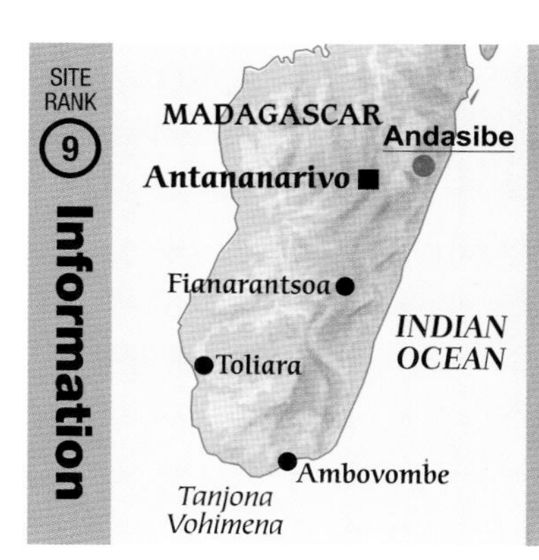

HABITAT Mid-altitude rainforest, marsh, villages

KEY SPECIES Nuthatch Vanga, Pitta-like, Scaly, Rufous-headed and Short-legged Ground-rollers, Cuckoo-roller, Madagascar Wood Rail, Velvet Asity, Common Sunbird-Asity, Madagascar Owl, Madagascar Blue Pigeon

TIME OF YEAR All year round

A good many people have their first taste of wild Madagascar here, in this set of reserves under the umbrella epithet of Andasibe National Park. None of them is ever likely to forget the experience. Analamazaotra Special Reserve, better known as Perinet, plus the newly opened Mantadia National Park, and the wilderness of Maromizaha, combine to protect some truly prime quality mid-altitude rainforest on Madagascar's eastern escarpment, and all this just four hours' drive from the capital city, Antananarivo. It is possible, therefore, to visit on a long day trip. Although most famous for its mammals, including the incomparable, tailless, forest-serenading Indri, the national park is home to a remarkable 70 species of endemic Malagasy birds, and the forests simply brim with all kinds of unusual plants and animals. There is nowhere better to become giddily immersed in the delights of birding in Madagascar.

Even the short walk from the village to the entrance to the special reserve guarantees birds that most people only dream about. The common garden or forest birds are mostly endemics, at least to the Indian Ocean Islands, and include species such as the Malagasy Bulbul, Common Newtonia, Souimanga Sunbird and Madagascar Wagtail. The smart and rather regal-looking Madagascar Blue Pigeon is likely to fly over or perch in a treetop, and before long you are bound to bump into one of the flocks of forest birds that are

Below: Common Sunbird-Asity is a member of a small family that is endemic to Madagascar; it feeds on nectar.

Below: the colourful Pitta-like Ground-roller is the easiest of its family to locate at Andasibe. Despite the evidence of this picture, it rarely leaves the ground.

very much a feature of Madagascar birding. These always include newtonias and bulbuls, along with such species as the Madagascar White-eye, Malagasy Paradise Flycatcher and Common Jery, and if you work hard on these flocks they will probably offer you your first glimpse of one of the endemic families for which the Malagasy region is renowned, the vangas. These mysterious shrike-like birds, usually placed in their own family, show a remarkable radiation of colour, form and habitat throughout the island, somewhat equivalent to the Hawaiian honeycreepers of the Pacific. You will probably see the rather small Red-tailed Vanga first, as it is a common component of flocks here, but the area is also good for the striking Blue Vanga, blue above and white below, with a conspicuous, staring yellow eye.

One of the most famous residents of the area, often found within the Analamazaotra Special Reserve, is another member of the family that joins these flocks, the Nuthatch Vanga. Surprisingly hard to find elsewhere, even in similar habitat, the Nuthatch Vanga, as its name implies, fills a similar ecological niche to the nuthatches or treecreepers, which are absent from Madagascar. It spends its time in the upper levels of the forest, hugging the branches and creeping along with a jerky motion, always working its way up, not down. This splendid bird is largely dark grey-blue in colour, with a black mask and a peculiar pink bill. It feeds on invertebrates, but does not probe into cracks in the same way that a treecreeper does.

Another of the families at the top of anyone's list are the asities, of which two species occur at Andasibe. The jewel-like Common Sunbird-Asity, with its frenetic character and long, sharply down-curved bill is remarkably similar to a sunbird, and similarly sips nectar, yet, in common with all the members of the family, it is now known to be a primitive songbird related to the broadbills, with a limited set of muscles making up the sound-producing syrinx. At least the Velvet Asity, a rather plump, short-tailed, slightly languid forest bird, looks a bit like a broadbill. It is the only complete vegetarian of the forest understorey and, to add to its uniqueness, also forms dispersed leks, with male birds holding compact territories and competing for females within sight of each other. Both asities have colourful wattles, green in the largely dark-plumaged Velvet Asity, blue in the largely yellow Common Sunbird-Asity; these colours are produced by arrangements of collagen fibres in a way unique in the animal kingdom. Incidentally, the Yellow-bellied Sunbird-Asity, which tends to replace the commoner bird above altitudes of about 1,600 m on the eastern escarpment, is only doubtfully recorded here, mainly at Maromizaha; the mid-altitude forest here peaks at 1,500 m.

The fabulous ground-rollers are the other 'big' family at Andasibe. With a little effort, and provided you visit from September to November when the birds are giving their low booming calls, it is possible to find all four rainforest species at Andasibe. The colourful Pitta-like Ground-roller and the Scaly Ground-roller are fairly easy to see, while the Rufous-headed Ground-roller, although something of a speciality, is found mainly in rugged ravines, and the Short-legged Ground-roller, which alone habitually perches well above ground, should make you work very hard. If you wish to see all four species, you need time and a lot of patience, but as you search you should be rewarded with views of other forest dwellers, such as Madagascar Wood Rail, the wonderful Madagascar Ibis, Red-breasted Coua and White-throated Oxylabes.

Andasibe isn't all forest, and indeed some of its best specialities are waterbirds. Lac Vert, in the middle of Analamazaotra Reserve, is one of the few reliable sites for the Madagascar Grebe, while the Malagasy Kingfisher is common here. There are several marshes in the vicinity, too, such as Torotorofotsy, 15 km west of the town, and these allow you to look for such species as Madagascar Rail and Grey Emutail. Who knows, if you get so exhausted that you are delirious and delusional, you might coax the almost mythically rare and elusive Slender-billed Flufftail on to your list, too.

There are so many specialities in this superb area that it is impossible to name them all. The splendid Cuckoo-roller is another member of a family endemic to the region, and is quite a common sight and sound (a rising whistle) in the forests; it often feeds on chameleons. At night, the entrance to Analamazaotra is an excellent roost-site for the Madagascar Owl, while Rainforest Scops Owl and Collared Nightjar also occur. Everywhere you look, indeed, and at any time, there is something to enjoy in this fantastic place.

Right: its habit of calling with its deep voice from mid-levels of the forest makes the Short-legged Ground-roller exceedingly hard to find.

Below: one of the world's great wildlife experiences is to hear a group if Indris in full voice; these large, tailless lemurs are a speciality of Andasibe.

The Spiny Forest

SITE RANK 1 Information

HABITAT A unique sub-arid scrub and woodland community

KEY SPECIES Long-tailed Ground-roller, Subdesert Mesite, couas including Crested, Red-capped, Giant, Running and Verreaux's, vangas including Sickle-billed, Lafresnaye's, Chabert's and Red-tailed, Archbold's Newtonia

TIME OF YEAR All year round

Is there any birding spot in the world that can compare to this? Its unique and strange birds, and the almost surreal scenery, combine to make the spiny forest of southern Madagascar one of the most exciting places in the world for birdwatching.

Many places in the world can be described as unique, but the spiny forest is dramatically different in every way to anywhere else on earth. Growing on sandy soils with a rainfall restricted to a maximum of 610 mm a year, this sub-arid woodland and scrub is dominated by some singularly peculiar plants, most of them unique to Madagascar and to this narrow coastal strip. The most famous are the baobabs, tall deciduous trees with bizarrely bulbous trunks and spindly-looking branches, which look amusingly like normal trees that have been planted upside down; six species grow in Madagascar, while just one occurs on the nearby African continent. These lumbering giants are scattered across the landscape, while the rest of the spiny forest is dominated by woody euphorbias, and trees and shrubs that look rather like cacti, with thick rows of spines joining forces with small deciduous leaves along their branches. These are the Didieraceae, whose growing habit has given rise to the name 'Octopus Tree' for one species, *Didierea madagascariensis*, which effectively describes its somewhat ramshackle, spreading branches that flail out in all directions from close to the base of the

Right: parties of the wonderful Sickle-billed Vanga are a relatively common sight in the Spiny Forest.

Opposite: huge Didierea *trees sprout from the sandy soil, giving an unearthly feel to the place.*

Above: the smallest and most arboreal of its family, the Verreaux's Coua is easily overlooked.

tree. The *Didierea* trees give this woodland its spiny appearance, removing any softness from the edges of the trees, and the thin, skeletal outlines of the sticking-up branches give the forest its surreal horizon.

It is hardly surprising, given the nature of the habitat, that the spiny forest hosts some unique birds, and they, too, sometimes veer on the side of weirdness. One species, in particular, is distinctly odd, the marvellous Subdesert Mesite. The small family of mesites, which is restricted to Madagascar, is something of a puzzle anyway, being placed somewhere near the rails, but having been associated with doves or pheasants in the past. The truth is that these birds are quite unlike anything else, with their small size (about that of a thrush), reptilian movements, small wings and heavy tails. When walking, the Subdesert Mesite bobs its head like a chicken and flicks its tail, and it never seems to be in a hurry. It feeds on invertebrates and a little fruit and spends most of its time on the ground. Lacking a true collarbone it does not fly well, and mesites usually prefer to clamber rather than fly up to their nests. These birds live in groups and, should you be fortunate enough to stumble upon them, they will fly up to a branch and just sit there, peering at you. The Subdesert Mesite has a long, decurved bill and its pale breast is spotted with black and, in the male, streaked with pink.

Mesites are extremely difficult to find and, if you visit the spiny forest in search of them, you are advised to enlist the help of a guide; several work near the hotels in Ifaty, near Toliara. The habitat is rough, dry and thorny, and the birds exceedingly elusive, so a little local experience will make a difference. Many of the local people follow the birds by checking for their tracks in the sand.

A local guide is also advisable if you want to find the region's other major five-star speciality, the exquisite Long-tailed Ground Roller. This bird, the only representative of its endemic family to be found outside rainforest, is rather similar in shape to a roadrunner, and not dissimilar in pattern, with cryptic brown speckling on the upperparts, including the long, graduated tail. However, this terrestrial species is also marked with perfect powdery sky-blue on its wings and outer tail-feathers, and a bold white submoustachial streak that cuts through its brown ear-coverts and sides of the throat. It is somewhat more languid in its hunting, too, alternating long periods of standing still with brief runs after invertebrates such as ants, worms and Madagascar's famous cockroaches (which make hissing sounds). Long-tailed Ground-rollers move effortlessly over this difficult ground, making tracking them down something of a challenge.

Another group of birds that define the spiny forest are the couas, strong-legged, long-tailed and soft-plumaged cuckoos, all of which have colourful patches behind the eye and make loud, far-carrying calls. The Crested Coua in particular, a greyish species with an elegant mop, makes a chorus of descending *koa-koa* calls in the twilight. No fewer than five species occur in the spiny forest, of which Crested and Verreaux's are arboreal while the Red-capped, Giant and Running Couas are all primarily ground-dwelling. These birds are predatory and eat chameleons and snails as well as insects.

The vangas represent the greatest example of adaptive radiation among passerines in Madagascar, and several species of these superficially shrike-like birds occur in the spiny forest. The most distinctive is the splendid Sickle-billed Vanga, pure white but for jet-black wings and tail; its remarkably long, blue, curved bill is used for probing into cracks and holes for invertebrates, and the birds live in noisy groups of up to 25 individuals. The rare Lafresnaye's Vanga has a massive, shovel-like bill used for gripping animal prey, and its overall shape and plumage pattern, with black cap and pale underparts, recalls the butcherbirds of Australasia. The White-headed Vanga, often found alongside the Sickle-billed and similar in plumage, although short and thick of bill, often joins mixed bird parties with the black-and-white Chabert's Vanga. The tiny Red-tailed Vanga, not much larger than a warbler and with a smart black face-mask and yellow eye, hardly looks like a related species.

These are just some of the characters found in the extraordinary spiny forest. The supporting cast includes a number of interesting species, such as the chameleon-eating Banded Kestrel and the Madagascar Hoopoe, as well as a number of great rarities, including the small brown Archbold's Newtonia and the Subdesert Brush Warbler. Owing to their rarity, these birds would top the bill in most areas of the world, but here, in this most astonishing of places, they hardly raise an eyebrow.

The Seychelles

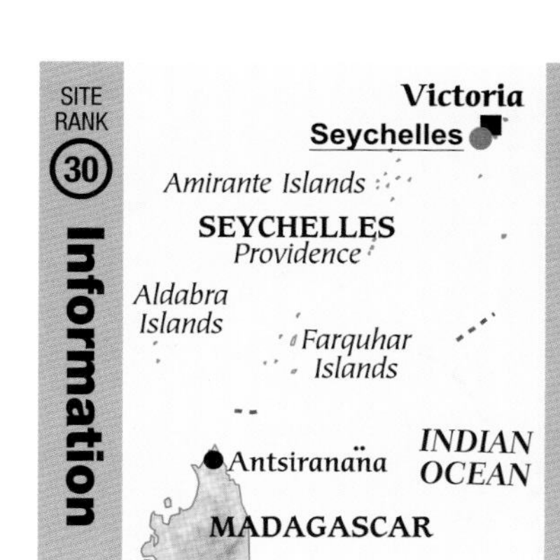

HABITAT Islands, forest, cultivation, mangroves

KEY SPECIES Eleven endemic landbirds including Seychelles Magpie-Robin and Seychelles Paradise Flycatcher, and a host of tropical seabirds including White Tern

TIME OF YEAR July to October is best, when the seabirds are breeding

Right: the incredible Seychelles Paradise Flycatcher is found only on the island of La Digue; this is a male.

Below: the Seychelles combine great birding with a picture-postcard tropical landscape.

The Seychelles is a group of small islands 1,500 km off the eastern coast of Africa, in the Indian Ocean. It is a luxury tourist destination, offering the romantic tropical island experience with sun-drenched beaches, palm trees, coral reefs and top-class accommodation. To wildlife enthusiasts and conservationists, however, they are something much more: an example of successful habitat restoration and the recovery of bird populations that at one time seemed doomed to extinction.

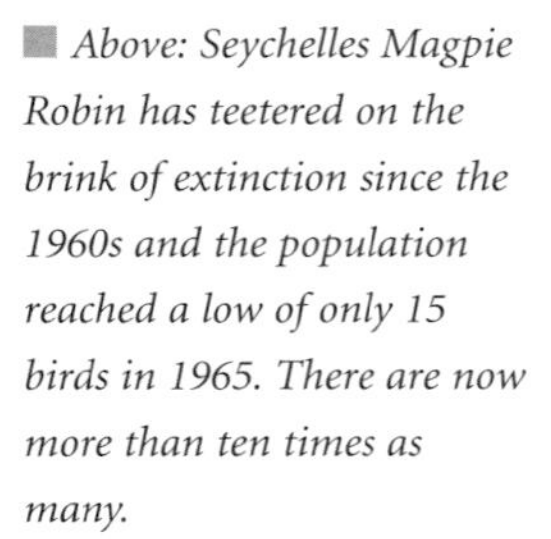

■ *Above: Seychelles Magpie Robin has teetered on the brink of extinction since the 1960s and the population reached a low of only 15 birds in 1965. There are now more than ten times as many.*

■ *Above right: another species back from the brink is the Seychelles Scops Owl; declared extinct in 1958, it is now known that a healthy population occurs on Mahe.*

The Seychelles include about 40 granite islands and more than 100 coral or sand islets. The larger islands have hilly centres covered with montane forest and, not surprisingly, it is these islands that hold most of the endemic bird species. Indeed, Mahe, the largest island, is home to seven of them and one, the Seychelles Scops Owl, occurs nowhere else. This small reddish-brown owl was discovered in 1880 but declared extinct in 1958, having been recorded just once in the meantime, in 1940. Following up on reports of strange croaking calls being heard at night, however, it was rediscovered only a year later, when reports tentatively suggested the existence of 20 birds. It is now known, however, that there are about 360 birds, mostly found in the Morne Seychellois National Park, and it has now been intensively studied, the very first nest having been discovered in 2000.

This back-from-the-brink story is not the only one to feature among the birds of the Seychelles. On the nearby island of Fregate, for example, the Seychelles Magpie-Robin has faced a more genuine struggle for survival. Its problems are common to many Seychelles birds, and to many species of birds on small islands generally. Once widespread on most of the major islands, this attractive black chat with a broad white wing-bar fell victim to a catalogue of ills; foraging mainly on the ground, it was frequently preyed upon by introduced cats and its tree-hole nest was vulnerable to rats. This, together with habitat destruction and degradation, eventually limited the magpie-robin to just the one 2.19-sq-km island, and by 1965 there were only 15 individuals remaining, perhaps making it the rarest bird in the world.

Then the conservation agencies stepped in, yet their first attempts at helping the magpie-robin failed. Some birds were translocated to other islands, but died out; cats were eradicated from Fregate in 1982, but still the population hovered in the danger zone. It was not until 1990, when the habitat on Fregate began to be managed especially for them, that the birds at last began to recover, reaching a total of 85 individuals by 1999. Birds were translocated to Aride, Cousin and Cousine islands and the present total from the four populations is well above 150 birds. In 2005, to richly deserved fanfare, the Seychelles Magpie-Robin was downgraded from Critical status to Endangered by BirdLife International.

Meanwhile, the island of Cousin was witnessing its own drama. At the same time that the magpie-robin was faltering on Fregate, the Seychelles Warbler population declined to 29 individuals on Cousin, and it was in severe trouble. However, in 1968 Cousin Island came up for sale and became the first island in the world to be owned by an international conservation organization (the International Council for Bird Preservation, forerunner of BirdLife International). The island was promptly purged of its coconut palms and the original scrubby vegetation regenerated. The warbler recovered

Above: not all Seychelles birds are endangered, although this White Tern chick does look precariously balanced on the branch of a tree.

and its population is now over 2,000 individuals, including birds translocated to Aride and Cousine. Studies have shown that this species lays only one egg per clutch, exceptionally low for a member of its family.

The next species that was expected to reach dangerously low populations was the Seychelles White-eye. Until recently it was thought to occur only in Mahe, where in 1996 there were thought to be only 35 individuals surviving, all dangerously vulnerable to cats and to nest-predation by rats and introduced Common Mynas. With no hope of eradicating these threats on such a comparatively large (30 km long) island, things looked bleak until, miraculously, a previously unknown, thriving population of more than 250 was discovered in 1997 on the predator-free islet of Conception. For once, the bird managed its own recovery.

The other Seychelles endemics have spared the conservationists too much trauma, but several are vulnerable to catastrophic events. The magnificent Seychelles Paradise Flycatcher, for example, the male of which is entirely sooty black with a long trailing tail, occurs only on the island of La Digue, where its population numbers about 230 birds. The gecko-eating Seychelles Kestrel numbers about 450 birds, mostly on Mahe, while the Seychelles Swiftlet has a larger population, but many of its breeding caves are still to be discovered. Happily, the Seychelles Blue Pigeon, the Seychelles Bulbul and the Seychelles Fody are all secure, and the Seychelles Sunbird seems to have positively benefited from man's alteration of the local environment. It can be seen everywhere.

Thus it is that a birder visiting the Seychelles can still see a good range of endemic birds. The other main attraction is the seabirds, and the island of Aride has some of the most important colonies in the Indian Ocean, totalling about one million birds of ten species. There are up to 360,000 pairs of Sooty Terns, 200,000 pairs of Sooty Noddies (Seychelles has all the world's largest colonies of this bird) and 72,000 Audubon's Shearwaters. Bird Island, an isolated speck 100 km north of Mahe, may hold a million Sooty Terns in the breeding season, of which up to 600,000 pairs may actually be nesting. This island also provides habitat for 10,000 Brown Noddies, and several other species of tern. Meanwhile, the population of the exotic-looking White Tern, with its snowy plumage and large dark eye, numbers about 14,000 pairs throughout all the islands. Famously, it lays its single egg precariously on a fork in a branch from where, on occasion, it may be deliberately tipped by a badly behaved Seychelles Fody.

These two last-named sum up the Seychelles quite well, bird-wise. The fabulous White Tern adorns the tourism brochures, but it is birds like the fody that provide lasting and uplifting memories of the Seychelles.

Australasia

If there was an accolade for the region with the world's most distinctive avifauna, a strong contender would be the Australasian region, and especially Australia itself. Visiting birders from elsewhere often take a long time to get used to such groups as fairywrens, butcherbirds, woodswallows, honeyeaters and thornbills, finding few equivalents elsewhere. Australia's count of endemic birds, some 330, represents more than 50 per cent of all regularly occurring species.

Australia is famously dry and dominated by arid landscapes. A high number of its bird species live a nomadic lifestyle in the interior, following food supplies in no particular direction, and sometimes returning to an area that was formerly known to be in their range after many years' absence. Other birds are specialists, highly adapted to stay put in such biomes as desert grasslands, which are often dominated by spinifex grass, or in the acacia shrubland known as mulga. A few even occur on the stony desert plain known as gibber, the sort of habitat one would expect to be completely empty of any life, let alone birds. Many of these tough inhabitants are highly sought-after, not least the charismatic grasswrens, which include some of the hardest birds to see in the world.

Of course, Australia is not exclusively a continent of deserts. Indeed, the dominant plants, eucalypts, are found throughout in various forms. The various forests hold a number of endemic families, including the famous lyrebirds. Australia also has some tropical forest, mainly in the north-east, and this is actually by far the richest part of the continent for birds. The very north-east tip of Australia is a mere 160 km from its nearest neighbour, the island of New Guinea. This island is wild and mountainous, and its most famous inhabitants are the incomparable birds-of-paradise, a family which would be endemic but for four species occurring in Australia.

More than 2,000 km off Australia's south-eastern flank lies New Zealand, almost a continent in its own right in terms of its unique avifauna. Three families, the kiwis, New Zealand wrens and wattlebirds occur nowhere else, while the islands are also rich in seabirds and waders. The rest of Australasia comprises the Pacific islands, including New Caledonia and Fiji. Each group has large numbers of endemic species, and families such as parrots and pigeons are well represented.

Right: Black Swans in Australia

The Strzelecki Track

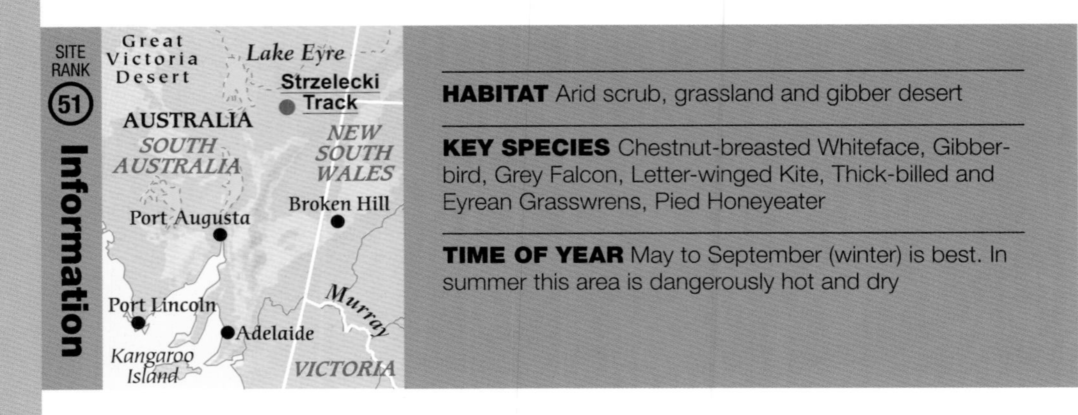

HABITAT Arid scrub, grassland and gibber desert

KEY SPECIES Chestnut-breasted Whiteface, Gibberbird, Grey Falcon, Letter-winged Kite, Thick-billed and Eyrean Grasswrens, Pied Honeyeater

TIME OF YEAR May to September (winter) is best. In summer this area is dangerously hot and dry

Right: a rare and localized species that favours the Australian interior, Letter-winged Kite is exceptional among raptors in that it prefers to hunt in near-darkness.

While visitors to Australia are content to absorb such icons as Laughing Kookaburra, Galah and Superb Lyrebird, the hard-core local birders take their holidays here, in the harsh arid deserts of South Australia. The area is home to some of the most difficult birds to track down anywhere in the world, let alone Australia, and this wilderness is frighteningly inhospitable. In some ways, this makes it irresistible, and the keenest birders will find plenty of remarkable species. Such outback birding is also dangerous, and if you succumb here you will not, tragically, be the first. Prepare well.

The Strzelecki Track, named after the Polish Count Edmund de Strzelecki (pronounced 'stress-lecky'), was originally laid out by one of Australia's famed outlaws, a certain Harry Redford, otherwise known as Captain Starlight. In 1870 he audaciously stole 1,000 head of cattle in Queensland and drove them south through the Strzelecki Desert to Adelaide. He was eventually tried for this heinous crime, but the court found him not guilty on the grounds of his public service in opening up this route through hitherto impenetrable country. Presumably the victim was less than impressed, but that's Australia for you!

The track is about 500 km long and runs from Innamincka, in remote north-eastern South Australia, to Lyndhurst, a small town 590 km north of Adelaide. In the good old days it was strictly four-wheel drive and sleep-under-the-stars territory, but nowadays, especially with the development of the gas field at Moomba, the road is all-vehicle and some motels have sprung up along the way. Nevertheless it is still a very remote spot.

Birders with little time invariably begin their exploration near Lyndhurst because, despite being at one end of the track, this settlement is within easy reach of two of Australia's rarest and least-known birds, the Thick-billed Grasswren and the Chestnut-breasted Whiteface. These occur about 30 km along the track from Lyndhurst, in saltbush and blue bush flats, without a tree from one horizon to another. The whiteface is exceedingly rare, and this is just about the only place in the world where you can guarantee seeing it. Its nest and eggs were only found for the first time in 1968.

Above: the dauntingly dry country plays host to some of Australia's least known birds.

While the whiteface is tricky to find, the grasswren belongs to a group of birds that are notorious for giving birders the run-around – literally. The Thick-billed is typical in that its habitat consists of sparse clumps of vegetation surrounded by bare ground, and what these birds do so well is to run headlong from one to the other with amazing speed – as if they were trying to keep down amid volleys of gunfire – and will sometimes zigzag to put you, the assailant, off the scent. The Thick-billed Grasswren is quite a sedentary species and lives in matched habitat to the whiteface, although it has a much broader distribution.

Further up the track, after the Strzelecki Crossing, the rolling sand hills hold another grasswren. If you haven't been driven mad by the Thick-billed, you can try your luck with an even more awkward species, the near-legendary Eyrean Grasswren. This is an odd little bird with the same long legs, streaked back and long cocked-up tail of the other grasswrens (they look a bit like miniature roadrunners), but it is a small species, no larger than a standard fairy-wren, and has a thick, stubby bill. It lives entirely in tussocks of sand hill cane-grass, and it seems

Right: Chestnut-breasted Whiteface is endemic to north-eastern parts of South Australia, where it is restricted to certain dry semi-desert habitats.

that this grasswren's bill is specially adapted to cope with the rather large seeds of this plant (it also takes insects). This little pest moves even faster than its counterpart down south, not just running from tussock to tussock, but actually flying between long hops, so to speak, a bit like an athlete attempting a triple-jump. If you fail to see the bird here, at least you'll be in good company. First discovered in 1874, this species then went missing until 1961, making it one of Australia's celebrated 'lost birds'.

A modern-day version of the lost bird, at least in the view of many Australians, is the enigmatic Grey Falcon. Although it is recorded every year from somewhere in the vast interior, the number of encounters is almost unbelievably small. There is nothing obviously remarkable about its ecology; it seems to specialize in high-speed pursuits of birds such as parrots, much as a Peregrine might, yet it appears to be a rare and unpredictable bird that is very thinly spread. If you see this species along the Strzelecki, you can count yourself exceptionally fortunate.

Visitors are much more likely to bump into another enigmatic raptor, the Letter-winged Kite, especially just south of the Strzelecki Crossing, where it can sometimes be seen roosting in dead trees. This is the world's most nocturnal bird of prey and it is typically seen hovering over the desert at dusk or by moonlight. Its primary food is the Long-haired Rat (*Rattus villosissimus*), a rodent that explodes in population when heavy rains soak the interior. When a rat plague strikes, the kites bring up a succession of broods and then, once the locality dries out, they move away, often not to return for many years. The Letter-winged Kite is thus a classic nomadic bird, moving from place to place as prey dictates, breeding well in good years and probably suspending reproduction when times are lean.

There are, as might be expected, many other nomadic birds here. In the last few years, for example, Pied Honeyeaters have made several appearances along the track. This is one of the so-called 'blossom-nomads', a honeyeater spending its life moving between localized blooms of favourite desert flowers, drinking their nectar when they flower.

One remarkable non-nomad commonly found along the track is the Gibberbird, a small yellowish, pipit-like bird that walks along the ground. This species occurs all year round on the desperately exposed, treeless, apparently lifeless gibber plains, which are basically rocky desert. No Australian bird is such a complete desert specialist, in a continent of deserts. Most of the Gibberbird's habitat contains no water whatsoever, and how it survives here is not yet understood. It is, in many ways, the epitome of toughness in this unyielding landscape.

Lamington

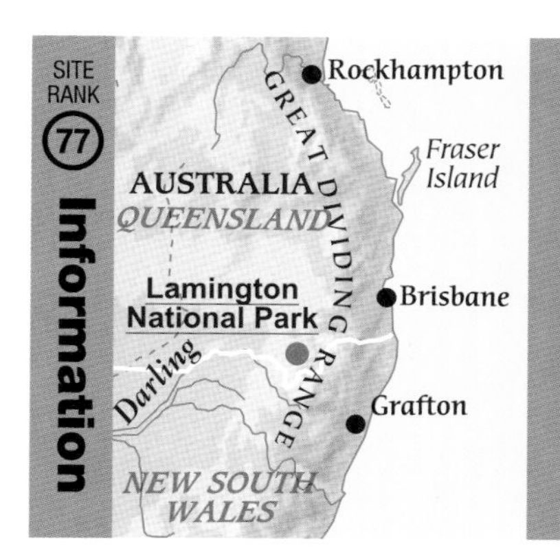

HABITAT Subtropical and temperate rainforest, sclerophyll forest

KEY SPECIES Albert's Lyrebird, Noisy Pitta, Wonga Pigeon, Yellow-throated Scrubwren, Rufous Scrub-bird, Satin and Regent Bowerbirds, Paradise Riflebird

TIME OF YEAR All year, but November and December are best for the breeding birds

Above right: the male Satin Bowerbird is a common sight in Lamington National Park. This bird is attending its bower on the ground.

This 205-sq-km national park in the mountains, just inland from Brisbane in Queensland, holds a treasure-chest of spectacular and remarkable birds, some of which are very hard to see anywhere else. It is not easy birding country, but the rewards are great, and at the end of a long day in the field there are one or two places where you can raise a glass in thanks for the wonders you have managed to find.

Lamington is essentially a forested national park. The primary habitat for the best birds is rainforest, but in areas above 1,000 m there is also some *Nothophagus* (Antarctic Beech) temperate woodland, a good site for one of the trickiest birds to see here (or anywhere), the Rufous Scrub-bird. The presence of this type of woodland suggests to scientists that the Lamington Plateau was once much cooler.

A really good way to start your birding is to pay a visit to O'Reilly's Rainforest Retreat, where you can either stay in the lodge or camp. The lodge has been here since the 1930s, and the birds have become used to the provision of food at the various stations set up for them. Just outside the dining room, such forest birds as the gorgeous Regent Bowerbird and the scarlet-breasted Australian King Parrot will come unconcernedly almost within touching distance, and the red-and-blue Crimson Rosellas will take food from the hand. Australian Brushturkeys consider the grounds their own, and they will wander around like peacocks on a country estate, joined by normally shy, ground-dwelling Wonga Pigeons, with their grey plumage and black-spotted white breasts. Meanwhile, from the open parts of the restaurant you can watch Grey Goshawks soaring over the forest. A proportion of these amazing bird-specialists are entirely pure white, making it easy to confuse them with cockatoos – which is probably what evolution intended.

This rich introduction to Lamington will probably lull you into thinking that the whole exercise will be easy, and you will happily skip along one of the rainforest trails unaware of your immediate fate. An hour later, having seen precisely nothing, you will then be suitably conditioned and ready for some hard birding work. Seeing the

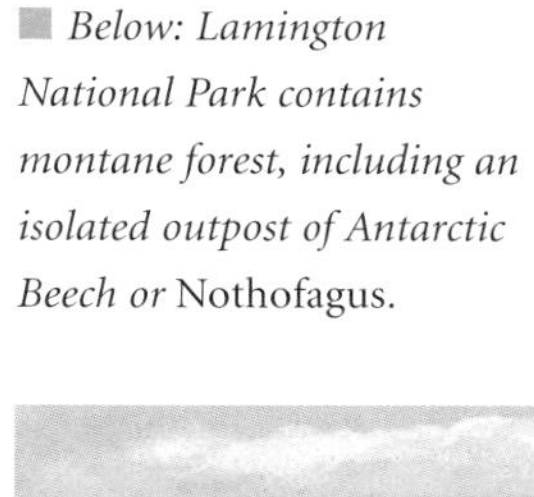

Below: Lamington National Park contains montane forest, including an isolated outpost of Antarctic Beech or Nothofagus.

park's most celebrated species, for example, takes some doing: it is the size of a pheasant, has a long, trailing tail and is very noisy, but Albert's Lyrebird is very shy and lives in thick undergrowth. You will need to be patient.

There are two species of lyrebird in the world, of which the best known, the Superb Lyrebird, is quite common in south-eastern Australia. Albert's Lyrebird, which is slightly smaller and lacks the wide, spatule-ended tail feathers of its relative, is confined to a mere 100-km section of montane forest in south-east Queensland and north-east New South Wales, entirely above 300 m. The world population is only about 10,000 and it is decreasing. Its skulking habits make much of its behaviour a mystery, including details of its diet and mating behaviour. The male sings its complex song, which includes about 70 per cent mimicry (mainly of the Satin Bowerbird) from an arena, which suggests that it is probably polygynous like the Superb Lyrebird. If you cannot hear the song, try listening for scratching sounds as the bird searches the leaf litter for food.

But the lyrebird is not the only leaf-litter brusher in the deeply shaded ground layer. Others include the Australian Logrunner, a small and rather tame thrush-like bird with three white wing-bars; the Noisy Pitta, which is colourful and sings a song that sounds like a squelch; the Yellow-throated Scrubwren, a small brown bird with a smart black face-mask and a passing resemblance to an American yellowthroat; and the two species of very similar scaly-plumaged thrush, the Bassian Thrush and the Russet-tailed Thrush. The Bassian Thrush has recently been shown to shift the leaf-litter in a unique way that you would not believe unless you read it in a learned journal. Apparently it squats down and effectively farts, evidently dislodging insect prey.

If you are very fortunate, or have a guide, you may come across a bower on the rainforest floor. Both the Satin Bowerbird and the Regent Bowerbird make bowers of the 'avenue' kind, which are essentially two rows of woven sticks and grass stems looking like hurdles on a racecourse, in the middle of which the male sings its song. The male Regent Bowerbird is such a stunner, with sensational black and golden-yellow plumage, that this alone is enough to impress the female. The more subtly plumaged Satin Bowerbird, on the other hand, has to decorate its bower with dozens of 'ornaments', which are scattered about at the end of the avenue. Most of these are blue, and may include feathers, straws, bottle-tops, clothes pegs – anything it can find of the right colour. This species also goes about some DIY, 'painting' its bower with chewed vegetable matter and saliva.

If you take your eyes off the ground briefly, and respond to a loud *yass* call, you may spot one of the forest's most elusive birds up in the canopy. The Paradise Riflebird is, as its name suggests, one of Australia's four species in the bird-of-paradise family. It is not especially spectacular to look at, with mainly black plumage and iridescent purple and green throat, but it does go in for the usual eye-catching display, perching on a stump and holding its wings out and arching them, as if to envelop any female standing in front. You would need to be fortunate indeed to witness this sight.

Talking of which, the hardest bird of all to see in Lamington National Park is none of those so far mentioned, but another great rarity, the Rufous Scrub-bird. It requires a 12-km hike to get to the best spot, and when you arrive you are likely to find the bird almost unbelievably hard to see. It sings loudly, but is a remarkable ventriloquist, so that won't help much. It is also the king of skulkers – it doesn't just hide within the litter, it sometimes hides under it, pushing forwards and downwards with its bill. It craves the dampest and thickest undergrowth, in the gloomy understorey, and it is also cryptically coloured. If you see it, you deserve to buy yourself a beer at O'Reilly's. Good luck!

■ *Below: some of Lamington's specialities, including the spectacular Regent Bowerbird, come to bird feeders at O'Reilly's and Binna-Burra Lodges.*

New South Wales pelagic

HABITAT Pelagic (open ocean)

KEY BIRDS: More than 20 albatross taxa including Gibson's Wandering and Campbell Black-bowed Albatrosses, other 'tubenoses' (notably Providence and Great-winged Petrels)

TIME OF YEAR All year, but the austral winter (June to October) is best

Nowhere in the world can you see as many species of ocean-going seabirds as on the waters off New South Wales, eastern Australia. This area is a seabirders' dream. Pelagic trips here have been running since 1985 and in the interim nearly 110 species of seabirds have been logged, including an amazing 45 species of Procellariidae (shearwaters and gadfly petrels), almost half the world's total.

Albatrosses also figure prominently, and on a recent trip in August 2006 no fewer than nine species or subspecies (albatross taxonomy has been a source of conjecture in recent years) of these majestic birds could be seen from a boat at the same time.

Several factors combine to make this an exceptional place for ocean trips. Underwater currents, possibly originating from the Great Barrier Reef to the north, bring high concentrations of marine organisms to the area, providing plenty of food for the birds. Secondly, the edge of the continental shelf, where the juxtaposition of shallow ocean deepwater brings further mixing of currents and high productivity, is only 30 km offshore, just a few hours' boat ride from port. Finally, this coast faces the mighty Pacific, so that seabirds can fly in unencumbered from the tropics, from the Antarctic or from anywhere to the east.

In order to tap into this seabird wonder, boat trips run out from three New South Wales ports – Sydney, Wollongong and Eden – once a month, each on a different

Right: the delightful White-faced Storm Petrel is regularly seen off New South Wales, especially on summer trips. It is the only storm petrel that breeds in Australian waters.

■ *Above: Providence Petrels breed nearly 1,000 km away on Lord Howe Island but wandering birds are regular off the New South Wales coast.*

■ *Opposite: Black-browed Albatross is one of 20 albatross taxa that have been seen on pelagic trips out of Sydney and Wollongong and it is also one of the commonest.*

weekend. All the sites record similar birds, but Sydney offers the chance to leave in style from its world-famous harbour, and if you leave from Wollongong you may witness some of the birds actually being captured and ringed by the Southern Oceans Seabird Study Association (SOSSA). Whichever your port of departure, however, you will almost certainly catch up with some localized species of seabirds, and every trip brings the chance of encountering some mouth-watering rarity. Part of the thrill is in the unpredictability; it is anyone's guess what will appear next.

The great delight and satisfaction of all pelagic trips, of course, is that they bring you into the birds' habitat, so you do not have to strain your eyes too hard trying to make out distant dots. Within a few minutes of leaving port you will be surrounded by such species as Short-tailed, Wedge-tailed and Flesh-footed Shearwaters passing by at close quarters. These all-dark species are rather similar and, by observing their different shapes and manner of flight, you will soon get your eye in. In these comparatively shallow waters they may be joined by various terns (Swift Tern is the most common, although 14 species have been recorded in all), by a few gulls and skuas and the ubiquitous Australasian Gannet.

It is not until the boat reaches the shelf edge, however, at about the 100-fathom depth line, that the more sought-after birds begin to put in an appearance, and the birding regulars start to get genuinely interested. It is here that you and they first meet petrels of the genus *Pterodroma*. These birds are famed for the way in which they typically travel over the oceans, rising up high by flying into the wind and then using gravity to glide down again, often moving along very fast in the process (dynamic soaring). Many of the species in this large genus are scarce and localized, and all *Pterodroma* sightings are highly prized by pelagic enthusiasts. Here, though, in the right season you would be hard pressed not to see plenty of Great-winged and Providence Petrels, while other regulars are Tahiti (which is not classified as a *Pterodroma*), White-necked and Gould's Petrels.

Another interesting group that are hard to find elsewhere are the prions. These small tubenoses with snowy-white undersides are filter-feeders, mainly consuming copepods that they snatch from the surface. Several species have been recorded in these waters, of which by far the most abundant is the Fairy Prion, a bird that may appear in hundreds or thousands between October and March.

The trip organizers will have prepared a concoction of offal and fish-oil known variously as 'chum' or, as here, 'berley', which will be thrown over the side to attract birds close to the vessel. This is the oceanic equivalent of putting food out on a bird table, and while the boat drifts (watch out for seasickness at this point) there will be high expectations about what might be drawn in. Almost invariably a few storm petrels will appear, either Wilson's or the delightful White-faced Storm Petrel, a bird that has the habit of making repeated two-footed hops on the water surface several times in succession, for reasons that are uncertain. Recent studies suggest that Wilson's Storm Petrels can follow a scent trail to find food, zigzagging their way to the source.

Despite the delights of all these birds, however, it is the albatrosses that are most likely to steal the show. On winter trips there may be dozens of these birds around the vessel, some in the water, others gliding within touching distance. These birds, so regal and magical when seen effortlessly sailing the winds, show their less dignified side when feeding, paddling on the surface like ducks and scrapping irritably among themselves over the offal. However, even at their worst they are captivating birds, full of charisma.

Recent changes in albatross taxonomy have greatly inflated the number of species, so that the list for New South Wales now exceeds 20. The most likely to be seen are the Black-browed and Campbell [Black-browed] Albatrosses, the [Indian] Yellow-nosed Albatross, the Shy [White-capped] Albatross and the Gibson's Albatross, the local form of the mighty Wandering Albatross. It will be no easy matter to identify all these different forms, and some birds may well baffle even the experts. This, of course, is merely a reminder of how much is still unknown about the identification and life-histories of albatrosses, and seabirds generally. Pelagic trips like these travel enticingly to the frontiers of our knowledge.

Kakadu

SITE RANK 33 **Information**

Melville Is.
Arafura Sea
Cape York
Darwin
Kakadu NP
Arnhem Land
Katherine
Gulf of Carpentaria
AUSTRALIA
NORTHERN TERRITORY
Tennant Creek

HABITAT Tropical savanna, floodplain wetland, sandstone cliff and patches of rainforest

KEY SPECIES Magpie Goose, Brolga, Partridge Pigeon, Chestnut-quilled Rock Pigeon, Banded Fruit Dove, White-throated Grasswren, Sandstone Shrikethrush

TIME OF YEAR All year, but best between April and October (dry season)

Situated in the 'top end' of the Northern Territory, Kakadu is the largest national park in Australia, encompassing 20,000 sq km of prime wilderness with high biological and cultural significance. It is partly owned by the Aboriginal peoples' Gagudju Association and is particularly famed for its fabulous rock art sites, such as Ubirr Rock and Nourlangie. These include depictions of prehistoric megafauna, such as the Marsupial Tapir, indicating that the area has been occupied by humans for at least 60,000 years. Impressive though this is, the adjacent escarpment of Arnhem Land, a 500-km-long sandstone plateau that dominates the skyline of the floodplain beneath, is thought to date back two billion years. It was there when all the continents were joined together, and before life emerged from the sea. At times in Kakadu, the feeling of antiquity is almost tangible.

These days the national park consists of four main habitats: the sandstone cliffs and outcrops, the floodplain of the South and East Alligator Rivers, the vast area of tropical savanna, and much smaller patches of rainforest. All have their typical birds, making this national park an exceptional place for the birder. About 300 species have been recorded in all.

The sandstone areas hold two birds that are endemic to the Arnhem Land area, and that will be high on the wish-list of any visiting birder. The best place to see them is probably Gunlom Falls, towards the south-west

Right: as floodwaters recede, waterbirds become more concentrated and easier to see. Australian Pelican, Royal Spoonbill, Green Pygmy Goose and Plumed Whistling-Duck are on show here.

Opposite: with a population numbering in the hundreds of thousands, Kakadu is the world's most important site for the peculiar Magpie Goose.

corner of the park. The handsome White-throated Grasswren, the largest of its charismatic group (see also the Strzelecki Track account), is quite easy to find here, especially in the very early morning. A chocolate-brown species, with white streaks on its upperparts and a brilliant white throat and upper breast, it is found in bare rocky areas with some spinifex growth. Like other grasswrens it is reluctant to fly, instead running with head and long tail lowered from bush to bush or into a crevice, never to emerge again. By contrast, in the rocky cliffs it is easy to catch up with Chestnut-quilled Rock Pigeon, a large pigeon whose dull brown appearance is transformed when it takes flight and shows its bright chestnut primaries. These birds usually occur in groups and fly off with loud wing-beats; they feed among rocks on the ground. Two other specialities of the area include the Sandstone Shrikethrush, a thrush-sized bird with a melodious song, whose plumage matches its habitat, and the black-and-white Banded Fruit Dove, here at its only Australian outpost. It occurs in patches of rainforest in the sandstone gorges and ridges, feeding mainly on the fruits of sandstone fig.

The rivers in Kakadu National Park may be named after alligators, but it is crocodiles that occur here, in large numbers. Most of these are the extremely dangerous Estuarine Crocodiles, known locally as 'salties'; they make birding in wetland areas a potentially hazardous business. Fortunately, you can see the birds while remaining safe by going on a cruise by the permanent billabong known as Yellow Water, in the centre of the park. Book early and take the earliest departure of the day, and you could well be treated to a couple of hours' birding ecstasy. The Comb-crested Jacana is easy to see on the abundant water-lily pads, often in company with the Green Pygmy Goose, one of the world's smallest wildfowl species. Herons are represented by such species as Rufous Night Heron and by enormous numbers of Intermediate Egrets (up to 90,000 in the park). The Brolga, an Australian member of the crane family, forages for the tubers of sedges, while the Black-necked Stork, almost as tall, wades in shallows in search of fish. On vegetation around the billabong perch two small kingfishers, the Azure Kingfisher, with orange breast, and the Little Kingfisher, deep blue with a white breast. The Little usually perches lower than the more common Azure, and takes smaller fish.

Below: the Black-necked Stork, known locally as the 'jabiru' (which is not to be confused with the South American stork species of the same name), often chases after fish with rapid steps and wings outspread, although these two are looking more leisurely.

Another species you can hardly fail to see in the wetlands is the Magpie Goose, for which Kakadu is the most important site in the world; an average of 1.6 million of these birds spends the dry season in the park. This peculiar long-necked, long-legged waterfowl has only partially webbed feet, spending much of its time on land digging for tubers and bulbs, its main food items. These are large birds, and their bold black-and-white plumage makes them stand out easily; they make loud, resonant honking noises, aided by an extraordinary elongated trachea. When breeding, they usually occur in trios, one male with two females laying eggs into the same nest.

Kakadu is a place of extreme seasons; the local Aborigines actually recognize six of them, giving each a different name. The dry season begins in April and lasts until October; as time passes the floodplain gradually dries out until much of the wetland area is caked mud. The high tourist season coincides with the austral winter in July and August, when it is easy to get about and many of the water birds are concentrated into spectacular gatherings. There is then the 'build-up' season, when it becomes increasingly humid and thunderstorms bring the first rains. Kakadu has more thunderstorms than anywhere else on earth, with 80 lightning strikes a day at peak times, and these continue into the true wet season from January to March. The monsoon rains can make roads in the park impassable.

The best flowering season happens at the end of the wet, when the first water-lilies adorn the rich waterways of the rivers and billabongs. In the sandstone country and in the savanna, the eucalyptus and grevillea bloom. The latter is a particular delicacy for honeyeaters, especially Little and Silver-crowned Friarbirds, two grey-brown species that often occur together, albeit with much competitive tension. Two colourful lorikeets also feed on the plenty, the rather local Varied Lorikeet and the local race of Rainbow Lorikeet, which is known as Red-collared Lorikeet. Flowers are available in the park for quite a few months, with these and other birds following the blooming of different plants in nomadic style. Fortunately, in the vast area of this magnificent national park, they have plenty of space in which to move.

Queensland Wet Tropics

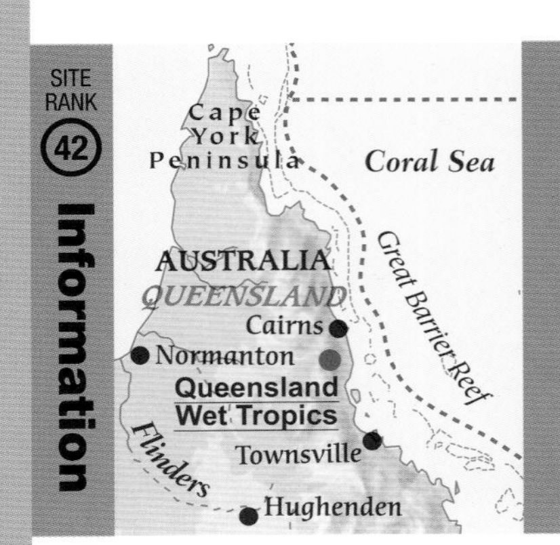

HABITAT Mainly rainforest, with some eucalyptus woodland, mangroves, farmland and wetland

KEY SPECIES Southern Cassowary, Australian Brush-turkey, Orange-footed Scrubfowl, Victoria's Riflebird, Golden Bowerbird, Tooth-billed Bowerbird, Atherton Scrubwren, Buff-breasted Paradise Kingfisher, Fern-wren, Chowchilla

TIME OF YEAR All year round

Above right: the Chowchilla is a forest floor forager which is restricted to the Queensland Wet Tropics. Its close relative, the Logrunner, occurs further south along the east coast of Australia

The Queensland Wet Tropics represent the richest part of Australia in terms of biodiversity. No fewer than 48 per cent of the continent's bird species have been recorded here, along the coastal strip between Cooktown in the north and Townsville in the south, among them a small selection of endemics and some others shared only with nearby Papua New Guinea. The area also holds 58 per cent of Australia's butterflies and one-third of all the marsupial mammals, as well as being extremely important floristically. Of the 19 families of flowering plants regarded as evolutionarily primitive, 13 occur in the Queensland Wet Tropics, by far the highest concentration in the world. The presence of such ancient plants, including the forerunners of iconic Australian plants such as banksias, grevilleas and probably even eucalypts, is indicative of how long this ancient land has been separated from the rest of the world.

The birds here are superlative, for not only are there a great many species in the 8,940 sq km of the QWT World Heritage Area, but a good many of these are of great character and interest, representing some of the most unusual of the continent's species. Most of the best birds occur in the rainforests, especially those on the various mountain ranges and highlands, such as the Atherton Tablelands close to Cairns, much of which lies at or above 800 m altitude. Any birdwatcher who visits the area is guaranteed to see some species with limited range and peculiar behaviour.

These rainforests are, in many ways, typical of many throughout the world, with tangles of close-growing undergrowth below huge, tall trees that have won the race for the light of the canopy. These trees are often festooned with creepers and epiphytes (a great way to see these forests is to take the cable car from Cairns to Kuranda, giving you a unique 8-km-long bird's-eye view). One particular characteristic, however, is the relative lack of ground predators, allowing a rich avifauna to flourish on the forest floor. Many of the special birds of the QWT are terrestrial.

A signature Australasian family, the megapodes, are represented by two species, both of which are common. Both build large mounds and incubate their eggs by making use of the heat generated by fermenting litter. Each male Australian Brushturkey builds a mound for its exclusive use, to which it attracts females every so often, while pairs of the Orange-footed Scrubfowl actually build the mound together and may even share it with other pairs. So while the scrubfowl – which, with its short crest and tail, looks as though it has recently had a

■ Above: a Southern Cassowary can be dangerous if threatened, but its numbers are declining fast due to factors such as road casualties.

■ Opposite: birders may find one of Australia's birds-of-paradise, Victoria's Riflebird, in this area, but only in rainforest at altitudes above 500 m.

severe haircut – is the smallest megapode, it actually builds the largest mound of any in the family. That is saying something: Australian Brushturkey mounds are often 1 m high and 4 m across, while scrubfowl mounds 18 m long and 3 m high have been recorded. The brush-turkey, with its dark plumage, long tail and orange wreath below its bare, red neck, is an Australian endemic that occurs right down the east coast, while the scrub-fowl is one of those species shared with New Guinea.

Many smaller species also inhabit the forest floor, from the relatively widespread Noisy Pitta to the rare and localized Chowchilla. The Chowchilla, endemic to the area, is a plump, thrush-like bird with dark brown upperparts and bold white underparts, stained orange on the female's chin and breast. It patrols the forest floor in groups consisting of a pair with young, plus some extra adults, and spends much time grubbing around in the leaf-litter. The name comes from its very loud call, which is performed at dawn by members of the family in turn, and tends to dominate all other bird sounds.

Another great speciality, the Fernwren, often follows Chowchillas around, feeding on insects disturbed by them. In common with the majority of scarce forest-dwellers of the QWT, it can be exceedingly difficult to see, and even goes so far as to search under the litter itself and out of sight from birders. A small, dark brown bird, the male with two bold white stripes above and below the eye, the Fernwren shares these forests with some close relatives, the very rare Atherton Scrubwren, and the rather widespread Large-billed Scrubwren, both of which feed mainly above ground, the latter usually in small parties.

Two of the most remarkable of the QWT birds are not terrestrial as such, but use the ground to construct extraordinary nuptial constructions known as bowers. These bowers, in tandem with elaborate displays, are used to impress the females. A highly restricted endemic, the Tooth-billed Bowerbird, builds one of the sim-

■ *Above: a common forest species, the Australian Brushturkey, like other members of the megapode family, incubates its eggs using the heat from an outside source – in this case fermenting leaf-litter.*

plest, effectively clearing a surface on the forest floor and decorating it with leaves. Each male uses its notched bill to cut fresh green leaves off their stalks, and then lovingly scatters between 30 and 90 of them on to its court. The leaves are replaced when they shrivel.

These efforts seem modest when compared with those of the male Golden Bowerbird. This species is a 'maypole' builder, going to extraordinary lengths to collect sticks and twigs, which are leant against or twisted around a sapling to make a 'tower' up to 3 m high. In fact, the Golden Bowerbird often builds two unequal towers on saplings 1 m or so apart, which it then decorates with flowers or lichen. The bird spends much of its day sitting on a horizontal perch between the two towers, waiting for a female to come along to admire its handiwork.

By far the largest bird of the forest floor is the imposing Southern Cassowary, a huge flightless bird with hairy dark plumage, scarlet and blue skin on its neck and a large casque protruding from the top of its head – which possibly helps it to push its way through the forest tangle. With its enormous, long, strong legs and alarmingly elongated sharp claw on the middle toe, which may be 17 cm long, the cassowary is one of very few species of bird in the world that is genuinely dangerous to humans. When cornered it can use the razor-like claw like a knife, and it is perfectly capable of disembowelling a human being; indeed, people are quite frequently killed by cassowaries in New Guinea and there has also been one fatality in Australia, in 1926.

Nevertheless, Southern Cassowaries are extremely important to the rainforests because, in the absence of the usual large grazers or monkeys that inhabit rainforests throughout most of the rest of the world, these large birds are the only dispersers for at least 70 species of trees – the fruits are too large for any other animal to eat. Unfortunately, cassowaries are already very difficult to see (Cassowary House, near Kuranda, is a reliable spot), and the population is fast declining owing to habitat reduction, attacks by dogs and road casualties. With only about 1,500 adults remaining in the QWT, the situation is serious. Without these birds, the make-up of these forests is set to change for the worse in the future.

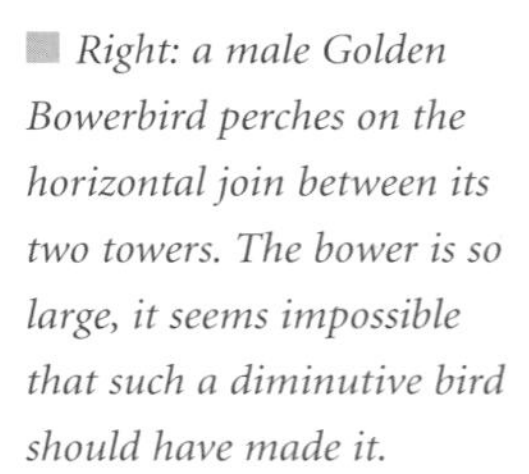

■ *Right: a male Golden Bowerbird perches on the horizontal join between its two towers. The bower is so large, it seems impossible that such a diminutive bird should have made it.*

Rivière Bleue

HABITAT Lowland rainforest

KEY SPECIES Kagu, Crow Honeyeater, Horned Parakeet, Cloven-feathered Dove, New Caledonian Friarbird, New Caledonian Cuckooshrike, New Caledonian Crow

TIME OF YEAR Access is easiest outside the rainy season (December to April), but the best birds are present all year round

New Caledonia is a long, thin island in the South Pacific 1,200 km off the eastern coast of Australia. In contrast to most of the other islands in the region it is not volcanic in origin, but actually a small splinter from the original ancient southern continent of Gondwanaland. Evidently it left Australia 85 million years ago and has been in isolation ever since, as evidenced by its unique fauna and flora – it would certainly need all that time to produce a bird like the Kagu.

The Kagu is one of the world's most peculiar and singular birds. It looks odd, sounds odd and acts odd. It is the sort of bird that, when you see it running towards you across the forest floor, you half expect it to stop and speak to you and confirm that you are merely dreaming. About the size of a large domestic chicken, it is generally reckoned to look like a cross between a rail and a heron, which is about right. It has a long coral-red bill that is designed to dig into the soil (it has plates to cover the nostrils), and long red legs that allow it to move quickly over all sorts of terrain, which it does in an upright stance. Its wings are rounded and fully developed, but the muscles powering them cannot lift it into the air; all it can do is glide while running downhill, so

Below: The Kagu moves along with a stop-start walk, rather like a plover.

it is effectively flightless. Its plumage is astonishing, all-over silver-grey, and about as camouflaged as a crow on an iceberg. It has a bizarre, wispy crest that normally rests on its back, but can be raised, hoopoe-style, when the bird is excited, which it frequently is. Large, red eyes are sited close to the bill to afford excellent binocular vision while the Kagu collects its favourite beetles, millipedes and lizards from the shady forest floor. When foraging it spends much time standing still, plover-like, before running towards prey. When alarmed or excited it occasionally hisses and, each dawn, pairs perform a territorial call that may be heard from 2 km away and is best described by suggesting that, if cockerels barked rather than crowed, they would sound like a Kagu.

The utterly surreal impression given by this unique bird, which is in its own family with no close relatives, is magnified when you find it at its world headquarters, the Rivière Bleue Provincial Park, not far south of New Caledonia's capital, Noumea. Rather than being an inaccessible wilderness, Rivière Bleue is a pleasant weekend retreat for many locals, with opening times, access roads, toilets and attendant tourists, with some well-worn forest tracks and a lake formed by damming. While you are eating your packed lunch at one of the picnic tables, the Kagus not infrequently breeze past you, strutting around

Right: Barred Honeyeater is another of the New Caledonia endemics that may be encountered.

Opposite: the Kagu spends much of its time digging into the soil with its bill, searching for a variety of animal prey.

Opposite below: as well as being the world's most important site for Kagu, the lowland rainforests harbour a remarkable wider avifauna.

like peacocks at a country estate. It all adds to the weirdness.

However, the perilous state of the Kagu population should shake you out of any dream, for this is a seriously rare bird. It is confined to the 400-km-long main island of New Caledonia (Grande Terre), and it is only within the 90 sq km of Rivière Bleue that it is secure or increasing, and that in itself is partly helped by a captive breeding programme. Elsewhere in New Caledonia, Kagus are threatened by their ever-dwindling habitats (partly to make way for the island's many nickel mines), predation of their eggs and chicks by introduced predators such as rats, and by direct killing by feral dogs, which are common in the countryside. There are fewer than 1,000 of these glorious birds remaining, and possibly no more than 500.

While you are searching for the Kagu (most birders seek out 'Mr Kagu' himself, Yves Lettocart, to do this – he is the warden and chief preserver of the species), Rivière Bleu has a selection of other species to enjoy in these tall, undergrowth-dense lowland forests dominated by the mighty Kauri trees (*Agathis*). Indeed, all but one of the island's endemics (the New Caledonian Thicketbird) occurs here. Rivière Bleue is a top site for the very rare Crow Honeyeater, for example, an outsize member of its family with mainly black plumage, orange-yellow face wattles and a pleasing song, as well as the New Caledonian Friarbird and the smart New Caledonian Cuckooshrike, almost all sooty dark-grey with pink undertail coverts. Up in the trees you might glimpse the rare Horned Parakeet, with its two wiry head plumes, or the dark-plumaged Goliath Imperial Pigeon, the largest and plumpest of the world's arboreal pigeons. With considerable searching, you might be fortunate enough to come across the rare Cloven-feathered Dove, a portly green pigeon that appears to wear woolly white leggings.

It would be tempting to say that the Kagu is the most famous of the inhabitants of Rivière Bleue, but this is probably not the case. In recent years another New Caledonian endemic, the New Caledonian Crow, has stolen headlines for its extraordinary levels of intelligence. Not only is it one of the select bands of birds that use tools to obtain food – it will take a stick to poke out grubs and insects from a hole, for example – but even more extraordinarily, it will actually fashion tools for itself. It has been known to strip a vine of all but its final thorn, for instance, and to make various hook-like implements. In the laboratory, New Caledonian Crows have been observed tailor-making instruments for specific jobs, suggesting that they possess exceptional cognitive skills. Clearly, over all those 85 million years, something very special happened on this small island.

Stewart Island

SITE RANK 22 **Information**

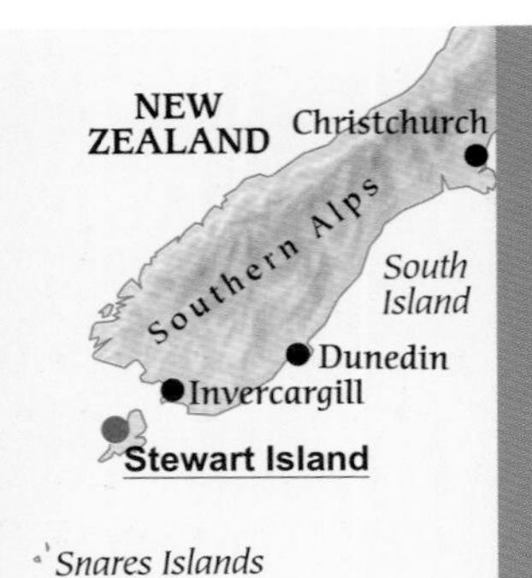

HABITAT Forest, scrub, beach and ocean

KEY SPECIES Southern Brown Kiwi [Stewart Island Tokoeka], Fiordland and Yellow-eyed Penguins, Albatrosses, Weka, New Zealand Pigeon, New Zealand Kaka, Saddleback (introduced), Rifleman (introduced), Tui

TIME OF YEAR All year round for kiwis; spring is said to be the best time for general birding

Most visitors to New Zealand know about the kiwi but never get close to seeing one in the wild. Stewart Island (or Rakiura), at the very southern tip of the country, is arguably the best place in the world to see them.

Stewart Island lies 24 km off the coast of South Island, across the sometimes tempestuous Foveaux Strait. It is quite large (1,680 sq km), very sparsely populated, and has only 25 km of roads. About 80 per cent of the island has been designated as Rakiura National Park. It rains here for 275 days a year on average, with

Above: the Weka is a confident member of the rail family, with omnivorous habits that stretch to predating frogs, small mammals and young birds.

Opposite: kiwi is top of every birder's wish list when visiting New Zealand. Stewart Island is one of the best places to find them, and recent taxonomic research has show that the birds of the island are a distinct species, which is known as Stewart Island Tokoeka.

the result that much of the land is covered in luxuriant forest dominated by podocarps, New Zealand's characteristic conifers, but there is also some scrub (known in New Zealand as 'shrubland') and grassland. Birding on the island itself can be done along the island's 245 km of tracks, and sometimes a hike will reveal a kiwi sighting. Uniquely on this island, the birds sometimes come out to feed during the day, mostly in mid- to late afternoon.

However, there are much more reliable ways of seeing a kiwi here. The local birds have developed the very unusual habit of leaving the forest at night to forage in the open on the tide line of beaches, probing the flotsam for isopods and other goodies. This remarkable behaviour has spawned a busy kiwi-spotting business. A tour boat leaves the only settlement on the island, Oban, every other night, right through the year, and docks near to Ocean Beach, where a 20-minute walk brings the tourists (15 are taken each night) into kiwi territory. The birds are successfully seen on almost every trip, so it would be an unlucky birder indeed who missed them. The island's densest kiwi population lives around Mason Bay, on the other side of the island. Getting there from Oban involves either a two-day hike (each way), a one-hour water-taxi trip and a four-hour hike (each way) or a 20-minute flight by a plane that lands on the Mason Bay beach at low tide.

Kiwi are, of course, remarkable birds. They look rather like small mammals, and have a habit, shared by many mammals, of reading much of their environment by smell. The nostrils are located at the end of the kiwi's long, slightly decurved bill, and the birds hunt by probing into the substrate, searching out earthworms, insects and plant material. The birds also seem to detect people by scent; if you flush a bird it will often run away to a safe distance and then raise its bill, apparently smelling the wind.

Since the 1980s kiwi taxonomy has been turned upside down. Genetic studies have revealed that, among others, the birds of Stewart Island were very different from those elsewhere, enough for them to be afforded specific status. Modern DNA testing has now determined that there are five species, and more testing could differentiate these even further. The 'new' kiwi of Stewart Island was christened the Stewart Island Tokoeka, a Maori name. As well as its tendency to feed in the open on beaches, this bird bucks the kiwi trend by living in family groups rather than being solitary.

The name 'Tokoeka' has a distinct meaning, related to another of the distinctive birds found here: the Weka, a species of rail. The Weka is rather similar in colour and shape to the kiwi, but has a short, thick bill for grabbing animal prey such as frogs, young birds and invertebrates. The Weka is quite common on Stewart Island, and the Maoris named the kiwi the Tokoeka to distinguish it; the name means 'Weka with a stick'.

The presence of the Wekas is just one illustration that Stewart Island is not just about kiwi-watching. In fact it is probably the best single place for birding in the whole of New Zealand. Native landbirds here include the sweet-singing Tui, the splendidly colourful New Zealand Pigeon, the New Zealand Kaka, a large pinkish parrot that lives in the forest canopy, and a number of smaller birds including the euphonious New Zealand Bellbird, the elusive New Zealand Fernbird, the Grey Gerygone and the Pipipi. Most of these can be seen with a little effort along the many tracks.

For those with little time to spare, however, a good option is to visit the excellent open bird sanctuary on the island of Ulva, a short ride by water-taxi from een

■ *Right: Yellow-eyed Penguin is the world's rarest penguin species. It breeds only in New Zealand and its population may be lower than 5,000 individuals.*

Bay, which is very close to Oban. The forest here holds many of the species mentioned above, crammed into only 2.6 sq km. The sanctuary is also particularly noted for its programme of rare New Zealand introductions.

The island has been free of introduced predators since 1997 and, in May 2000, 30 Saddlebacks were released, followed not long afterwards by New Zealand Robin (of the Stewart Island race *rakiura*), Riflemen and Yellowheads. Saddlebacks, which are renowned for their exceptionally melodious songs, are members of the New Zealand wattlebird family, a group of medium-sized, short-winged forest birds that includes the now extinct Huia; Saddlebacks are threatened on the mainland. The aptly named Yellowhead, a small insectivorous species related to the thornbills of Australia, nests in tree cavities where, on the mainland, it is continually compromised by predation by introduced Stoats. Here on Ulva Island it has the chance to establish itself without their malevolent presence.

Stewart Island is close to the prolific seabird breeding grounds of the Subantarctic Islands, which means that the seas here are good for watching pelagic species, including at least five species of albatross. Some of these can be seen from the one-hour ferry ride from Bluff on the mainland, but in recent years all-day pelagic trips have been run from Oban with the express purpose of searching these waters. Already they have turned up an impressive list of birds, including Grey-headed and Buller's Albatrosses, the rare Mottled Petrel, Southern Fulmar and Common Diving Petrel.

The seas and beaches here also provide habitat for Stewart Island's penguins, of which there are three species. The Blue Penguin is reasonably common, but both the Yellow-eyed Penguin (the rarest in the world) and the Fiordland Penguin require a little more work. However, guides on the birding cruises know the location of their nesting sites, so it's worth joining one of these.

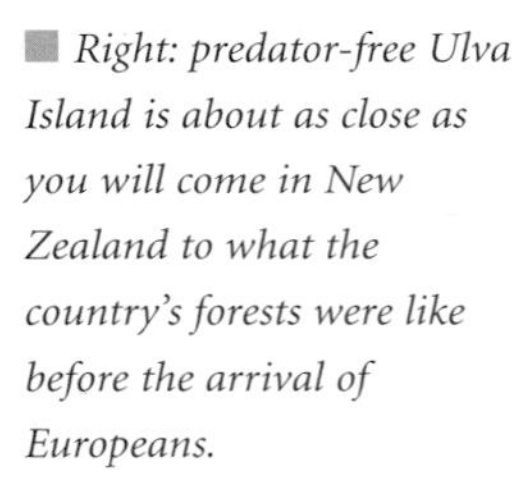

■ *Right: predator-free Ulva Island is about as close as you will come in New Zealand to what the country's forests were like before the arrival of Europeans.*

Tari Valley

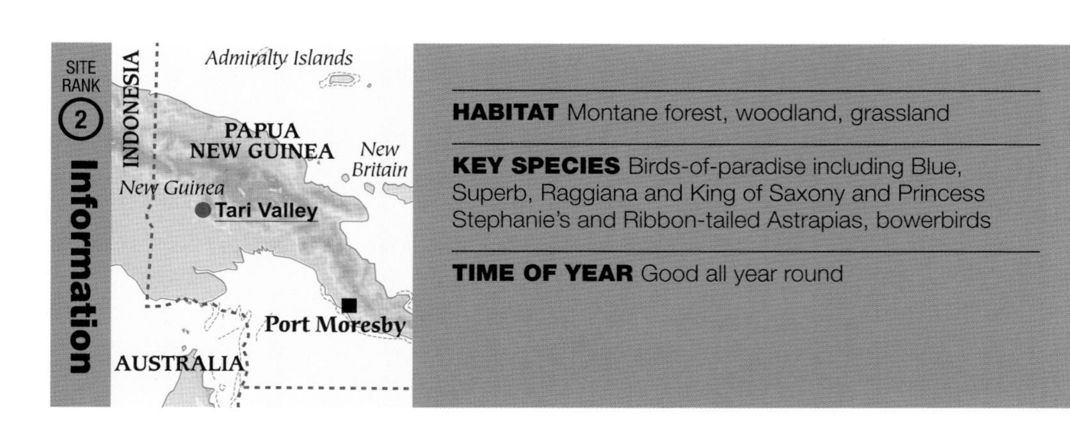

HABITAT Montane forest, woodland, grassland

KEY SPECIES Birds-of-paradise including Blue, Superb, Raggiana and King of Saxony and Princess Stephanie's and Ribbon-tailed Astrapias, bowerbirds

TIME OF YEAR Good all year round

Right: the extravagant head plumes of the male King of Saxony Bird-of-Paradise can be twice as long as the bird's body.

A site that holds more species of birds-of-paradise than anywhere else on earth certainly has a very special claim to fame, and this mantle currently belongs to the Tari Valley, in the Central Highlands of Papua New Guinea. The montane forests on the valley sides regularly hold ten species or more of these famed and incomparably weird and wonderful beauties, along with a superb set of other mouth-watering New Guinea birds.

The main attraction may be birds-of-paradise, but this is not, in fact, birding in paradise. In fact, it can be hard and frustrating work. There are frequent lulls when the forests are deathly quiet; the large, moss- and epiphyte-covered trees hide the birds well; most species stick to the high canopy or hide away in the undergrowth; and the adult male birds-of-paradise, which are inevitably the stars, are vastly outnumbered by females and sub-adult males and can be difficult to track down. This is all to do with these birds' social system. Most species are polygynous, small numbers of usually mature males monopolizing display sites and copulations, while the rest look on. It can take as long as seven years for a male to become a court-holder, with the fabulous plumage to match, and some never make it at all. The females, which for the most part perform all the breeding duties alone, are generally dowdy in comparison with their mates, but are also easier to see. Every sighting of a male is the result of hard slog.

The birding is tough, but the rewards are at times sensational. There really is no such thing as an ordinary male bird-of-paradise, after all. Even the more soberly plumaged compensate by being a bit weird, such as the Short-tailed Paradigalla, a small black bird with bright, light blue and yellow wattles by its eye, and a tail so short as to be almost indecent. Meanwhile, the spectacular ones are just incredible, such as the Tari's Valley's twin stars, the Blue Bird-of-Paradise and the Black Sicklebill, two species listed as Vulnerable by BirdLife International. The Blue Bird-of-Paradise is a medium-sized bird with a bold black back and breast, save for two half-moons of brick-red on the lower belly. The wings, however, are bright blue and most of the tail feathers are equally blue ostrich-like plumes, giving the impression of a crow wearing a blue tutu. The tail also has two long streamers that are slightly swollen at the end, like antennae. When a female approaches a lek, the males suddenly embark on a bizarre courtship display, in which they hang upside down from a branch and sway from side to side, blue plumes shimmering, and the metallic hum they make could, fancifully, be accredited to static being picked up by the antennae. Not many visitors are able to witness this astonishing sight, but those who do come away as changed people.

The male Black Sicklebill, meanwhile, is a far more

Above: adult male birds-of-paradise, like these Raggiana Birds-of-Paradise, can be hard to see, since they make up only a small proportion of the population of each species.

imposing bird than the Blue Bird-of-Paradise. It is much larger, with mainly dark, glossy plumage and its long, decurved bill gives it a rather primitive look. Its shape, however, is dominated by its sabre-like tail, which can be 1 m in length. There is a specially favoured ridge for Sicklebills near Ambua Lodge in the valley, and here it is occasionally possible to see a male displaying: iridescent blue fans either side of the breast are puffed out to frame the bird's head, while the huge tail is also fanned and the bird crouches down, almost horizontal.

The forests in the Tari Valley are rich in fruit and insects, and are remarkably free of predators and competitors for the birds. There are no squirrels or monkeys, nor any terrestrial terrors such as cats or mustelids. This benign scenario is thought to have enabled the birds-of-paradise to evolve outrageous plumage and display routines without compromising their ability to live and feed. Thus the species mentioned above are only part of the pageant. There is also the famous Raggiana Bird-of-Paradise which, like the Blue Bird-of-Paradise, performs communally in a lek; these species are largely reddish-brown, with bold yellow head and green throat, but when they display they uncover from their flanks a silky orange fan of plumes, which shimmer over their backs, making them look as though they are ablaze, half way up their forest tree. Two astrapias, Princess Stephanie's and Ribbon-tailed, display dark plumage and absurdly long, ornate tails; the Superb Bird-of-Paradise shows off harlequin plumage, and Lawes's Parotia is adorned with six hair-like plumes tipped with spatules, three on each side of the head. The extraordinary King of Saxony Bird-of-Paradise manages to trump even these; from each side of

■ *Right: a Crested Berrypecker lives up to its name. It is a relatively common and conspicuous member of the forest community.*

its head spring quills that are more than twice as long as the bird itself. They are powder-blue in colour, and only have clumps of barbs on one side, making them look like the old writing quills.

The birds-of-paradise, indeed, show a greater diversity of feather structure than any other family of birds in the world, and this fact, it turns out, did not escape the attention of the New Guinea people. The indigenous tribe here is the Huli, and it is famous for two reasons. First, the people were living a stone-age existence, completely unknown to the outside world, until the 1930s, when gold-seeking Australians first stumbled this way. Second, they live up to their nickname of 'Wigmen' by adorning their hair with flowers and feathers from the forests – and not for tourists; they wear this traditional dress routinely. Feathers from birds-of-paradise and other species are culturally and economically important, and have been widely traded in these parts for centuries. Meanwhile, the Huli traditional dances mimic the various display routines of their revered avian neighbours.

■ *Below: all but one of the world's nine owlet-nightjar species are found on New Guinea; this is a Barred Owlet-Nightjar.*

The Huli are not the only other inhabitants to admire bird-of-paradise feathers. So, it appears, does another bird of the forests, Archbold's Bowerbird. This essentially dull-coloured bird has forsaken the acquisition of extravagant plumes in favour of constructing an elaborate bower, which involves the placing of much material, such as orchid blooms, in the trees, and also includes a remarkable mossy mat on the ground, on which it arranges those astonishing blue plumes of the King of Saxony Bird-of-Paradise in an unusual case of one bird using another's plumes to impress a potential mate.

The bowerbirds of the Tari are no less impressive, in their way, than the birds-of-paradise. Macgregor's Bowerbird, for example, is famous for constructing a 'maypole' bower, in which a slender sapling is surrounded by short sticks, arranged like spokes, to make a conical tower up to 3 m tall. These may be structures rather than plumage, but visitors fortunate enough to see them can be just as impressed as they are with the more gaudy forest inhabitants.

One other quirk of Papua New Guinea is the presence of the world's only poisonous birds, which have toxins in their plumage. Several species, such as the Black Pitohui and the Ifrit, are so endowed; perhaps these amazing forests are not so benign after all.

Taveuni

SITE RANK 44 **Information**

HABITAT Tropical hill forest, reef, agricultural

KEY SPECIES Orange Fruit Dove, Silktail, Maroon Shining Parrot, Collared Lory, Many-coloured Fruit Dove, Island Thrush, Black-throated Shrikebill, Giant Forest Honeyeater

TIME OF YEAR Any time of year outside the rainy season (December to February)

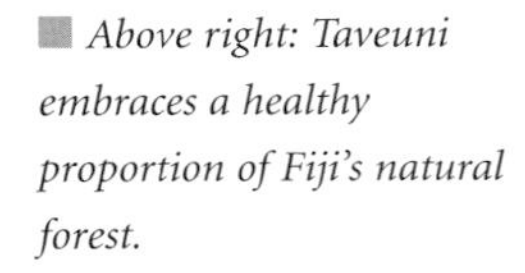

Above right: Taveuni embraces a healthy proportion of Fiji's natural forest.

If you are the sort of birder who likes to suffer for your craft – to brave fierce wind and lashing rain to do a count of seaducks, perhaps – then Taveuni is not the place for you. It would be hard to watch birds in a more comfortable and idyllic spot: a tropical island with palm-fringed, empty beaches, a coral reef offshore, lush jungle in the interior, and waterfalls plunging down high jagged mountains. There are some more than pleasant resort-style hotels dotted about and, just to cap it all, Fiji also happens to be populated by some of the friendliest and happiest people in the world (an officially attested fact). You might have a daydream about birding here.

You might dream, but is it an illusion? There are, it must be said, plenty of tropical island wonderlands in the world that look gorgeous, but where the underlying truth is an ecological disaster of introduced rats and deforestation – and extinctions of native species. That is not the case on Taveuni; four-fifths of the island, which is 42 km long and 15 km wide, is a national park (Bouma National Heritage Park), and 60 per cent of the original forest remains. It is not all perfect – there are some introduced plants and birds – but the island is sufficiently pristine to be currently under consideration as a World Heritage Site.

Above: *Collared Lory is a Fijian endemic that has adapted to thrive in artificial habitats such as gardens and farmland.*

There are plenty of birds here. Fiji as a whole boasts at least 25 endemic species, plus a number of others with restricted distribution in the South Pacific. Of these, 15 occur on Taveuni, several of which are hard to find elsewhere, although the island has no endemics of its own. Together with more widespread native species, a few introductions (including, bizarrely, the Australian Magpie, which was introduced to help control worms in farmland) and some waders and seabirds (including Lesser Frigatebird, Red-footed and Brown Boobies and Brown Noddy), the list for the island is around 100 species, which is quite impressive for such an isolated part of the world. The area around Des Voeux Peak (1,195 m), a marvellous forest dominated by feathery tree ferns, has the richest diversity of birds on Fiji.

It is not the statistics that make Taveuni special, though. The wonder of the place, and the main reason for including it in this book, is in the specific birds that live here. They are like the cast in a movie, an eclectic bunch of good-looking characters that lend a special appeal, individually and collectively, making birding here a unique experience.

The star of the show is a bird called the Orange Fruit Dove and it occurs on the forested slopes in the centre of the island. It is a small, compact, somewhat globular dove. The female, it should be admitted, is a bit green and unexciting, but the male is extraordinary, clad in a brilliant, shining, almost glowing, orange colour, with an olive-yellow head. At first sight, feeding on the trees, it really does look like a piece of fruit – an orange, no less – but, on closer inspection, its plumage has a somewhat hairy texture. To cap its singular appearance, this dove also has the oddest of calls, almost exactly like the ticking of a clock. A peculiar bird indeed!

Another star is a very small forest mite known as the Silktail, which is best looked for at Des Voeux Peak and on the Vidawa Rainforest Trail. The affinities of this forest species have long been debated, and it has even been placed in the bird-of-paradise family owing to the iridescent deep velvety blue-black colouration on the head and upperparts, dotted with sequin-like spots. What is really striking about it, however, is its brilliant white rump, which contrasts strongly with the rest of the plumage. The colour spreads up onto the back and down to cover most of the short tail, which is rounded and with a narrow black tip. This is a highly active sprite which uses its fairly long, slightly hooked bill to tear bark or to thrust leaf litter aside, all the while cocking and flicking its tail. It creeps along mossy branches like a nuthatch, and also makes sallies into the air, betraying its probable relationship to the monarch flycatchers.

One of the commonest birds on the island is another Fiji endemic, the Collared Lory. As well as being found in the forest, this gem of a parrot occurs in villages and plantations, where it can indulge its passion for drinking nectar from the blooms of coconut palms. Despite its dreary name, it is one of the most colourful of the lories (which is saying something): brilliant crimson on most of the underparts, offset by deep purple on the cap and belly, with deep green wings. The much larger Maroon Shining Parrot, a two-colour parrot with deep maroon underparts and bright green upperparts, also manages to cope outside the forest core, although coconut palms are not to its liking.

Amid all the glamour there are character actors here, too. One of the conspicuous species of the forest and edge is the Streaked Fantail, which, in common with the other members of its family, is a small insectivorous bird that makes spectacular aerial sallies after flying insects, in this case usually within the shaded canopy. It perches in a rather horizontal manner, with wings drooped, often rocking from side to side on its perch and flicking open its long tail, making tinkling calls and often coming close to an observer to have a look. Another of similar ilk is the equally inquisitive black-and-orange Vanikoro Flycatcher, which also snatches insects in flight, even around the roofs of buildings and in gardens.

Taveuni also has a clown, or at least a bird dressed in spectacular harlequin colours. Of all the birds in Fiji, perhaps the male Many-coloured Fruit Dove has the most improbable plumage pattern. It looks surreal, a mixture of pink, cream, green and yellow; the sort of pattern a child might invent for an outlandish tropical bird. Even here in Taveuni, this is one bird whose plumage is a bit rich.

Right: *Taveuni is blessed with a number of staggeringly colourful birds, not least its fruit doves. The astonishing male Orange Fruit Dove (left) seems to glow in the canopy, and just about manages to outdo the improbably plumaged Many-coloured Fruit-Dove (right).*

Antarctica

It is easy to forget that Antarctica is a continent, so isolated does it seem from the rest of the world and, in the ornithological sense, so birdless. It nonetheless covers an area of 14.3 million sq km, which is theoretically potential habitat. However, much of this area is covered in an ice sheet, and a good deal is on a plateau at altitudes of more than 2,000 m, with the highest peak 5,140 m above sea level. The lowest recorded temperature on Earth, -89.2 °C, was measured here and, to make matters worse, most parts of the continent are incredibly dry and are, in effect, desert.

The saving grace of the Antarctic region is, however, the surrounding sea, which is incredibly rich in marine life. This is partly because of the tempestuous currents, which mix and bring food from the bottom to the surface in the form of upwellings. The Antarctic Convergence is where northward flowing Antarctic surface water sinks below the southward flowing subantarctic water, while the Antarctic Divergence occurs where currents flowing eastward from the continent mix with westward-flowing currents from the Antarctic Ocean. The upwellings support some of the most abundant marine life on earth, in turn providing food for huge populations of birds. Thus, groups such as albatrosses, penguins, petrels, shearwaters and storm petrels thrive. Indeed, such is the richness that the Arctic Tern, a breeding bird of the northern Holarctic, flies all the way here for the austral summer.

The hard ground of the Antarctic continent, plus the various subantarctic islands such as South Georgia, Heard Island and, indeed, parts of southern South America, provide breeding grounds for many of these species. The vast colonies of penguins found here, in particular, are among the most spectacular ornithological sights in the world.

Besides all the seabirds mentioned above, there is also one family of birds, the sheathbills, which breeds only in this region. Not surprisingly, its two members can often be found feeding among the colonies of seabirds and seals.

Right: Emperor Penguins on the Antarctic Peninsula.

South Georgia

HABITAT Rocky islands, cliffs, tussock-grass, seashore, ocean

KEY SPECIES Penguins, albatrosses, petrels, storm petrels, diving petrels, South Georgia Shag, [South Georgia] Yellow-billed Pintail, South Georgia Pipit

TIME OF YEAR Cruise ships pass the island from November to February

Situated at nearly 55°S, in the South Atlantic Ocean some 2,150 km east of the continent of South America, the British-administered sub-Antarctic island of South Georgia is a pretty remote and unforgiving place. Half covered with glaciers, its peaks covered with permanent snow, a land frequently lashed by violent gales stirred up by ocean and unfettered by landmasses, this 160-km-long island is one of the planet's most fiercely inhospitable and challenging outposts. Not many people get the chance to venture here.

Yet its very loneliness and isolation contribute to South Georgia's allure for naturalist and aesthete alike. Seen from far off, its high cliffs and snow-capped mountains rise like magic from the turbulent seas. Close to, its high cliffs, deep bays and soaring fjords are majestic, and its combination of black rock, bright snow, dull-green tussock grassland and grey-brown beaches makes for a curiously satisfying mix of subtle shades. Furthermore the island's position, close to the Antarctic Convergence and plunged into some of the most biologically productive waters in the world, makes it the perfect breeding base for literally millions of seabirds and seals. For a few months of the year, far from being lonely, the island is heaving with masses of lazy pinnipeds and energetic birds.

Some of the figures for breeding seabirds give an indication of the richness of South Georgia and the whole Antarctic region. There are, for example, an estimated 22

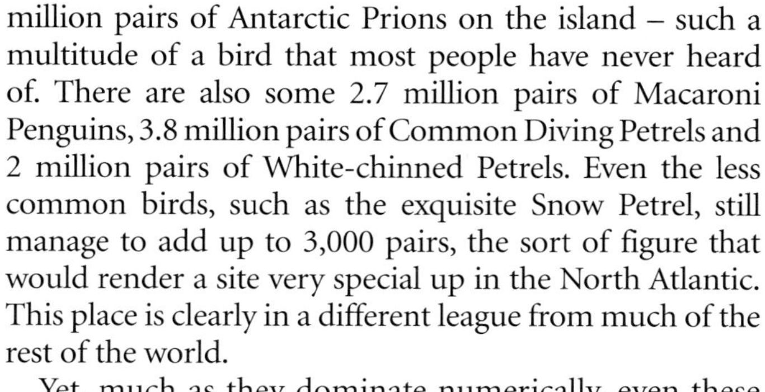

million pairs of Antarctic Prions on the island – such a multitude of a bird that most people have never heard of. There are also some 2.7 million pairs of Macaroni Penguins, 3.8 million pairs of Common Diving Petrels and 2 million pairs of White-chinned Petrels. Even the less common birds, such as the exquisite Snow Petrel, still manage to add up to 3,000 pairs, the sort of figure that would render a site very special up in the North Atlantic. This place is clearly in a different league from much of the rest of the world.

Yet, much as they dominate numerically, even these birds don't necessarily play to the gallery. Instead there are two birds here of complete star quality and it is they that usually dominate the punch-drunk post-trip reminiscences. The first is that supreme seabird, the Wandering Albatross, with its aerodynamic 4-m wingspan in the air and intimidating waddling bulk on land. There are about 4,000 pairs on South Georgia, mainly on Bird Island, a relatively small island off the north-west tip, with others on Albatross Island and Prion Island. Currently it is only on the latter that visitors are able to set foot among these colonies, no doubt hoping for a glimpse of this bird's incomparable display, when it points its bill to the sky, opens its wings as if about to catch a large beach ball, and makes its clanging call. Just the sight of these birds, eyeball to eyeball, tends to be a

■ *Right: a King Penguin surfs into shore; the South Georgia population of this species is close to half a million individuals, including young birds.*

■ *Below: two Wandering Albatrosses alight; these birds may not breed until they are 11 years old.*

Above: the highly pelagic Light-mantled Sooty Albatross obtains much of its food by diving underwater.

profoundly moving experience for most visitors. The encounter is tinged with regret these days, because this fantastic bird is in serious decline, mainly owing to the practice of long-line fishing. Its numbers are dropping by 4 per cent every year and have declined by 50 per cent since the mid-1990s.

The Wandering Albatross is not the only albatross here; in fact, four species breed. The Black-browed Albatross is the most common, with 100,000 pairs, closely followed by the Grey-headed, with 80,000, the latter figure representing nearly half the world population. The Black-browed breeds every year, while the Grey-headed breeds once every two years; the reason for the difference is that the Grey-headed's fledging period is longer (averaging 141 days to the Black-browed's 116), mainly because it seeks food over deeper water, further away from the island, with more squid and less krill. The rare Light-mantled Albatross also maintains a population of up to 8,000 pairs.

The other big box-office star on South Georgia is the King Penguin, the second-largest penguin in the world. There are about 400,000 breeding birds on the island, of which 100,000 are found at St Andrews Bay, on the north-east coast. This immediate area is claimed to have the greatest density of wildlife on earth, with all its penguins mixed in with Southern Elephant Seals and Antarctic Fur Seals, and who would care to argue? The sight of these fine birds, impeccably clad in white, grey, black and orange in contrast to their shabby, overweight, woolly brown chicks, is as amusing as it is impressive. The King Penguin breeds only once every two years, as does the Wandering Albatross, and the chicks similarly have to cope with the harsh Antarctic winter before fledging. No wonder they look a little the worse for wear.

There are other penguins on South Georgia besides the King and Macaroni Penguins. The Gentoo Penguin maintains a population of 105,000 pairs and the Chinstrap one of about 6,000 pairs, while the odd pair of Adelie Penguins occasionally breeds, too.

Alongside the seabirds, South Georgia also has a few oddities. Foremost among these is a passerine, the endemic South Georgia Pipit. It is thought that the ancestors of this species arrived on the island some 1.5 million years ago, and evolved into this strongly streaked speciality, which is often found foraging on the shoreline. Its main habitat is actually tussock grassland by streams, but predation by introduced Brown Rats has now restricted the breeding population to about 20 offshore islands. This small bird tends to feed behind clumps of thick grass to keep out of the wind, and its well-lined nest is partly domed. The population today is about 4,000 pairs.

Another oddity is a duck, the South Georgia Pintail, a distinctive darker brown race of the widespread South American Yellow-billed Pintail. This bird, with a population of about 2,000, feeds mainly in the inter-

■ *Above: part of the King Penguin colony at St Andrews Bay. Each individual breeds only once every two years, but presumably not to escape the crowds.*

tidal zone and, in contrast to most of its dabbling-duck congeners, dives freely.

These are just some of the highlights of this amazing place. Visitors will also be greeted by those arch scavengers the Snowy Sheathbills, members of the only bird family confined to the Antarctic region. Subantarctic Skuas, and two species of giant petrel, Northern and Southern, also breed and give predatory hassle to many of the rest, forcing smaller breeding seabirds such as the prions, Blue Petrels and Wilson's and Black-bellied Storm Petrels to visit their colonies only at night. South Georgia is also one of the few places in the world to host two species of diving petrel, the Common and the South Georgia Diving Petrel, the latter highly localized and maintaining a population of 50,000 pairs. There is also the South Georgia Shag, a blue-eyed black-and-white species that breeds on outlying offshore islands, plus colonies of Antarctic Terns. What more could you want to see in such a magnificent setting, in a place that few ever have the good fortune to visit?

South America

South America is known as 'the bird continent'. It is home to nearly one third of all the world's known bird species, meaning that it has by far the richest continental avifauna and is about 800 species ahead of its nearest rival, Africa. It is also here, for the last 40 years, that the most new species have been discovered. South America's biodiversity is enhanced by two significant geographical features: the lowland (Amazonian) rainforest and the Andean mountain chain. Both are extraordinarily rich in species, and where they meet can be found the most diverse avifaunas on earth with, for example, more than 1,000 species recorded at Manu, Peru.

The rainforest is not a single ecosystem, of course, but actually has many different components, depending on soil, altitude and water level. Different species live in each of these components, while others are widespread. Within the various strata of the forest there are other distinctions, among canopy species and understorey species, for example.

The Andean chain is the main reason for South America's astounding biodiversity. Mountains act as isolating agents, separating bird populations that cannot fly over them, and the sheer length and breadth of the Andes ensures that here are dozens of refugia where unique species have evolved. Furthermore, climatic variations from west to east (the western slopes are in the rainshadow of easterly winds, and are thus much drier than their counterparts) and from low altitude to high altitude, multiply that diversity.

Apart from forest, South America has a host of other key birding habitats. There is the world's driest desert, the Atacama, along the Pacific coast, plus grasslands in the north (Llanos) and south (Pampas), plus areas of scrubland (caatinga) and savanna (cerrado). South America holds the world's largest wetland (the Pantanal), and there are seabirds such as penguins along the west and south coasts.

South America has a number of endemic families, including rheas, seedsnipe and trumpeters. It is also exceptionally rich in such groups as puffbirds, toucans, woodcreepers, furnariids, antbirds, cotingas, manakins, hummingbirds and tyrant flycatchers. Many of these would be endemic were they not shared with Central America.

Right: Stripe-headed Antpitta in Peru.

Manu

SITE RANK 3 **Information**

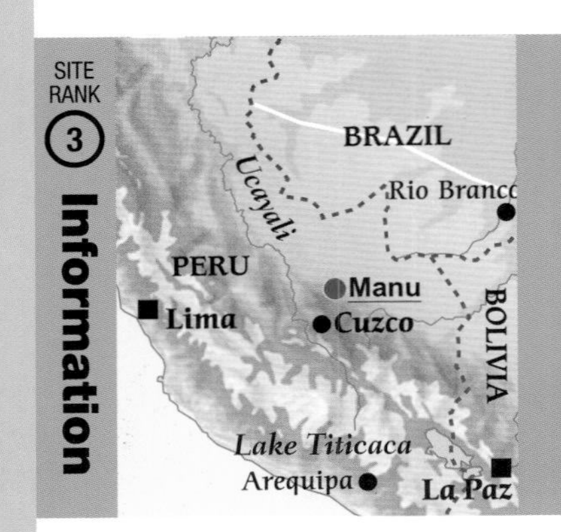

HABITAT Various forest types including humid tropical, humid upper tropical, humid subtropical and humid temperate forest; river, lakes, varzea woodland

KEY SPECIES Antbirds, hummingbirds, tanagers, Andean Cock-of-the-rock, macaws and other parrots, Manu Antbird

TIME OF YEAR All year round

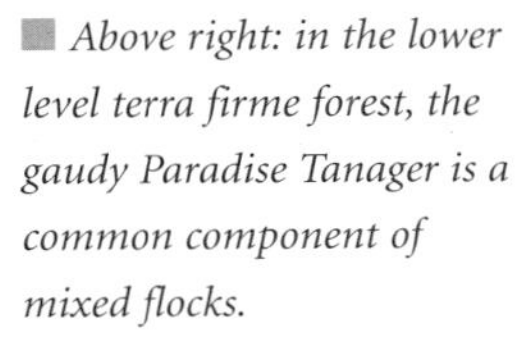

Above right: in the lower level terra firme forest, the gaudy Paradise Tanager is a common component of mixed flocks.

If you were ever curious as to which is the richest birding locality on earth, this is it. The bird list for Manu National Park stands at around the 1,000 species mark. Such is its sumptuous richness that about one-third of all South America's superabundant avifauna has been recorded here, and no less than one-tenth of all the bird species in the world, all within a single, albeit large, protected area in a single country.

Manu lies in the south-eastern part of Peru, to the north-east of the famous tourist hot-spots of Cuzco and Machu Picchu and west of the bustling frontier town of Puerto Maldonado. Its species richness stems partly from

■ Above: up to 100 Red-and green Macaws may gather at the Branquillo clay lick.

■ Opposite: a male Andean Cock-of-the-rock bows in display. A well-watched lek of these birds is situated close to the lodge of that name, although the species is fairly common in suitable habitat at Manu.

the fact that its area encompasses the complete altitudinal range of moist east Andean slopes, from 4,000 m altitude down to 365 m, with all their varied birdlife; and partly from its biogeographical position, where three endemic bird areas meet: the south-eastern Peruvian lowlands, the eastern Andes of Peru and the western Andes of Peru, with their combined list of more than 50 restricted range species. Of course, the national park is also rather large, encompassing 15,000 sq km of core area, together with a 'Reserved Zone' of 2,570 sq km and an outer 'Cultural Zone' of 9,129 sq km. This is enough to hold almost the entire watershed of the Manu River and most of the tributaries of the Alto Madre de Dios. Nevertheless, many other parks of comparable size elsewhere in the tropics would struggle to reach even half the number of bird species recorded here. Incidentally, it is also the site where two great ornithologists, Scott Robinson and the late Ted Parker, managed to record 331 species in 24 hours in September 1986.

With such an amazing diversity of birds, there are obviously plenty of highlights. Perhaps, though, the Manu Lodge and Manu Wildlife Centre, which are in the Reserved Zone to the east of the park proper, offer the best introduction to pure biodiversity. Located on the Rio Alto Madre de Dios, within a huge area of pristine *terra firme* forest, with scatterings of *varzea* (seasonally flooded) forest and transition floodplain forest, the immediate area of the wildlife centre had a bird-list of 556 to the end of 2006, with a great number still to come. It is the sort of place you could stay for months and still see new species almost every day. Kitted out for surprisingly comfortable birding, this lodge is to wildlife watchers what an amusement park is to a child: packed with an array of thrilling treats. One of the most famous is the Blanquillo clay lick, where simply hundreds of parrots visit every day to eat the mineral-rich clay on the exposed bank. The most prominent visitors are the spectacular Red-and-green Macaws, of which there may be up to 100 or so, while a procession of smaller species are regular visitors: Blue-headed Parrot, Mealy and Yellow-crowned Amazon, the splendid Orange-cheeked Parrot and White-eyed and Cobalt-winged Parakeets. Nobody is entirely sure why these birds risk their lives to visit the lick (which is exposed to the view of predators), since they are all primarily herbivores, subsisting on fruit and seeds. It has been suggested that the minerals may help to ameliorate the effects of plant toxins contained within their food. Whatever the reason, just watching these spectacular gatherings from floating blinds on the river would alone be quite enough reason to visit Manu. Yet the Blanquillo lick is not the only one hereabouts. Another area of clay deep in the forest is a favourite spot for large mammals, including tapirs and peccaries. It is also good for the larger forest birds, especially cracids and yet more parrots.

A good way to appreciate the forest around Manu Wildlife Centre is to climb the lodge's own canopy tower, which is built into a giant *Ceiba* tree. Here, a few hours at dawn will reveal a procession of forest canopy jewels, often with 70 or more species in a single mixed flock, notably those absurdly colourful tanagers, such as

Above: Manu's forests are the richest on earth for bird species.

Below: the forest near Amazonia Lodge is a hot-spot for antbirds – this is a Great Antshrike, which is one of the easiest species to see.

the almost too opulent Paradise and Opal-rumped Tanagers. A completely different experience would be to venture into the *Guadua* bamboo thickets, that shady world where the forest harbours its inner secrets. Here the specialists include the recently described Manu Antbird, the White-cheeked Tody-Flycatcher and the curious, if frustrating, Peruvian Recurvebill. The trip to the different sites along the river is a wonder in itself, encountering large numbers of equally specialized birds, such as Yellow-billed and Large-billed Terns, Orinoco Goose and Sand-coloured Nighthawk, the world's most diurnal nightjar.

All these birds are found only in a small part of Manu National Park. Chief among the other attractions are the transitional forests at Amazonia Lodge at an elevation of 500 m, and Cock-of-the-rock Lodge, in the south-west of the park, at 1,600 m. The former site, in the zone between tropical and subtropical forest, has a bird-list of about 550 species, including many species characteristic of slightly higher elevations, such as Blue-headed and Military Macaws, Koepcke's Hermit and Gould's Jewelfront. The diversity of antbirds here could possibly be one of the highest in the world, with Great Antshrike and such gems as Spot-backed Antbird and Stripe-chested Antwren among the dozens of species. Amazonia Lodge is, of course, famous for its lek of Andean Cock-of-the-rocks, near the accommodation, which can be seen from permanent blinds and watched at leisure. The many highland birds here include Booted Racket-tail, a spectacular hummingbird, plus both Golden-headed and Crested Quetzals and such delightful Andean staples as Streaked Tuftedcheek, Pearled Treerunner, Hooded Mountain Tanager and Capped Conebill.

There is yet more to be seen at the highest elevations, where the elfin forest grows and there are open areas of paramo. Birds here include various tinamous and specialized high-altitude species such as Puna Thistletail, Tit-like Dacnis and Moustached Flowerpiercer. As yet, there has been little ornithological exploration in this inaccessible part of the park. Most likely it will be here that the species-counter will click on into the 1,000s, cementing Manu's position as the indisputable top place in the world for bird species.

Abra Patricia

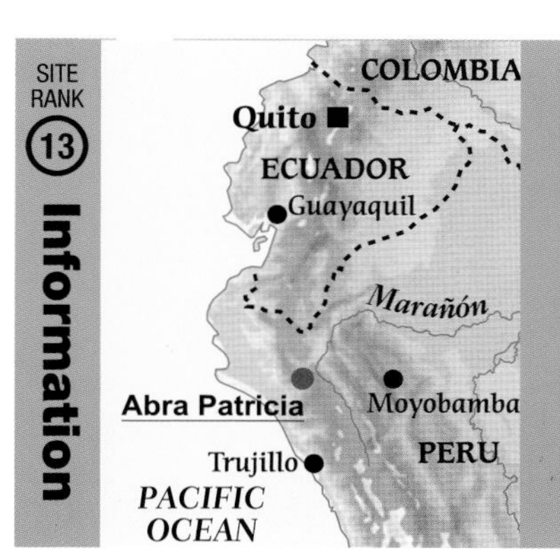

HABITAT Various forest types on east Andean slopes, including humid montane and elfin forest

KEY SPECIES Marvellous Spatuletail, Royal Sunangel, Long-whiskered Owlet, Lulu's Tody-Flycatcher, dozens of species of tanagers

TIME OF YEAR All year round

Any birder visiting Abra Patricia is travelling to the frontiers of ornithological knowledge. More new bird species have been discovered in this corner of northern Peru in the last three decades than anywhere else on earth. The area is also home to what must be the world's most spectacular hummingbird, the Marvellous Spatuletail, and it is the sort of place where such incredible species as Andean Cock-of-the-rock and Amazonian Umbrellabird simply make up the supporting cast. The bromeliad-laden cloud-forests teem with an array of colourful tanagers and smart furnariids, and the area itself, with its undulating pattern of plunging ravines and densely forested slopes, makes for a stunning backdrop to some of the world's most exciting and challenging birding.

Birders planning to work the road that travels over Abra Patricia inevitably begin in the area around Pomacocha (Florida) and it is here, on the hills nearby, that the Marvellous Spatuletail occurs. This species is as rare as it is astonishing to look at, more or less confined to just a 100-km stretch along the valley of the Rio Utcubamba. It is a medium-sized hummingbird and is unique in its family in having just four tail feathers, instead of the usual ten. In the adult male, two of these – which are as long as the bird itself – stick straight out, while the other two, which are mainly bare shaft, angle out and then across each other beyond the end of the straight part of the tail, eventually ending in rackets, or spatules. When the bird flies, these rackets move independently in all directions, and it is thought that their odd movements serve to confuse predators – it certainly works with bird-watchers. The main food plant of this hummingbird appears to be a red-flowered lily, which grows among thickets between 2,100 and 2,900 m.

Below: a male Marvellous Spatuletail shows off its astonishing tail feathers.

From Pomacocha a newly-paved highway winds up towards the pass of Abra Patricia (2,400 m) and then plunges down the slope, zigzagging wildly, until it eventually heads to the distant lowland forests of Amazonia. The most famous stretch for birds is by Alta Nieve, a few km beyond the pass and at about 2,000 m altitude, where the moist elfin forest becomes stunted and the bushes are richly covered by lichen, moss and other epiphytes, with patches of bamboo, palms and tree ferns. It was here, in 1976, that the remarkable Long-whiskered Owlet was discovered one wet morning, caught in a researcher's mist-net. This tiny owl, with its dark hair-like whiskers reaching from the base of the bill to beyond the side of the head, is still completely unknown in the wild state, although five specimens have been captured. The nature of the captures has set off a strong rumour that it might even be flightless which, if it were true, would be truly remarkable in an owl. As yet, the race to be the first to observe this enigma in the field has not yet been won, so no one can confirm or deny how it gets around, catches food or reproduces.

The owlet is not the only enigma here. Barely better known is the Ochre-fronted Antpitta, another bird discovered here in 1976 and only recently seen wild for the first time. It lives in the dense understorey of these forests, and is no more confiding than most other terrestrial antpittas, so the details of its life history are completely unknown. Birders here should, however, have a better chance to see some of the other recently described species for themselves. The delightful Royal Sunangel (described in

The year 1979 was memorable because both Bar-winged Wood Wren (above) and Royal Sunangel (right) were discovered at Abra Patricia and described for the first time as species new to science.

1979) appears to prefer areas where the stunted forest is juxtaposed with taller forest. The male, which is royal blue all over, is highly territorial and reluctant to leave stands of its favourite bush, the red tubular-flowered *Brachyotum quinquenerve,* while females seem to occur lower downhill and feed on different plants. The Bar-winged Wood Wren (1979), looking remarkably like a small antbird, can usually be found by careful searching of the stunted forest undergrowth, while the Cinnamon-breasted Tody-Tyrant (1979) draws attention to itself by making short aerial forays after insects. The gorgeous Lulu's Tody-Flycatcher (2001) has a broader distribution in the bamboo thickets of the cloud-forest proper, and it is a wonder that it escaped detection until after the more cryptic discoveries. With its bright orange-red head, brilliant yellow underparts and grey nape, this little bird is now known to be quite common in the area as a whole.

These headline-grabbers are but a tiny fraction of the bird community in the cloud-forest and elfin forest of Abra Patricia. Unless you have been to such a habitat it is almost impossible to convey its sumptuous richness, both in the density of vegetation and the thrilling multi-coloured pageantry of its birds. The typical currency of dawn in a cloud-forest is a mixed flock of outrageously vivid tanagers, with names to match the visual feast: White-capped, Yellow-scarfed, Beryl-spangled, Grass-green and Blue-winged Mountain Tanagers. Buzzing to and fro there will also be hummingbirds, conveying the same sense of exuberant wonder in their names: Rainbow Starfrontlet, Emerald-bellied Puffleg and Blue-fronted Lancebill. Almost all of these birds have specific altitudinal preferences, so that a mixed feeding flock at 2,100 m will be quite different from a flock found at 1,500 m above sea level.

Indeed, a whole new set of birds appears once you have reached the village of Afluentes. Here the forest is taller, and humid montane forest species, such as the Andean Cock-of-the-rock, are common. Although only about 20 km from the pass, you almost have to start working all over again if you are to identify the components of the mixed flocks that bustle through the canopy here.

Despite its wonders, the forest on the hills here is under threat. Local people have been moving into the forests and clearing them at an alarming rate, and until recently all the richest parts, including the localities for Marvellous Spatuletail had no official protection at all. However, since 2004 1,820 sq km of forest east of the pass have been given protection, and the Abra Patricia Bird Reserve created. Hopefully, this will be just the beginning, and significant tracts of these teeming forests will be saved for future generations to enjoy.

Opposite: Cloud-forest near Abra Patricia; the epiphyte-laden canopy is typical of the habitat.

Santa Marta

SITE RANK 26 **Information**

HABITAT Various forest types from lowland to cloud-forest and elfin forest; paramo grassland; cultivations

KEY SPECIES Santa Marta Parakeet, Blossomcrown, Bearded Helmetcrest, Santa Marta and White-lored Warblers, Santa Marta Brush Finch

TIME OF YEAR All year round, although May is generally considered the best birding month. The wet season runs from September to December

Above right: the Santa Marta Brush Finch is one of the easiest endemics to see.

Below: the Santa Marta Bush Tyrant is found in the temperate forest zone, between 2,100 and 2,900 m above sea level.

For many years now Colombia has been famous as the country with more bird species than anywhere else in the world. The current total is just over 1,800 species, but continues to rise; a new hummingbird, for example, was described for the first time in 2007. Such a staggering total is partly explained by the remarkable variety of habitats, which include a western extension of the Llanos of Venezuela, some dry, almost desert-like vegetation in the north, plenty of lowland rainforest in the interior, and many mountain ranges; and partly by the shape of the Andean chain, which bulges in Colombia and breaks into three north–south cordilleras, which are separated by wide valleys. The ranges are sufficiently isolated to hold distinctive sets of birds, including some of their own endemic species, and the valleys are similarly full of localized inhabitants. It is this potent mix that has led one recent writer to enthuse about Colombia's 'megadiversity'.

However, there is a further quirk to Colombia's topography that lies outside the Andes. It could be argued that the boost provided by this area is the one that actually pushes the country's bird-list out of reach of its nearest rivals, Peru and Brazil. That quirk is the Sierra Nevada de Santa Marta, an isolated mountain massif on the northern coast, far removed from the Andes and surrounded on all sides by lowlands covered with dry, shrubby and thorny vegetation, with its own self-contained climate and natural history. It contributes many species to the Colombian list including, remarkably, almost 20 endemics.

This triangular massif is highly unusual topographically, because the land rises up to 5,775 m, just 45 km from the coast, so that the range's snow-capped peaks loom over the tropical beaches far below, and making the Sierra Nevada de Santa Marta the highest coastal range in the world. With the land rising so spectacularly, it is easy to imagine how different habitats can be crammed into a relatively small area. Indeed, dry, thorny scrubland cloaks the foot of the range, while the tops are dominated by permanent ice and, just below this, cold, wind-blown paramo grassland. In between, where people have not cultivated the land, various types of forest grow, from wet lowland forests up to about 1,000 m, then subtropical and temperate cloud-forest and finally, at about 3,000 m and above, elfin forest that covers the ridges below the paramo.

In all, about 630 species occur in the range, if you include the dry, thorny forest of the foothills. Most of the endemic species occur at altitude, however, and it is these that most birding visitors will be seeking. The Santa Marta Brush Finch is usually the first to appear as one ascends, and has been recorded as low as 600 m. This splendid bird, with grey-black upperparts, brilliant yellow underparts, a reddish eye and silvery ear-coverts, occurs abundantly in various edge habitats, including cultivations and secondary growth. Two hummingbirds

Above: not all the best birds here are endemics; the White-tipped Quetzal is locally fairly common at altitudes below 2,000 m.

are also found on the lower slopes, the Coppery Emerald, which is copper-bronze all over in the male, and the extremely pretty Blossomcrown, a primarily green hummer with a subtle patch of rose-pink at the back of the crown and a spot of white behind the eye. Neither species is common, and the latter is mainly found in lightly forested areas.

Higher up, the endemics proliferate, and include three delightful New World warblers. The perky Yellow-crowned Whitestart is a consummate member of the cloud-forest mixed-species flocks, moving around actively in pairs, calling *chip, chip* constantly and wagging its tail. This yellow, black-faced bird often occurs alongside the Santa Marta Warbler which, in complete contrast, is a shy, skulking species of bamboo thickets. Once seen, it has a most distinctive appearance with a black-and-white face pattern on a green and yellow body. The White-lored Warbler, a yellowish bird with a grey head, is found mainly on the edge of the forest.

The cloud-forest is rich in species although, as everywhere in this habitat, the birds come in fits and starts. Among the other endemics are the rather dull-coloured Santa Marta Antpitta, which is considered to be slightly easier to see than most other antpittas, often appearing in clearings in the early morning; and the Santa Marta Tapaculo, a predominantly grey species that often feeds on the ground. Two spinetails, the Rusty-headed and the Streak-capped, often join mixed species flocks, the former usually feeding lower down than the latter. Other endemics in this zone include the Santa Marta Bush Tyrant, the very rare Santa Marta Parakeet and the Santa Marta Mountain Tanager.

There are other specialities to be found in the elfin forest and paramo zone, several of which are hummingbirds. The spectacular, shy Santa Marta Sabrewing, mainly iridescent green with a deep blue throat and breast in the male, regularly occurs in this zone, along with the Black-backed Thornbill, a rather dark species with a very thin and, for a hummingbird, short bill. These two occur alongside the spectacular Bearded Helmetcrest, very much a bird of the high tops, which often perches to feed and has even been observed walking on mats of grass, picking up insects. This stunning species, which also occurs in Venezuela, has a showy black-and-white crest, and the male has a deep, purplish beard that hangs down to the upper breast. It makes quite a sight, especially among the spectacular mountain scenery of the Santa Marta range.

Birders visiting the range have recently been surprised to find a small owl in the cloud-forests. Of screech-owl type, this species has not been recorded before, and it is highly likely that it is new to science.

Galapagos Islands

SITE RANK 5

Information

PACIFIC OCEAN

Galapagos

Archipiélago de Colón Islas Galápagos (Ecuador)

HABITAT Volcanic islands, with cliffs, beaches, forest and semi-desert.

KEY SPECIES 'Darwin's' finches including Woodpecker Finch, mockingbirds, Galapagos Penguin, Flightless Cormorant, Waved Albatross, Swallow-tailed and Lava Gulls and other seabirds.

TIME OF YEAR All year, seabird colonies busiest between February-September.

Above right: on the whole the famed Darwin's finches are disappointing as a tourist attraction – this is a Medium Ground Finch, which occurs only on Floreana.

Below: typical Galapagos habitat – the islands are volcanic and many of them appear bleak and dry.

These islands, which lie in the Pacific Ocean 1,000 km west of Ecuador, on the Equator, are among the most famous destinations in the world for watching wildlife. Their fame is partly aesthetic and partly scientific. In the first place they are remote, starkly beautiful and relatively unspoilt, with many islands, even quite large ones such as Española, still uninhabited by people. Their wildlife is tame, abundant and often spectacular. Secondly, these islands were visited in 1835 by the pioneering English naturalist Charles Darwin, and in the course of making observations here, most famously on the native finches, he was able to ruminate on what would become his theory of evolution, one of the great landmarks of modern science.

Following Darwin's footsteps is a thrill that many thousands of visitors cannot resist and, whether or not the islands are over-hyped (for example, Darwin actually made very little reference to the archipelago's finches in his writings, and their importance to his thinking is disputed) scarcely matters to those that come here for the memorable sights and sounds. The main point is that much of what occurs on the Galapagos is unique: the Giant Tortoises, the Marine Iguanas and, of course, the birds, of which there about thirty endemic species. What you can see here, often at close quarters, cannot be seen anywhere else.

Oddly enough, those celebrated finches, nowadays often collectively known as Darwin's finches, are pretty dull to look at and extremely difficult to identify, so to a

Right: the rare Flightless Cormorant, with a vulnerable world population of only 900 individuals, is found only on two islands.

casual visitor they can prove to be pretty disappointing. They are basically small, dark (to blend in with the volcanic soil) and streaky, with the females browner than the males. Where the various species differ is primarily in their bill shape, which is closely correlated to their feeding behaviour and ultimately the ecology of the islands. However, don't expect these distinctions to be obvious in the field, because often they are not: the bill of the Large Cactus Finch is not very different from that of a Large Ground Finch, except in diagrams in textbooks.

However subtle, the differences between the finches are truly fascinating. Marvel, for example, at how the bill of the Large Ground Finch is so hefty that this bird can tackle much larger seeds than the Medium Ground Finch or the Small Ground Finch. The distinction, furthermore, between the two arboreal insectivorous tree finches is also intricate: the Large Tree Finch has a thick, hooked bill for tearing bark, whereas the bill of the Small Tree Finch is narrower and used mainly for plucking insects from the surface. One species you will have no trouble at all in recognizing is the Warbler Finch, because its bill is much thinner and is used for probing into cracks and among leaves. It has, however, recently been split into two species, the Olive Warbler Finch and the Dusky Warbler Finch, presumably to keep birders on their toes.

Some of the finches exhibit distinctly odd behaviour. For example, the Woodpecker Finch is a tool-user which plucks twigs from herbaceous bushes (or the spines off prickly pear cacti) and uses them to pry termites and insect larvae from bark or wood cavities. The behaviour of the Sharp-beaked Ground Finch is even more strange; on the island of Tower, where there are huge numbers of breeding seabirds, it will casually approach the sitting Red-footed Boobies and peck at the base of their feathers; that will be enough to draw blood, which the finch drinks!

It seems that unusual forms of behaviour are rather commonplace here. The Wedge-rumped Storm Petrel, for example, is the only member of its family to occupy its breeding areas during daylight hours rather than during the night. This has presumably arisen in the relative absence of predators, since there are no skuas here and the gulls, on the whole, are well behaved; the squid-eating Swallow-tailed Gull, indeed, is virtually nocturnal as a feeder itself, unlike most other gulls. The storm petrels, however, suffer from predation by Short-eared Owls. Another behavioural quirk has recently been discovered in the Galapagos Hawk. Despite being large and broad-winged, this raptor is apparently highly adept at hunting small landbirds, rather than the larger prey or carrion you might expect.

Aside from the finches, the mockingbirds are perhaps the most sought-after of the islands' inhabitants, at least by birders. The Galapagos Mockingbird is widespread, but the other species all occupy tiny ranges. The San Cristóbal and Hood (Española) Mockingbirds occupy their respective islands, while the Floreana Mockingbird now occurs on only two small islets off its namesake, and

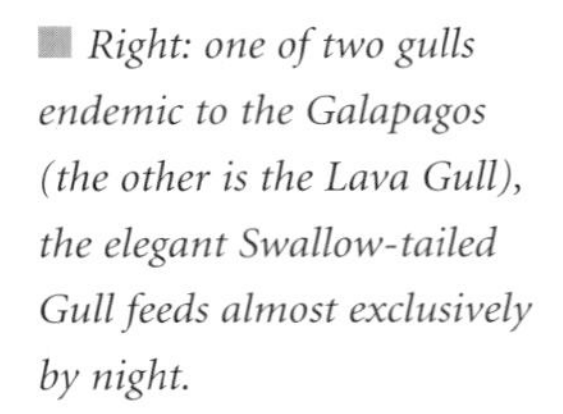

Above: Waved Albatrosses is a Galapagos speciality – the species breeds on Española.

has a world population of no more than 260 individuals. In common with many Galapagos animals, these mockingbirds have a habit of being inquisitive and utterly fearless, and some apparently cannot resist the urge to peck at visitors' shoelaces!

Although the landbirds are undoubtedly fascinating, the seabirds of the Galapagos provide much of the spectacular viewing. There are huge colonies of Red-footed and Blue-footed Boobies, together with smaller numbers of both Magnificent and Great Frigatebirds, dotted throughout the archipelago. One of the islands' most impressive spectacles is the colony of Waved Albatrosses on Española, which stands at about 18,000 pairs. These birds can often be watched performing their comical sky-pointing displays. Two very rare endemic species are also found here. The world's northernmost penguin, the Galapagos Penguin, occurs in small numbers on some of the western and central islands, while the Flightless Cormorant is found only on Isabela and Fernandina – it is the only member of its family to have lost the power of flight. The world populations of these vulnerable species are estimated at 1,200 and 900 respectively.

Of course, any species confined to a single island or island group is more likely than most to be vulnerable to extinction, and the birds of the Galapagos are no exception. In recent years, both the penguin and the cormorant have been affected by fluctuations in the El Niño current. One species of finch, the Mangrove Finch, has been decreasing for unknown reasons, and its population now numbers a shaky 100 individuals in just three sites on Isabela. In the 1990s, visiting boats are thought to have brought in parasitic flies that are now found in a high proportion of finch nests, reducing breeding success. Despite the impression that these islands give of being far distant and locked away from modern life, it appears that even they are not immune to threats from the outside world.

Right: one of two gulls endemic to the Galapagos (the other is the Lava Gull), the elegant Swallow-tailed Gull feeds almost exclusively by night.

Tandayapa

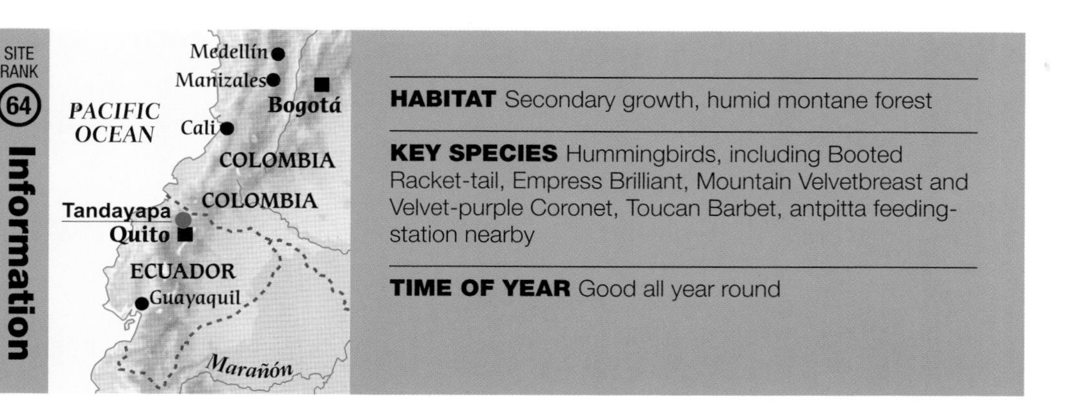

HABITAT Secondary growth, humid montane forest

KEY SPECIES Hummingbirds, including Booted Racket-tail, Empress Brilliant, Mountain Velvetbreast and Velvet-purple Coronet, Toucan Barbet, antpitta feeding-station nearby

TIME OF YEAR Good all year round

The bird lodge and environs of this small village in north-west Ecuador host what is probably the highest number of species of hummingbirds in the world. Up until 2007, an astonishing 31 species have been recorded from the deck of the lodge, representing almost 10 per cent of all the species in one of the most diverse families in the world (indeed, boasting some 330 species, the hummers currently lie second behind the tyrant flycatchers, which are way out in front with 400). Within a short distance of Tandayapa, another ten or more can be added to a hummer-hungry birder's list.

In various parts of South America, it is likely that some other spots must be strong competitors in the hummingbird stakes. However, with up to 20 species visiting the feeders or blooms every day, and with as many as 100 hummers possibly in view at once, the display at Tandayapa is nothing short of phenomenal.

The bird lodge lies in the humid tropical forest zone (cloud-forest) at 1,750 m. It was originally constructed on a denuded ridge, but the planting of 30,000 trees has allowed the area to be enclosed by secondary growth, and the forest birds have settled in. The reason for the hummingbird diversity lies not just in the location – right in the midst of the North Andes, where speciation has run riot within isolated valleys – but also in its altitude. Both high altitude species, such as the Mountain Velvetbreast, and visitors from low altitude, such as the White-necked Jacobin, make occasional forays to mix in with the regulars to whom the altitude is ideally suited.

With such diversity present in one place, it is relatively easy to get an idea of the different characters and characteristics of the hummers. For example, some species are present at the feeders throughout the day, whereas others make short appearances just once or twice a day, or even less frequently. These differences are related to the feeding habits of the birds, which tend towards one of two main strategies: the territory holders and the so-called 'trap-liners'. Territory-holders, as their name implies, devote much time defending a rich, reliable source from other birds; as territory-holders they also

Below: the aptly named Booted Racket-tail is a common visitor to the feeders at Tandayapa Bird Lodge.

try to attract females to the site. Thus, at Tandayapa, a species such as a Sparkling Violetear, a large hummer, with a royal-blue breast and tail and deep green upperparts, will be highly aggressive, trying to keep rivals of other species, as well as its own, away from the nectar source. Other territorial species include the Buff-tailed Coronet, a scaly green hummer with a buff shoulder tuft and pale belly, which never really seems to be satisfied until the whole feeder is completely clear. It is often a surprise to see just how aggressive and ruthless hummers can be.

The trap-lining strategy is to follow a beat around known nectar sources, stopping briefly at each one throughout the day. This way there is no need to try to defend a nectar source, although skirmishes are nonetheless frequent when lots of trap-liners meet and the feeders are busy. Thus most of the less frequently observed species are trap-liners, including Tandayapa's only hermit, the Tawny-bellied Hermit, which comes in for a few brief minutes each day. Other trap-liners include the abundant Andean Emerald, which is mainly iridescent green, and the much less numerous Violet-tailed Sylph, which is unusual among hummers for sometimes trap-lining in pairs.

Some hummingbirds are quite particular in their habits, which quickly marks them out from the rest. For example, the tiny woodstars have a very smooth, insect-like flight that lacks the jerky movements of the other

Above: the gaudy Toucan Barbet is a speciality of the site.

Opposite above: Violet-tailed Sylph is almost a Colombian endemic, but its range reaches just into this part of Ecuador.

Opposite below: the cloud-forest here is at an intermediate level of 1,750 m, and it receives occasional visits from species that are typical of both lower and higher elevations.

hummers, and when hovering they are far more horizontal in posture; the White-bellied Woodstar, with white underparts and purple throat, also makes an insect-like rattle with its wings, alerting a birder to its approach. The large, iridescent green Empress Brilliant, on the other hand, has an interesting quirk. In contrast to the other hummingbirds, it regularly uses the perches on the feeders to drink, instead of hovering; this mirrors its behaviour in the forest, where it perches to drink from the bracts of certain forest plants.

The Empress Brilliant is one of several species found here which are very restricted in range, and is thus one of the most sought-after species here. Others are simply stunning, or spectacular. Anyone could appreciate the splendid Violet-tailed Sylph, with its amazingly long forked tail, which takes up more than half the bird's entire length, or the Booted Racket-tail, with its absurd fluffy white booties. The latter is a common species, and visitors can sometimes see it in display, when it is fluffed up even more. Hummingbirds generally have sparkly, iridescent plumage and most are splendid to look at, but surely the Velvet-purple Coronet must be one of the world's finest. With a glittering green back, striking black-and-white tail, and underparts and head in various shades of deep purple or pink, this rather uncommon hummer is one to bring all the staff of the bird lodge out to watch and admire.

In a rich habitat such as this, there are, naturally enough, plenty of other things besides hummingbirds to keep visitors occupied. Just over 300 species of bird have been recorded in the valley, including such specialities as Toucan Barbet, Sickle-winged Guan, Scaled Fruiteater and Golden-headed Quetzal. The forest-floor hide is an excellent place for the jet-black Immaculate Antbird. Within easy reach there is also a private nature reserve, the Pas de las Aves Reserve, where a number of shy forest-floor birds, including Yellow-breasted, Moustached and Giant Antpittas have become accustomed to being fed with worms, and will approach to within a few metres of incredulous visitors. It is, however, the sight of the buzzing hummingbirds that will remain in the memories of those fortunate enough to come here.

The Pantanal

HABITAT Floodplain wetland, gallery forest, cerrado

KEY SPECIES Waterbirds, including Sunbitten, Jabiru and Boat-billed Heron, Greater Rhea, Hyacinth Macaw, Chestnut-bellied Guan, Southern Screamer, Scarlet-headed Blackbird

TIME OF YEAR Roads passable only from July to December; rains begin in October

Each year, between December and April, deep in the heart of South America, rains in the Mato Grosso cause the mighty Rio Paraguay and its many tributaries to burst their banks. As they do so they flood an enormous alluvial plain of some 140,000 sq km, spanning the border between Brazil, Bolivia and, to a lesser extent, Paraguay. This is the Pantanal, the world's largest freshwater wetland and one of the richest birding areas in South America.

Most of the Pantanal is thinly populated and dotted with large ranches or 'fazendas', which together graze about 8 million head of cattle on the flat grasslands. This practice has been continuing for more than 200 years, with little deleterious effect on the Pantanal's astoundingly abundant wildlife. The breeding populations of a good many waterbirds, including various herons, storks, ibises and wildfowl, run into the thousands or the tens of thousands, and the overall health of the area is illustrated by the high number of raptors – up to 20 species – that seem to be everywhere. The Pantanal is also a superb place for mammals, including the Jaguar, which is pretty choosy about where it lives.

The area is bisected by a dirt road, the Transpantaneira, which runs from the town of Poconé, near Cuiabá, 148 km south to Porto Jofre, on the banks of the Rio Cuiabá. Despite having more than 100 bridges across the various waterways, the Transpantaneira is largely impassable between January and May, after which the waters recede. As they do so, the birdlife becomes more and more concentrated until, by September, what wetlands are left will be thronged with thousands of birds of dozens of species. For the birder, this is the best time to visit.

The flooding of the plain also coincides with a phenomenon known as the 'piracema', the seasonal migration of many hundreds of species of fish. During the dry season (May to October) these fish move upriver towards the Pantanal area and then, as the waters rise in the river, they reproduce. Following invisible cues, they then rapidly move into the standing water of the floodlands where they feed until the end of the rains. Many then return to the rivers whence they came, but in their abundance they are susceptible to being stranded or ambushed in transit by eager fish-eating birds. This is what sustains the enormous numbers of birds such as Wood and Maguari Storks, Jabiru, Snowy and Great Egrets, Capped and Cocoi Herons, Neotropic Cormorant, Anhinga, Black Skimmer and Large-billed and Yellow-billed Terns, as well as predators such as Great Black and Black-collared Hawks.

There are also plenty of wetland species here that do not rely on fish. Species such as Limpkin and Snail Kite consume molluscs, Purple Gallinule and Southern Screamer eat vegetation, while the various ibises

Right: the area's most famous bird, the Hyacinth Macaw, is the world's largest parrot.

Opposite: getting around in the Pantanal is not always straightforward – the many wooden bridges frequently have to be rebuilt if they are destroyed during the wet season.

(including the Plumbeous Ibis, which is commoner here than anywhere else) consume small invertebrates. The great variety of wetland habitats on the Pantanal allows all these different species to find a niche and fit in. Besides the flooded grasslands, there are permanent marshes, swamps and ponds, together with seasonally flooded woodlands and gallery forests growing on the banks of the rivers. Thus, such species as Sunbittern, Rufescent Tiger Heron and Sungrebe can hunt on the quiet backwaters of rivers and streams, the Wattled Jacana can wander over water-lilies (including *Victoria amazonica*, the largest water-lily in the world), Ringed and Brazilian Teals can dabble in the shallows for seeds and the superb Scarlet-headed Blackbird, with its astonishingly bright red head and upper breast, feeds on insects in the tall, dense marshland above deeper water.

Perhaps surprisingly, in view of the abundance of all these wetland species, the Pantanal has gained just as much fame among birders for its variety of forest and scrub species as for its wealth of waterbirds. Indeed, the most celebrated of all its avian inhabitants is actually dependent on palm trees – the magnificent Hyacinth Macaw. This largest of all parrots is clad almost entirely in deep cobalt blue except for bare yellow skin at the base of the bill and around the eye. This regal bird, which flies with the typically imperious slow wing-beats of the larger macaws, is

■ *Above: all three South American species of stork occur in the Pantanal, including the mighty Jabiru.*

■ *Right: Capped Heron is generally rather scarce and not gregarious; it usually feeds alone at the edge of the water.*

becoming rarer and rarer in the wild, owing mainly to illegal collecting for the cage-bird trade. Numbers have dwindled to fewer than 10,000 birds at a handful of sites, of which the Pantanal is by far the most important.

The Hyacinth Macaw is, in fact, not the only sought-after macaw of the Pantanal. The Golden-collared Macaw, which actually has a smaller world range than the Hyacinth Macaw, is found in the *cerrado*, gallery woodland and ranch lands of the area, being more adaptable than its counterpart and thus not being reliant on palms. It is smaller than the Hyacinth Macaw but is still striking, with largely green plumage broken up by a stunning patch of golden-yellow on the nape, and a black forehead.

The presence of large birds such as macaws is typical of areas in the Neotropics that are free from the pressure from excessive hunting. Much the same applies to the cracids, and the drier areas of the Pantanal are excellent for these, too. Most birders will be seeking the Brazilian endemic Chestnut-bellied Guan, but other species include the terrestrial Bare-faced Curassow, Common Piping Guan and the noisy Chaco Chachalaca. The presence of such birds is typical of the richness of all the habitats on the Pantanal.

A visit towards the end of the dry season will often produce more than 100 species each day. A special advantage of the area is that, not only are there masses of birds in quantity and quality, but it is often very easy to see them, too. All of this adds up to making the Pantanal one of the most intoxicating birding areas in the world.

Serra da Canastra

HABITAT Savanna (cerrado), open grassland, gallery forest, streams and rivers

KEY SPECIES Brazilian Merganser, Brasilia Tapaculo, Cock-tailed and Streamer-tailed Tyrants, Helmeted Manakin, Red-winged Tinamou, Greater Rhea, Red-legged Seriema

TIME OF YEAR All year round

Although Brazil is most famous for its rainforests, there are plenty of other habitats within this vast country that are also important for wildlife. One of these is known as *cerrado*, which is tropical savanna with grassland and scattered trees. In the last 40 years nearly two-thirds of all Brazil's *cerrado* has been modified for agriculture or destroyed for development, and as a habitat it is, if anything, even more threatened than the celebrated rainforests of Brazil's interior. Not many large patches remain, but one of the best is within the Serra da Canastra National Park, in the state of Minas Gerais.

This large national park, which has an area of 2,000 sq km if you include the nominally protected buffer zone around the outside, consists of two high mountain plateaux with a wide, deep valley in between. It is an area full of rivers and fast running streams, among them the Rio São Francisco, which begins its 2,700-km course to the Atlantic Ocean here. There are dozens of waterfalls and precipitous cliffs, making it one of the most scenic national parks in the country.

Below: Serra da Canastra National Park's fast-flowing rivers are one of the last outposts for the critically endangered Brazilian Merganser.

The tumbling rivers are home to what, currently, is the Serra da Canastra's most famous bird, the frighteningly threatened Brazilian Merganser. This marvellous endemic sawbill, with its grey-brown body, bottle-green head and long, wispy occipital crest, has suffered a huge decline in population in the last 30 years, and there are now fewer than 250 individuals in the world. A major conservation initiative has been launched to save the species, but it may already be too late, since not all the reasons for its decline are fully known and the populations are highly fragmented. At one of its last strongholds, this species is usually seen loafing about on rocks or overhanging branches, but it can be exceedingly difficult to track down on the many rivers and streams.

It is easier to locate the specialities of the *cerrado* itself. The countryside of much of the reserve is strikingly similar to that of the more famous African savannas, with the long grass and scattered trees, so it is no surprise to find South America's equivalent of the Common Ostrich, the Greater Rhea, wandering around the savanna here. Rheas are smaller than ostriches, with duller, grey-brown plumage, but they still stand 1.5 m tall, making them South America's largest bird. In winter they often gather into flocks and graze on any leaves, fruits and seeds they can find. Another interesting parallel to the African savanna is the presence of a long-legged, snake-eating predator to take the place of the Secretarybird. In common with its counterpart, the Red-legged Seriema is primarily terrestrial, prowling over the grassland in search of snakes, as well as small mammals, large insects, frogs and young birds, breaking into a run when it has something edible in its sights. Seriemas are more closely related to cranes or bustards than to birds of prey, and often break their predatory mould by eating vegetable matter. They are very wary birds, and are sometimes tamed by locals to keep as lookouts for vulnerable stocks of poultry.

One species that is truly unique to this habitat is a pocket dynamo known as the Cock-tailed Tyrant. A small, upright-perching bird that feeds on insects and forages by launching into the air on sallies, or by hovering, the male has a quite extraordinary display that is one of the highlights of birding the *cerrado*. Using remarkably fast, insect-like wing-beats, it rises into the air in methodical, helicopter-like flight up to heights of between 5–100 m above ground, all the while raising and lowering its black, bushy tail as far as it will go, and making soft ticking noises. The display is typical of a bird that needs to be noticed in its wide-open habitat, and it certainly works. Cock-tailed Tyrant is among the many *cerrado* birds whose range is rapidly contracting as the habitat disappears.

■ Above: Brazilian Mergansers can be extremely difficult to track down among the rocks and vegetation in their turbulent habitat.

■ Below: one of the park's many grassland specialities is South America's largest bird, the Greater Rhea.

These delightful and distinctive mites are not threatened yet, but the Serra da Canastra is already one of the very best places to see them.

Another interesting species found here is the Campo Miner, a member of the ovenbird family. In common with the Cock-tailed Tyrant it works hard to get noticed, having a tendency to perch atop the many tall termite mounds that dot the landscape of the *cerrado*. In display it will also frequently launch into the air, singing and flapping with slow wing-beats, showing off its attractive chestnut-brown wing-bars. As with all the miners, it nests below ground, in this case usually down an armadillo burrow. One of the curious features of its ecology is its preference for recently burned areas. Regular burning is one of the necessities for this landscape, keeping the growth of bushes down and maintaining open spaces. When fires occur the Campo Miners mysteriously appear, apparently from nowhere, and may begin to inspect holes while the ground is still smoking.

Another important habitat within the national park is gallery forest, which grows around the larger rivers and adds greatly to the overall biodiversity of the area. This forest is home to another of the area's great rarities, the skulking Brasilia Tapaculo. For a *Scytalopus* tapaculo it is almost daringly colourful, with white on the breast, a bluish wash to the upperparts and a little patch of ochre on the flanks. But it reverts entirely to type when birdwatchers are searching for it, being noisy but more or less impossible to see. Other good species in the forests include the rare Grey-eyed Greenlet, Helmeted Manakin and White-striped Warbler.

It is really the rolling savanna and more open grassland (*campo*) that defines this magnificent area. Besides birds, the many attractions include the rare Maned Wolf and the Giant Anteater, which patrol the plains in search of rodents and ants respectively. The superb Black-chested Buzzard-Eagle often flies overhead in search of cavies, rabbits and viscachas, and among the many other specialities are Red-winged Tinamou, White-tailed Hawk, Hellmayr's and Ochre-breasted Pipits and an excellent assortment of finches and hummingbirds. This truly is Neotropical grassland at its very best.

Manaus

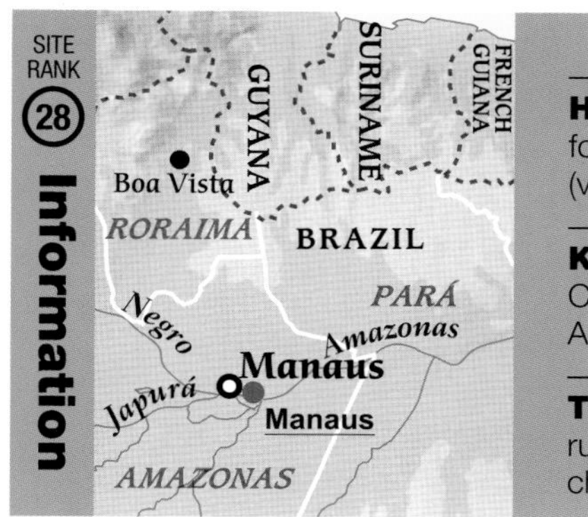

HABITAT Lowland terra firme rainforest, white-sand forest (campinas), permanently (igapó) and seasonally (várzea) flooded forest

KEY SPECIES White-winged Potoo, Dusky Purpletuft, Crimson Fruitcrow, Pelzeln's Tody-Tyrant, Klages's Antwren, Scaled Spinetail

TIME OF YEAR All year round, but the rainy season runs from November to April, when birding can be a challenge

Below: jacamars are one of many families to be well represented in the area around Manaus. Each fills a slightly different niche with, for example, the Yellow-billed Jacamar usually favouring the forest understorey.

Everyone knows about the riches of Amazonia; it is generally reckoned to be the world's greatest centre of biodiversity, especially on its western edge. However, if you asked the average birder what species they would most like to see should they go there, many would struggle for an answer. Apart from the well-known parrots and hummingbirds, the birds are almost as obscure as they are diverse, and thus the Amazon Basin is something of a specialist destination. What birds they are, though.

Located some 2,250 km from the Atlantic Ocean, Manaus is a large city at the very heart of Amazonia. It is a major port at the confluence of the two mightiest rivers on earth, the Amazon itself, which in this part of Brazil is called the Rio Solimões, and the Rio Negro, which drains most of the northern Amazon Basin. These two rivers famously converge, and for a short distance their differently coloured waters flow alongside each other, muddy-white on one side and dark on the other. They also form a major ecological boundary, with the birds of the southern bank of the Amazon showing many differences from those of the Rio Negro above Manaus itself. The distributions of antbirds, for example, which tend to be shade-loving birds unwilling to cross large, open areas, is often delineated by large rivers.

Not surprisingly, with the immense volume of water involved, the levels of the rivers can vary considerably, by up to 12 m, and it is this that has proved to be the saviour of the region's huge area of forest, since it keeps urban development and cultivation in check. Thus, happily, the forest here still stretches as far as the eye can see, with all its secrets and wonders, and one of those wonders is that not all forest is the same. At least four major forest types occur near Manaus: terra firme forest, the classic 'tropical rainforest', growing on permanently dry, well-drained soil; white-sand forest, or campina, which grows on sand and is stunted; on the islands, igapó, or permanently flooded forest; and várzea, which is seasonally flooded and boasts taller trees.

Each of these forest types has its own special bird species – often a whole suite of them – and this variety ensures that the Manaus area is one of the most species-rich accessible areas on earth. Over 600 species have been recorded, including dozens of rare and little-known jewels.

Just 20 km from Manaus, Ducke Forest Reserve holds some superb terra firme forest and, happily, boasts a 42-m meteorological tower from which birders can watch the canopy. The many species here include a high proportion of so-called Guianan specialities, which tend to thrive on regions with sandy soil. Birds often seen in this area include Red-fan Parrot, Guianan Toucanet, Guianan Red Cotinga, Wing-banded Wren, the canopy-dwelling Racket-tailed Coquette and the ground-dwelling Ferruginous-backed Antbird. Further north, covering the same habitat, there is

an even better tower (the INPA Tower) with a big reputation in birding circles. Harpy Eagles can sometimes be seen from here, and other great birds include Dusky Purpletuft, Crimson Fruitcrow, Yellow-billed Jacamar, Red-billed Pied Tanager and both Red-and-green and Blue-and-yellow Macaws. This general area of terra firme forest is also often referred to, somewhat jokingly, as the 'potoo capital of the world'; this is one of the few places where the very rare White-winged Potoo and the Rufous Potoo can be seen with relative ease.

The campina forests occur in patches and boast a highly specialized avifauna. The Pelzeln's Tody-Tyrant, known from a single individual and not seen for 160 years, was rediscovered in this area in 1992, and other rarities include Bronzy Jacamar, Spotted Puffbird and Saffron-crested Neopelma.

To many, however, the real delight of the Manaus area lies in the forests directly influenced by the rivers themselves, along the banks or on the numerous islands in the river channels. Different birds can be found on different types of islands, or even on different parts of the same island. The latter case is superbly demonstrated on Marchantaria Island, a large island in the Rio Solimões not far from Manaus. Marchantaria is being continuously modified by the flow of water of the white river; land is eroded at the western end, while mud is deposited at the eastern end. This ensures that the downstream end remains in an early state of vegetational succession, while the forest at the upstream end is very much older and taller. Most of the specialities occur at the younger end, and include such poorly-known birds as Olive-spotted Hummingbird, Pearly-breasted Conebill and a marvellous selection of spinetails, including Dark-breasted, Red-and-white and Parker's. The more copious growth at the older end attracts Scaled Spinetail, Castelnau's Antshrike and Zimmer's Woodcreeper, among others, and the distinction between the two parts of the island is wondrously clear to see.

Five hours upriver from Manaus, on the much less turbulent, millpond-like waters of the Rio Negro, lies one of the world's largest collections of river islands, the Anavilhanas Archipelago. These islands offer a superb chance to appreciate the glorious subtleties of ecological differentiation in this part of the Amazon Basin. The permanently flooded igapó forest holds specialities such as Blackish-grey Antshrike, Snethlage's Tody-Tyrant and Leaden and Klages's Antwrens, while such birds as Cherrie's Antwren, Grey-chested Greenlet and the splendid Amazonian Umbrellabird are mostly or exclusively found in the taller, seasonally flooded várzea. Admittedly, finding and identifying these species is a matter of considerable expertise; indeed, you could argue that the birding guides here are just as specialized as the birds themselves!

Nevertheless, the delight is in the detail, and a week or two's hard birding in this area will offer a fantastic insight into the intricate and fine ecological requirements of the different species. It is not for everyone, but for the specialists it can be heaven.

Right: despite its huge size the Harpy Eagle can be notoriously difficult to track down as it is frequently reluctant to leave cover of the forest canopy; the INPA Tower near Manaus has a good reputation for sightings of this species.

Opposite: the stunning Crimson Topaz is one of the area's largest and most striking hummingbirds. It is especially fond of open, flowery places near blackwater rivers and streams, where it will also hawk for flying insects.

Iguassu Falls

SITE RANK 69

Information

HABITAT Waterfalls, river, humid subtropical forest, town and garden

KEY SPECIES Great Dusky Swift, Toco Toucan, Black-throated Piping Guan, Black-collared Swallow, Spot-billed Toucanet, hummingbirds

TIME OF YEAR Any time of year

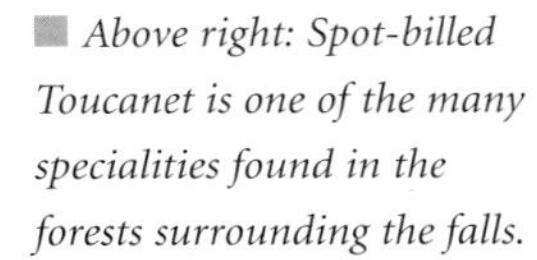

Above right: Spot-billed Toucanet is one of the many specialities found in the forests surrounding the falls.

The Iguassu Falls are one of South America's major tourist attractions, and no wonder. They are highly impressive; there are said to be 270 different falls along a 2.7-km stretch of the mighty Iguassu River, and just about every one of them is higher than Niagara (the highest are 82 m). Thus, arguably, Iguassu is the largest waterfall system in the world. However, it isn't the statistics that hit you; it's the atmosphere of the place. There are so many different falls, spread out over a wide area and mostly surrounded by lush, impenetrable subtropical forest, that you can get lost in wonder and entertain the irrational feeling that you are in a place no one has been to before. Also, you can get so close to the turbulent waters, either by taking a raised walkway, trail or boat, and you can become so immersed in their mist and spray, that you get the feeling that you've expe-

Right: an instantly recognizable member of its family, the Toco Toucan is common and easy to see at Iguassu. It occurs around the falls themselves and sometimes even in the gardens of the nearby hotels.

Opposite: an aerial view of the falls shows how spread out they are; when the water is flowing at full capaity there are said to be 270 separate chutes in total.

rienced everything except tumble over them yourself. Don't miss getting up close to the awesome *Garganta do Diablo* (Devil's Throat), a 700-m-long U-shaped cavern that allows you to creep along a walkway where the water thunders down past you on three sides. It's a white-knuckle encounter.

Since no one in their right mind will visit this corner of Argentina or Brazil (the falls are at the border) without visiting Iguassu, it is highly convenient that this spectacular place is also excellent for birds. It is not as rich as some of the more tropical parts of South America, but the 200-odd bird species that have been seen at the falls and in the national parks on both sides of the border include a number that are highly sought after. It actually makes for a pleasingly manageable introduction to South American birding.

One species that even the non-birders might well see is the Great Dusky Swift, a large sooty-brown swift that can usually be seen around the cliffs but gathers in swarms in late afternoon. It is actually attracted by the waterfalls themselves, rather than to the river or to the forests. This superb flier seems to spend half its life covered in spray. For much of its life it is a normal swift, patrolling over the forest in search of the insect layer that floats like plankton above the leaves. However, upon its return from feeding, the Great Dusky clings on to the moist cliffs around the falls, and has been known to roost and nest under an overhang, or even behind the permanent curtain of water. Could there possibly be a safer place to breed? These impressive birds nest in colonies on ledges, and mix their saliva with plucked mosses and liverworts to make themselves a small cup-shaped nest. The young, subject to continuous moist spray, are covered with warm insulating down.

Another waterfall junkie is the Black-collared Swallow, for which Iguassu is very much a southern outpost. This poorly-known species, an elegant, streamer-tailed swallow with royal blue upperparts and a thin dark breast-band across its white underparts, nests in rock crevices on the riverbank, and often perches over the water. Throughout its range it occurs only near the most turbulent waters and forages low over the surface, presumably specializing in the insects that fly around chutes and rapids.

As to the river itself, this attracts a wide range of various aquatic birds. The graceful White-winged Swallow patrols over much calmer, wider waters than its congener, and such long-legged birds as Green Ibis, Purple Gallinule and Great and Snowy Egrets wade by the riverbank. Green Kingfishers perch on foliage above the water or hover over the shallows, while rails, including the Blackish Rail, may be seen where marshy vegetation fringes the forest.

Some inhabitants of Iguassu are best classed as forest birds, but nonetheless depend on the presence of standing water. These include the secretive Slaty-breasted Wood Rail and the plain-plumaged but distinctive Riverbank Warbler. The latter bird is a real character, a bird that wanders over creeks and puddles with its body horizontal, in similar style to the waterthrushes of North America, constantly swinging its tail from side to side. This long-legged passerine, pale below and dark olive-green above, would be easy to miss were it not for its loud, exultant, liquid song and shouted '*chick*' call, easily heard above the roar of water.

The ecology of the rarest bird at Iguassu, the threat-

Above: huge numbers of Great Dusky Swifts breed and roost behind the walls of water; a good time to see them is late in the afternoon, when swarms gather above the falls.

Opposite: although declining over much of its range, the Andean Condor is still fairly common on the island of Tierra del Fuego.

ened Black-throated Piping Guan, is intriguing because, although it is a forest bird, this black-and-white guan with a large red and blue hanging throat-wattle is only ever found in strips of forest that are next to rivers. Iguassu is one of the very best places to see it, and occasionally it will even wander along the walkways by the falls.

The 660 sq km of national park surrounding the falls protects a good deal of excellent subtropical forest, although recent restrictions on visitors have made access rather difficult. Among the highlights are the improbable Red-ruffed Fruitcrow, with its scaly red-and-orange breast patch, the excellent Spot-billed Toucanet, the Blond-crested Woodpecker, with its blackish plumage offset by a cream-coloured head, and such gaudy creatures as the splendidly named Black-goggled Tanager and Plush-crested Jay. There is also a lek of White-bearded Manakins along the Macuca Trail.

Some of these forest birds enter the gardens in the area, especially those of the larger hotels, and these are an attraction in themselves. Many of the gardens have flowers and some have hummingbird feeders which, this being the Neotropics, attract a good range of half a dozen or so species. These include two fairly localized hermits, the Scale-throated and Planalto Hermits, plus the attractive Violet-capped Woodnymph. Pride of place, however, goes to the ultra-smart Black Jacobin. If you thought a black-and-white hummingbird would be a disappointment, you should set your eyes on one of these smart beauties.

Other birds of the town and garden at Iguassu include various species familiar in South America, such as Guira Cuckoo, Roadside Hawk, White-eyed Parakeet, Palm and Magpie Tanagers and Rufous-collared Sparrow. There is also one more character that you simply cannot miss and, in a way, it is one of the birds that really epitomize Iguassu. It is the Toco Toucan, the largest of its family, and the only one that is not a true forest bird. Groups of this iconic species are everywhere, flopping across boardwalks in single file, or feeding in the fruiting trees in hotel gardens. If Iguassu needed a little extra touch of the exotic, these birds provide it.

Ushuaia

HABITAT Andean-Patagonian forest, lakes and bogs, Patagonian steppe, seacoast and islands

KEY SPECIES Magellanic Plover, Magellanic Woodpecker, Andean Condor, Ruddy-headed Goose, White-throated Treerunner

TIME OF YEAR Austral summer is best, November to March

The town of Ushuaia, on the island of Tierra del Fuego that lies off tip of the South American mainland, has a good claim to be the southernmost town in the world. It is situated at 55° south, and is, appropriately enough, the port used by most cruise ships sailing the 800 km to Antarctica. Record-breaker or not, it certainly has a spectacular setting amid the forests, rolling hills and tempestuous skies of this remote region; the light railway that goes to the nearby Tierra del Fuego National Park is known as 'the railway at the end of the world'.

No birder working this area will break any records for numbers of species seen, but what characters are present tend to be distinctive and special. The Andean Condor, for example, is still found here in reasonable numbers, despite serious reductions over much of the rest of its range. Always a magnificent sight, with its huge size, improbably broad wings and smart black-and-white colour scheme, the Andean Condor here in the deep south often patrols sealion rookeries and other large gatherings of animals, searching for dead bodies. Other iconic birds include Magellanic Penguins breeding on islands in the Beagle Channel, Black-browed Albatrosses playing the ocean winds offshore, and marauding Chilean Skuas giving all and sundry a difficult time.

Yet even the less well-known birds are exciting, as a trip into the Tierra del Fuego National Park can testify. Here, among the airy, old growth Andean-Patagonian forests made up of tall, leafy Lenga trees, their trunks and branches covered with lichens and the understorey dominated by feathery ferns and spiny bushes, live such birds as the Magellanic Woodpecker and the delightful White-throated Treerunner. The woodpecker is one of the largest in the world, and acts like it. With its bold black plumage relieved by a red head (for the male) and white rump and underwings, this smart woodpecker hops heavily from branch to branch, pecking and probing and sometimes making deep excavations. Interestingly, it often moves around in trios of two red-headed males and one female, the latter with a wispy crest resembling a tuft of blow-dried hair. The treerunner, by contrast, roughly fills the niche of a nuthatch, hugging the trunks and branches, although never climbing down head-first. Despite belonging to the ovenbird family (Furnariidae), it has a remarkably nuthatch-like shape, with a long, straight, sharp bill, large head and plump body, although the tail, which supports the bird as its climbs, is more like that of a treecreeper. A clean, smart bird, the White-throated Treerunner is brilliant matt white on the breast and throat, with a pleasing mixture of cryptic browns on the back.

These forests contain a maximum of only six tree species, three of them 'southern beeches', or *Nothofagus*. There is the equivalent of a tit here in the form of another delightful furnariid, the Thorn-tailed Rayadito. Flocks of this small, richly brown species move around the forest, often hanging upside-down as they probe and glean. They have distinctive black eye-stripes and crowns, and 'serrated' tail edges, the feathers ending in 'thorns'. Hidden in the understorey, the Magellanic Tapaculo does what tapaculos do best – eludes birders. While Austral

Thrushes work the ground litter, Austral Parakeets, with their coppery-coloured tails, feed up in the Nirre trees, and flocks of Black-chinned Siskins, surprisingly similar to the Eurasian version, work the edges. These birds are preyed on by Austral Pygmy Owls.

The more open areas of the national park consist of lakes, bogs and sodden, grassy areas. Many waterbirds are to be seen here, in particular a unique trio of geese, all with slightly different habitat requirements. The Ashy-headed Goose forms something of a link between forest and open country, since it usually occurs in clearings and occasionally nests in hollow dead trees. It has an ash-grey head, with a reddish-brown band across the breast and upper back, and black-and-white barred flanks. The Upland Goose, larger, with a longer neck, barred breast and a white head for the male and brown for the female, is mainly a bird of the grasslands and marshes, while on the seashores, in the southern part of the park, the Kelp Goose grazes on seaweed. In this species the male is pure white, with a black bill and yellow legs, while the female is mainly very dark brown, with white barring on the underparts. Amazingly, a fourth species, the Ruddy-headed Goose, in which both sexes resemble a small female Upland Goose, occurs in the northern part of Tierra del Fuego island, where it tends to graze on pampas-like grassland.

Another common inhabitant of the marine shoreline is the Flightless Steamer Duck, so named because of its energetic, 'rowing' style of locomotion across the water when in a hurry, which resembles the turbulence made by old-style paddle-steamers. This huge diving duck, with its marbled brown-mauve plumage and bright yellow bill and feet, is thoroughly bad-tempered, avoided by all. It shares its habitat with such species as the Great Grebe and, on the shoreline itself, Magellanic and Blackish Oystercatchers.

A trip along the Beagle Channel from Ushuaia to Harberton brings a change of scene and a change of birds. There is a large colony of Magellanic Penguins at the eastern end of the channel, and keen birders sometimes pick up strays here from Antarctica, such as Gentoo or even King Penguin, among the masses. In a similar vein, Snowy Sheathbills also sometimes turn up. More regular breeding species along these rocky coasts are the Magellanic Diving Petrel, Chilean Skua and two excellent shags, the red-faced Rock and the larger, blue-eyed Imperial. Any trip to the Beagle Channel tends to be excellent for 'pelagic' seabirds, and these often include White-chinned and Southern Giant Petrels and Southern Fulmar.

No visit to Ushuaia is complete, however, without a pilgrimage to see a couple of Tierra del Fuego specialities. Hard work on the Martial Glacier above Ushuaia usually pays off with views of the rare White-bellied Seedsnipe (as well as Yellow-bridled Finch), while a trip up north, to the coast near Rio Grande, is needed to catch up with the peculiar Magellanic Plover. The latter, a soft-grey wader with a white belly and pinkish legs, is something of a taxonomic puzzle, but it is almost certainly not a plover. It often uses its short bill to turn over stones or other objects, and even uses its short legs to dig for food – hardly plover-like behaviour. There is some evidence, indeed, that it could be an offshoot of the sheathbills. These curious birds are quite easy to find on the steppe country on the northern coast of Tierra del Fuego.

Right: White-bellied Seedsnipe has been described as the Southern Hemisphere equivalent of a ptarmigan as it fills a very similar ecological niche.

Opposite: Magellanic Woodpeckers rely on old-growth forest with large trees in order to thrive. The male has a crimson head and the female a black head with a curled crest.

Below: Nothofagus forests help to give Tierra del Fuego National Park its distinctive character.

La Escalera

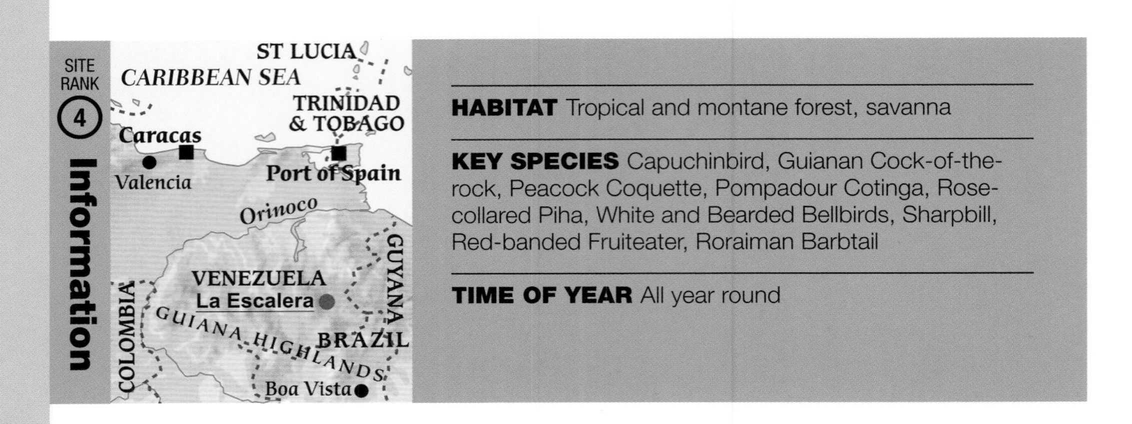

HABITAT Tropical and montane forest, savanna

KEY SPECIES Capuchinbird, Guianan Cock-of-the-rock, Peacock Coquette, Pompadour Cotinga, Rose-collared Piha, White and Bearded Bellbirds, Sharpbill, Red-banded Fruiteater, Roraiman Barbtail

TIME OF YEAR All year round

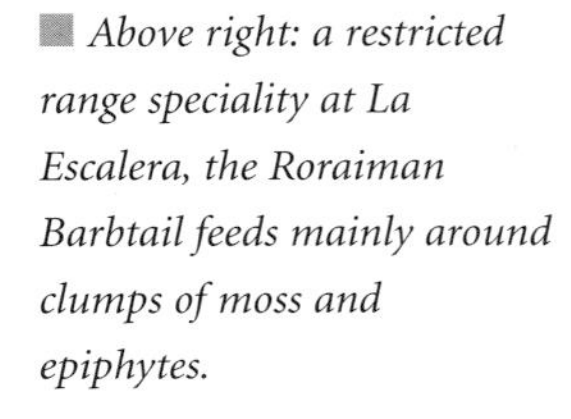

Above right: a restricted range speciality at La Escalera, the Roraiman Barbtail feeds mainly around clumps of moss and epiphytes.

In south-east Venezuela lies an area of spectacular table mountains known as the Tepuis. These great walls of rock, well spaced like an inland archipelago, rise up majestically and almost vertically from the surrounding rich lowland forest, reaching heights of 1,500–2,800 m, some sheer cliffs alone soaring up 1,000 m. The commanding flat summits are boggy plateaux studded with patches of savanna and bizarre elfin forest rich in orchids, mosses, lichens and bromeliads, cut off from the rest of the world. Rivers rise from some of these plateaux and tumble majestically off their edges; the celebrated Angel Falls, which drop from the Auyan-tepui, are the highest in the world.

With their imposing cliffs and dense skirting vegetation, the Tepuis are virtually inaccessible except to experienced and committed mountaineers, and some of their mist-enshrouded summits have never been explored. There may well be new ornithological discoveries to make here. It is, however, unlikely that any of the plateaux hide a whole world inhabited by ferocious dinosaurs, despite the idea behind Sir Arthur Conan Doyle's famous novel *The Lost World*, which was inspired by this very area. Even so, amid such an awesome landscape, it is not hard to imagine.

Dinosaurs may be lacking, but birds are not, and the region holds about 40 endemics. Most of these are found above 600 m in very dense cloud-forests which are almost impossible to reach except on a dedicated expedition. However, near the heart of the Tepui region there is an area where the land rises in less precipitous style on to a grassland plateau known as the Gran Sabana. A road follows the gradient up from the lowland forest and on

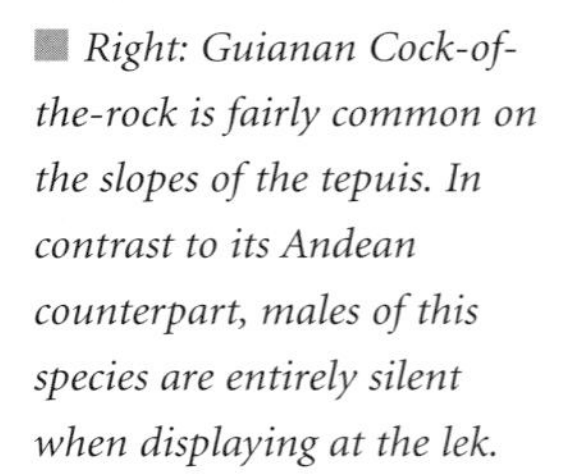

Right: Guianan Cock-of-the-rock is fairly common on the slopes of the tepuis. In contrast to its Andean counterpart, males of this species are entirely silent when displaying at the lek.

Right: the gorgeous Red-banded Fruiteater is attracted to fruiting trees, and in particular to members of the widespread Neotropical family Melastomataceae.

to the escarpment. Owing to its seemingly endless series of hairpin bends, this road is known as 'La Escalera' or 'The Staircase'. It may not be quite a stairway to heaven, but this road gives birders what is effectively their only chance to sample the ornithological treasures of the Tepuis.

Some of the inhabitants match their sumptuous surroundings. Notable among the endemics are the tiny Peacock Coquette, a mainly iridescent green hummingbird with long green tufts that stick out from the cheeks, making the bird look as though it is wearing a scarf; the tufts have large black eye-spots, hence the bird's name. Another gem is the Red-banded Fruiteater, a cotinga of which the male has neat, blue-grey plumage broken by a brilliant orange breast-band and golden-bronze wings, while the Rose-collared Piha would be a dull smoky-grey bird but for the male's astonishingly intense crimson-pink neck-ring and vent. More esoteric delights come from the very smart Roraiman Barbtail, a furnariid with a neatly brown-streaked front and chestnut supercilium, a bird that creeps up moss-laden branches in nuthatch-style (it is related to the treerunners in the same family); and from the likes of the Tepui Swift, a large species with intense rufous colouration on its head, neck and upper breast.

In common with the birds of cloud-forests generally, the birds here tend to gather in mixed-species flocks. These often move through the thick forest at dizzying speed, which means both that the birds can be difficult to pick out and that there can be long gaps between moments of drama. Thus it can take four or five days of hard birding before even a satisfactory number of endemics can be seen, and getting good views of them can be very difficult. To enjoy yourself here you need sharp birding skills. It will also help if you seek out flowering Melastome trees, as these will attract birds from a kilometre or more around.

Not all the birds of these cloud-forests are endemic to the Tepuis; indeed, two highly sought-after prizes are somewhat more widespread in their distribution. One of these is the Sharpbill, a small cotinga with oddly spotted plumage and staring orange eyes; a confirmed flock member, the Sharpbill has the unusual ability to unroll curled leaves to obtain arthropods inside, and will also hang upside down at the ends of branches. The other is the Guianan Cock-of-the-rock. This globular orange bird with a round sail for a crest is, like its better known Andean counterpart, a lek species; however, unlike the Andean Cock-of-the-rock, the males display on or just above the ground. When a female arrives, all the males flutter down as one, with a whirr of wings that clears extra debris from each of their forest-floor courts. Here on the Tepui foothills, the Guainan Cock-of-the-rock is usually seen near fast-flowing rivers.

This area is exceedingly rich in cotingas. Besides the Sharpbill, cock-of-the-rock, piha and fruiteater, several other superb species found here include the improbably plumaged velvet-coloured Pompadour Cotinga and two species of bellbird, the White Bellbird with its two-syllable ringing note and the Bearded Bellbird with its amazingly loud *tok* note, which can be heard 2 km

away. The sounds of these bellbirds give the forests a special atmosphere.

At the base of La Escalera, where the forest is lowland humid tropical, a trail leads to the lek of yet another cotinga, one that would, even without the other delights here, draw birding visitors to this fabulous part of South America. This is the weird Capuchinbird, a frankly rather ugly creature with loose-fitting rich reddish-brown plumage and a bald, blue-grey head. You can find this species by listening for a herd of cows mooing up in the treetops, for this is exactly what the Capuchinbirds sound like – the similarity is positively eerie. At the lek, birds display by perching bolt upright and fluffing out their orangey under-tail coverts, making them look still more unsightly. In common with most cotinga leks, just one male holds the key perch, and monopolizes copulations as much as it can, but this is an unusually tricky business. Every other male in the lek will constantly attempt to usurp the dominant male's position by force or by trickery. Subdominant males have a habit of joining forces for the whole breeding season and displaying as a team to intimidate the top bird; other males will pretend to be females (the plumages are alike) and thus gain access to the perch. Finally, females are aggressive to each other, and will pretend to be males in order to drive their rivals away. The system is probably ordered, but it does not, it must be said, give the appearance of being so.

Much as the Capuchinbird is more widespread than many of the other specialities of the Escalera, it is still the bird that leaves many birders with their fondest memories of this part of the Tepuis.

Los Llanos

HABITAT Savanna and grassland, some seasonally inundated; gallery forest

KEY SPECIES Herons including Zigzag and Agami, ibises including Scarlet, storks, Sunbittern, raptors including Snail Kite, and Black-collared Hawk

TIME OF YEAR Most productive time is the dry season (November to March) when water birds congregate in a few small areas

Los Llanos (the plains) is a vast area of flat savanna in northern South America, hemmed in by high mountains on three sides: the Andes to the west, the Venezuelan coastal mountains to the north and the Guianan shield to the south. It is about 1,200 km long from west to east, and covers 300,000 sq km, nearly one-third of Venezuela, as well as parts of neighbouring Colombia. Venezuelans have great affection for this hinterland, much of which is farmed for cattle. It is regarded as a romantic place where tough people eke out a living off the land, rather like the cowboys of the American west. It is certainly a place of open, flat ground and big skies.

Below: Black-collared Hawk is a familiar wetland predator in this part of Venezuela.

The main wildlife value of Los Llanos stems from its high rainfall. The region has disproportionate seasons: an extremely vigorous wet season between May and October, and a drought season from November to March. Ninety per cent of the 1.5 m of annual rain falls during the wet season, turning the plains into a series of shallow lakes, while the ground dries to a crust by March. There are several large rivers in the area, notably the Rio Apure, which sustain many of the wetland creatures when times are hard.

Los Llanos is undoubtedly among the best places in the world for waterbirds. Although there are dozens of species to be seen here, as you might expect – seven species of ibis and 20 species of herons, for example – it is the sheer numbers of each that are impressive. Huge flocks are a common sight as birds commute about the area, either seeking out feeding sites or going to roost. The best time to appreciate this abundance is, perhaps surprisingly, during the dry season; this is when feeding sites dwindle in number as the ground dries, and the birds become concentrated into vast crowds.

Among the most conspicuous of the birds of the Llanos are the large wading guild, such as herons, storks and ibises. As a rule these long-legged birds are easy to see, and their different feeding techniques and micro-habitats can quickly be appreciated. Take the ibises, for instance; Los Llanos is undoubtedly the very best place in the world to see members of this family. The commonest species are the improbably-plumaged Scarlet Ibis and the equally stark American White Ibis. Together with the Glossy Ibis, these species concentrate their feeding in standing water a minimum of 3 cm deep, often immersing head as well as bill when probing. All feed on small arthropods, such as crustaceans and aquatic insects, but the Glossy Ibis feeds, on average, on smaller items than the others, while the White Ibis tends to favour slightly shallower water than the Scarlet. Meanwhile three other species concentrate their efforts on moist rather than inundated ground. The Bare-faced Ibis, a black species with dark orange bill and legs, is found mainly in marshy ground entirely free of surface water; the Green Ibis, which is all-over bronzy-olive rather than green and has a bushy neck, is a specialist of shallow water within 2 m of a

Above: Los Llanos boasts an abundance of wildfowl – these are mainly Black-bellied Whistling Ducks.

lake shoreline; and the Sharp-tailed Ibis, a large dark species with a horizontal carriage and long tail, also feeds on moist soil, but probes to much greater depths than the Bare-faced Ibis, and takes larger prey. Finally, a seventh species, the Buff-necked Ibis, feeds mainly on dry ground, where it primarily takes insects.

This sort of division also separates the two storks of the area. Both feed in typical stork fashion, wading in the water and finding prey by sight, rather than probing, but the Maguari Stork, which resembles the well-

Right: the spectacular Scarlet Ibis is a common sight in Los Llanos.

Above: a Sunbittern lowers its wings and raises its tail during its remarkable display.

known White Stork of the Old World, eats smaller food than its giant relative, the Jabiru. The former captures frogs and fish up to the size of eels, while the latter, a white bird with black head, red throat sac and enormous scabbard-like bill, will even tackle young caimans. As to the herons, they occupy an even wider range of niches than the storks and ibises, but all catch their prey visually, usually by a patient wait in statuesque position hoping for prey to come into view. Among the more notable species in Los Llanos are the colourful but elusive Agami Heron and the tiny and almost mythical Zigzag Heron, probably the least known heron in the world and one of the hardest birds to see in the whole of South America. If you see one, make sure you pinch yourself to make sure it's not a dream.

With such a high volume of potential food, it is no surprise that Los Llanos is also an exceptional place for raptors, including several interesting wetland specialists. The dark grey Snail Kite uses its exceptionally curved upper mandible to slit the columellar muscle of large Apple Snails and thus extract them from their shells. Snail Kites hunt in flight over open marshland, swooping down to snatch their prey in one talon, while the closely related Slender-billed Kite hunts the same prey within riverside forest by dropping down from a perch. On the other hand, the rusty-coloured Black-collared Hawk eats fish, snatching them from the water's surface following a low approach; its talons are fitted with sharp spicules on the soles to give them grip on the slippery prey, as with the Osprey. The Great Black Hawk hunts a wide variety of wetland creatures, often simply by wading into shallow water. Finally, the Lesser Yellow-headed Vulture is the marshland scavenger; it is one of the three New World vulture species that can detect carcasses by their scent alone.

The best places for all these wetland birds are to be found in the southern part of this huge complex, where rainfall is highest. There are several cattle ranches in the area which cater for visiting birders, and have an eco-friendly outlook towards the management of their part of Los Llanos.

The northern parts contain more permanent grassland which is not seasonally inundated. Several other specialities occur in this area, including the Savanna Hawk and Double-striped Thick-knee. However, it is really the waterbirds that make this region such a lure for birdwatchers.

Lauca

SITE RANK 18

Information

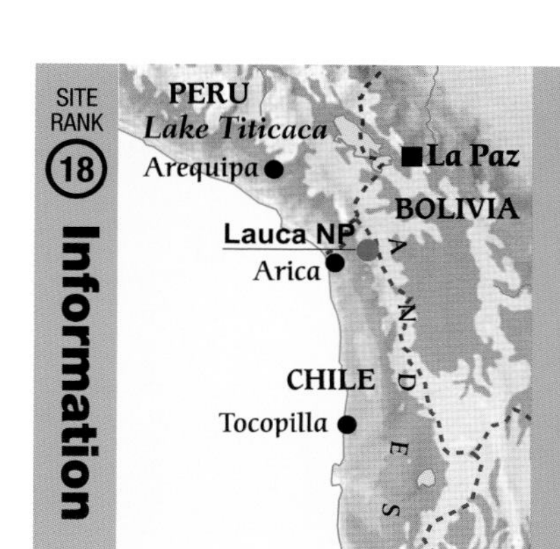

HABITAT High-altitude lakes and bogs, puna

KEY SPECIES Diademed Sandpiper-plover, Giant Coot, Silvery Grebe, White-fronted Ground Tyrant, James's, Andean and Chilean Flamingos, Darwin's Rhea, Grey-breasted and Rufous-bellied Seedsnipe

TIME OF YEAR Best visited in austral spring, October to November

Above right: Lauca's spectacular scenery includes smouldering volcanoes that are well in excess of 6,000 m.

A place for the daring birder, Lauca is one of the most spectacular national parks in the whole of the Americas. Most of its 1,379 sq km is on the *altiplano*, a high-altitude Andean plateau that runs down from southern Peru and through Bolivia, the remnants of what was once a vast inland sea. Most of the habitat is above 4,500 m, sometimes making for freezing cold, breathless birding in high winds, and it is not for the faint-hearted.

The rewards, however, can be high. Many tourists come simply for the scenery, the wide open expanses of the grey-brown *puna* – a type of grassland characterized by clumps of bunchgrass separated by bare soil – dwarfed by two enormous snow-capped volcanoes, Ponerape (6,342 m) and Parinacota (6,282 m) that together are known as the Payachatas. The water here is mainly from snowmelt, although there is also a brief wet season in summer. So, while the countryside is general-

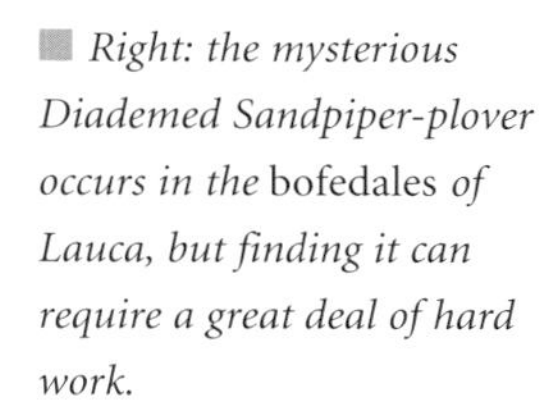

Right: the mysterious Diademed Sandpiper-plover occurs in the bofedales *of Lauca, but finding it can require a great deal of hard work.*

Right: a Giant Coot brings back material to its fortress-style nest.

Below: small groups of Darwin's Rheas roam Lauca; these altiplano rheas of the have slightly shorter legs than their lowland counterparts.

ly very dry, there are some wet areas, where most of the birds gather. These wetlands consist of bogs dominated by cushion-plants – a habitat known locally as *bofedale* – and shallow, highly saline lakes.

The *altiplano* is home to a highly distinctive set of

birds. Among the most dominant are the ground tyrants, of which several species occur together. Ground tyrants are plainly-marked, terrestrial passerines with long legs and a distinctive upright posture. Most, such as the Rufous-naped and Cinereous Ground Tyrants are found in the drier, rocky areas, but the highly localized White-fronted Ground Tyrant, the largest of the group, is entirely confined to the *bofedales*. Sharing its narrow niche is the splendid greyish White-winged Diuca Finch, which feeds on seeds from the cushion-plants rather than on insects.

The most sought-after birds on the *puna*, however, are somewhat larger in size than ground tyrants and finches. They include the distinctly portly Ornate and Puna Tinamous, of which the latter, oddly, is more ornate than the former, with smart black-and-white neck stripes and rufous belly and under-tail; plus two species of seedsnipe, the dove-like Grey-breasted Seedsnipe of the boggy areas, which is often quite easy to see, and the partridge-like Rufous-bellied Seedsnipe, which can be more tricky to find. Seedsnipe are odd birds, rather like a cross between a wader and a pigeon. They spend their lives nibbling the tips of vegetation, such as buds and leaves. Despite frequently living beside water, no seedsnipe has ever been recorded drinking.

The largest bird here, though, is Darwin's Rhea, which is often attracted to the boggy areas, making it relatively easy to find despite its inherent shyness. These splendid flightless birds, with their subtle grey-brown coloration and neat white speckles, which look like lumps of icing-sugar on their backs, live in small groups. In the austral summer, males acquire a small harem of females, each of

■ *Above: the Andean Flamingo, the largest of South America's flamingos, feeds on algae and is confined to lakes at high altitudes between 2,500 and 4,800 m.*

which contributes to a communal batch of up to 50 greenish eggs which are laid in a single nest made by the male. It is thought that the eggs, by mechanisms unknown, stimulate each other to hatch more or less at the same time. The youngsters are attended to entirely by the male.

There are a number of lakes in Lauca National Park, and one of these, Lake Chungara, is at 4,514 m altitude, said to be the highest in the world. Large and shallow, with copious waterweed but little emergent vegetation, it makes the perfect home for the Giant Coot: 10,000 have been counted here, making it the largest gathering in the world. Giant Coots are large and, in typical coot style, bad-tempered; fights over nest material are a matter of daily life. Giant Coots live permanently in a single territory and their nests are their castles. These structures are huge, up to 3 m long and 1.5 m tall at the rim, with a deep cup in which the chicks, in their wind-battered neighbourhood, spend a great deal of time. So large are the nests that a human could easily fit inside one.

Alongside the coots are Silvery Grebes of the black-and-white headed *altiplano* form *juninensis*, which are also here in their thousands. These colonial grebes occur in lakes with high concentrations of aquatic invertebrates, such as midge and bug larvae. Greyish in colour, and floating high on the water, they seem to gleam on the lake surface, much as their name suggests.

Up here at high altitude, evaporation from the lakes is rapid, so that most are highly saline. This makes them attractive to flamingos, including the Chilean Flamingo, which feeds on arthropods, and two *altiplano* specialists, the Andean and James's Flamingos, which feed on algae. The latter two species have a much more efficient filtering system in their bill than the Chilean, allowing them to take in this much smaller foodstuff. Alongside these long-legged birds are Andean Avocets, scything through the shallows for food similar to that of the Chilean Flamingo.

Giant Coots are fun and flamingos attract sightseers, but for the hard-core birder there is one species, way ahead of all the others, that they will be desperate to see at Lauca: the Diademed Sandpiper-plover. This bird has all the attributes to elevate it to almost mythical status: it is peculiar, poorly known, extremely attractive, very difficult to spot and extremely rare. Its precise ecological requirements are not certain, although it breeds in some of the boggy areas of the park. It is very small, not much larger than a small sandpiper, and has the infuriating habit of feeding in discrete corners of the bogs, often behind cushion plants and out of sight. It has a long, thin bill that probes the water, and which it uses to pick insects from plants. When it flies it makes rapid wing beats which barely rise above horizontal and, oddly, its path follows a slightly undulating course. Subtly smart, the Diademed Sandpiper-plover is marked below with delicate bars, and has a chestnut, white and black patterned head.

This Sandpiper-plover is pure box-office and, despite all the other wonders of this magical place, if a birding group is lucky and skilled enough to find it, this bird will inevitably top everybody's list, and dominate the excited bubbles of conversation on the long bus-ride back down to civilization. It is the jewel in Lauca National Park's well-laden crown.

Central America and the Caribbean

The Central American avifauna is very rich and is a composite of Nearctic and Neotropical influences, with more than 1,300 species if you include Mexico as a whole, but no endemic families (the closest is the silky flycatchers, one of which makes it into North America).

Much of the land bridge between the two big continents is mountainous. This simple feature swells Central America's biodiversity because the composite ranges are isolated from each other by lowlands, allowing new species to evolve in each. Furthermore, in the highlands one finds the same distinction between wet and dry slopes as seen in the South American Andes, with a vastly higher precipitation on the eastern slope caused by easterly trade winds, and correspondingly more luxuriant forests. The birds are very different from those on the dry, scrubby hillsides on the western slope. Together with concomitant gradations in altitude, the result is an impressive 340 endemic species in this small region. Among well represented families are birds of prey, hummingbirds, trogons, parrots and woodcreepers. Further north, Mexico holds an impressive selection of desert and arid country species, including thrashers, wrens and a number of hummingbirds.

Meanwhile, the Caribbean is quite a different region, with a distinctive avifauna that includes about 160 endemic species of its own. Needless to say, much of this is a result of it being an archipelago, although the island chain also straddles two continents. Caribbean endemics include two complete families, the todies and the Palmchat, the latter found only on Dominica. Well represented landbird families include hummingbirds, parrots and New World Sparrows. The Caribbean is also good for tropical seabirds such as frigatebirds, tropicbirds and terns.

Right: Keel-billed Toucan in Guatemala

Tikal

HABITAT Subtropical rainforest, open areas, small marsh

KEY SPECIES Ocellated Turkey, Great Curassow, Crested Guan, Orange-breasted Falcon, Ornate Hawk-Eagle, parrots including Mealy, Red-lored and White-fronted Amazons

TIME OF YEAR All year round

Not many places on earth combine great birding with ancient culture such as the great city of Tikal, in Guatemala. This is a place where even the most dedicated birder's eyes will be turned towards the archaeological wonders, and where the student of ancient history cannot fail to be distracted by the Ocellated Turkeys wandering between the ruins.

The archaeological remains of the Mayan civilization are completely surrounded by tall subtropical forest, giving them a more intimate feel than most other great archaeological sites. You cannot escape the history here; you can feel it. The five temples, with their thin, neatly stepped pyramid shape, speak eloquently of the lost culture; and while you wander around the site, finding mounds as yet unexcavated and still covered with thick vegetation, you can easily imagine that you are the first person to encounter them, their secrets hidden just beneath your feet.

Dawn is a particularly atmospheric time, when you can just make out the towering shapes of the temples in the half light, and the forest begins to wake up. The slightly melancholy, shivering note of the Great Tinamou is often one of the first voices to be heard, but it is soon joined by the loud shrieks of parrots. The morning fly-past of parrots at Tikal is one of the most impressive in Central America, and involves hundreds of birds of at least five species: Mealy, Red-lored and White-fronted Amazons, and White-crowned Parrots, plus the smaller Jamaican Parakeets. The birds leave their roosts and fly past the ruins on their way to foraging in the vastness of the forest; Tikal lies in the heart of the largest patch of forest in the whole of Central America.

Several predators can be attracted to these commuters, and in the last few years one in particular has delighted visiting birders. This is the very smart Orange-breasted Falcon, a large falcon with a wide distribution, but one that is inexplicably rare just about everywhere. Here at Tikal, however, it will sometimes sit on a temple top scanning for the medium-sized birds (its prey also includes pigeons and jays) that make up its diet. It often hunts at dawn and dusk, allowing observers to compare it with the smaller, slimmer and much commoner Bat Falcon, which

Opposite: sightings of the White Hawk, a snake and lizard hunter, are regular at Tikal.

Right: the brightly coloured Collared Araçari can be difficult to pick out in the forest canopy around Tikal, although some of the temples can afford a bird's eye view.

Opposite, below: the tops of the Mayan pyramids stand above the canopy of the cloaking forest.

also occurs here. The latter generally takes smaller prey, such as tanagers, and even some hummingbirds.

As the day begins to warm up, a good birding strategy is to climb to the top of one of the temples and watch over the forest. Temple IV, the highest of the five at 60 m high, is the best vantage point; built in 741AD, it was the tallest building in North or Central America before the skyscrapers of the Eastern Seaboard were built in the late 1800s. Raptors seen here regularly include the uncommon White Hawk, a canopy hunting specialist that preys on snakes and lizards, and Grey-headed and (in season) Swallow-tailed Kites. Ornate Hawk-Eagles are occasionally seen, but sadly the Harpy Eagle has not now been recorded for many years.

Lying within a national park of 576 sq km, and with a strict ban on hunting both observed and largely enforced, Tikal is one of the very best places in the Americas to see some of the larger forest birds that tend to disappear as soon as people move into an area with guns and traps. Perhaps the most impressive of these is the Great Curassow, a large cracid in which the male is almost all-black with a white vent, yellow cere and rather ludicrous crest that looks as though it has been wet combed and curled. It tends to eat fruit from the forest floor, and can often be seen wandering near the forest edges. Another species, the more arboreal Crested Guan, with its white-streaked dark brown plumage and bare red throat, is equally easy to see and appreciate in the open forest that surrounds the central 2 sq km of the site.

Probably the star of the show, however, is the Ocellated Turkey. The world's 'other' turkey, this species has a restricted distribution in south-east Mexico, Guatemala and nearby Belize. At Tikal it is both common and tame; indeed, it wanders around the ruins as if it were a tourist itself. Slightly smaller than its close relative the Wild Turkey, and with a blue head studded with red warts, large blue eye-spots (ocelli, hence the name) on orange tail coverts and a smart green sheen to its wing coverts, it is a more colourful bird than its better-known relative.

The temple tops are great for birding, or for simply enjoying the sight of a cloak of forest as far as the eye can see, but of course Tikal was not originally built as an observation post for birders; the temples are largely opulent burial mounds. There are probably many more birds here now than when Tikal was in its heyday as an independent city state between 250 and 900AD because at this time, during the Classic Period of the Mayan culture, the whole city, which is known to have covered 120 sq km in all, provided a home for a population approaching 90,000 people. It would not have been safe for large, tasty birds. Instead it would have been heaving with the business and commerce of this remarkable civilization, recounted so vividly in its glyphic texts written on the large ceremonial stones (the

■ Right: the lack of hunting at Tikal allows large birds such as Great Curassow to flourish; this is a male.

stelae) that dot the site. Among the site's most famous features are the celebrated ball courts, where a strange game resembling football was played, with marked-out goals. It was a game of high stakes – the losers were sometimes the victims of human sacrifices.

It is known that Tikal was occupied between about 800BC and 900AD but then, at the height of its powers, it was rapidly and completely abandoned by the Maya who built it. It is still a mystery why the city and its culture disappeared so rapidly, but theories abound. One thing is for sure, though. The rich forest that covers this site is a sober reminder to our present society that we, the people of the planet, are less significant and a little more temporary than we tend to realize.

■ Below: nobody can miss the Ocellated Turkeys of Tikal, which have free rein among the ruins.

Canopy Tower

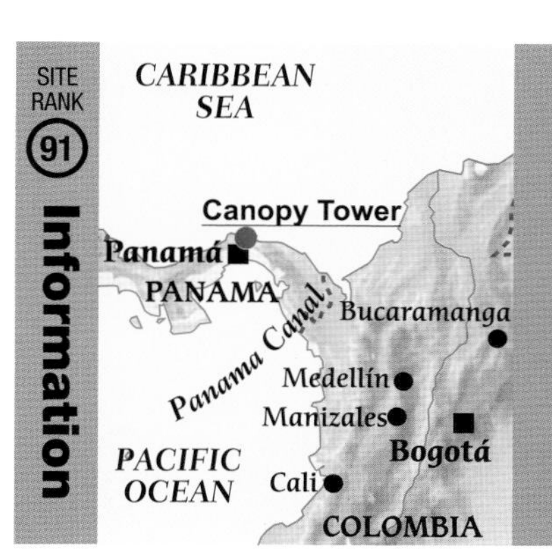

HABITAT Rainforest, second growth

KEY SPECIES Blue Cotinga, Ocellated Antbird, Brownish Twistwing, Great Tinamou, Collared Aracari, migrant raptors, hummingbirds

TIME OF YEAR All year round, with migrants January to March and September to November. Can be quiet in August.

Above right: Panama City is visible from the Canopy Tower.

Below: the highly-strung White-vented Plumeleteer is a common visitor to the Tower's hummingbird feeders.

Birding in rainforest can be a frustrating business, as the richest part of the habitat, the canopy, is inaccessible to ground-based observers, and viewable only by peering uncomfortably upwards for long periods of time. Any opportunity, therefore, to observe the canopy from a height comes as a welcome change. Few places in the world provide such a comfortable way of doing this as the Canopy Tower in Panama.

This building, perched atop a hill (elevation 300 m) not far from Panama City, was built by the United States Air Force in 1965 to house radar equipment, aiding the defence of the nearby Panama Canal. In 1996, no longer active, it was ceded back to the government of Panama and, thanks largely to the influence of a local businessman, Raul Arias de Para, has since been converted into an exclusive ecotourism lodge and observatory. Although it looks rather odd, with its round design and large geotangent dome on the top, it is suited to the task perfectly. The roof and upper floor allow a 360° view, overlooking the surrounding forest of Soberania National Park, with distant views to Panama City and the canal itself.

Birding here inevitably begins at dawn, when many of the forest's avian inhabitants become active. These include two noisy parrots, the Mealy Amazon and the Red-lored Amazon, whose piercing calls soon become a familiar part of the soundscape as they commute between their roosting and feeding sites. Meanwhile, in the treetops, the mixed feeding flocks that are such a feature of rainforest birding soon begin to form. In this forest tanagers predominate, the widespread Blue-grey and Palm Tanagers being joined by the more localized Plain-coloured and Golden-hooded Tanagers, and other motley members include the brilliantly blue Red-legged Honeycreeper and an assortment of tyrant-flycatchers and vireos. At the right time of year, October to March, these residents are joined by migrant birds from North America, including Scarlet and Summer Tanagers and a whole range of warblers, including Bay-breasted and Chestnut-sided. To visitors from the United States and Canada, therefore, these flocks contain an eclectic mix of the familiar and the unfamiliar.

Viewing from above the forest also allows visitors to see some of the more independent species of the canopy, which don't join the mixed flocks. Many of these feed on fruit, and they include a splendid trio of toucans – the Keel-billed and Chestnut-mandibled Toucans and the gorgeous Collared Araçari. However much toucans, pigeons and trogons might appeal, nothing causes quite such a buzz among the birders on the rooftops as the first daily sighting of the Blue Cotinga, a local speciality.

Right: Ocellated Antbird is usually found close to columns of army ants, where it picks off insects that have been disturbed.

The male is clad in dazzling iridescent turquoise-blue, with patches of dense purple on the throat and breast – a bird to take the breath away, even among the glittering array that resides here.

Later in the morning raptors appear, soaring on the thermals. These include an assortment of resident species such as the Great Black, Zone-tailed and Short-tailed Hawk, plus the occasional White Hawk. In the migration season, however, the skies can almost fill up with thousands of raptors, making their crossing from North America to South America along the narrow isthmus of Panama. In the autumn of 2004 a series of counts along the whole Panama Canal revealed almost three million passing raptors from October to mid-November, mainly Turkey Vultures (1.2 million), Broad-winged Hawks (one million) and Swainson's Hawks (750,000), the latter on their epic flight down to northern and eastern Argentina. Recent daily counts from the tower itself have reached the 150,000 mark, which suggests that, in the right conditions, this could be a major migration watchpoint. Besides the common species, Swallow-tailed and Mississippi Kites also come through.

No one remains in the tower for too long, because even in the idleness of this idyllic spot it is impossible not to be tempted into the forest below. However, as you reach the bottom, don't forget to check the hummingbird feeders, where aggressive White-necked Jacobins and White-vented Plumeleteers bully a succession of other species all day long.

Leaving the tower for the roads and tracks, you now have access to the low growth, where the birds are so different from the canopy that it is hard to believe you are in the same forest. In these shady areas antbirds often predominate in the feeding flocks. Much research has been done on Neotropical flocks, and it has been shown that the members share a common territory, which is defended from other flocks, although the respective species challenge only their own kind. Membership is restricted to just one singleton, pair or family per species, and each member breeds within the group boundary. Flocks have 'nuclear' species, which are permanent members, plus 'attendant' species which join from time to time. In the forests here the Dot-winged, Checker-throated and White-flanked Antwrens are always found together, and have been nicknamed the 'three amigos'.

If you are fortunate when walking in these forests you will eventually come across an army-ant column on the march, together with its attendant flocks of highly specialized ant followers. These birds don't usually eat the ants themselves, but instead mop up the many invertebrates (and small vertebrates) that flee from the advancing columns. Several of the species here are obligate ant-followers (they don't forage in any other way), of which the best known is the splendidly plumaged Ocellated Antbird. This smart bird, with its neat black scales and blue face mask, lives in small family parties consisting of a pair, their young and their youngsters' partners;

Above: Blue Cotinga is something of a local speciality.

if intruders threaten, these birds will defend their territory by perching together, shoulder to shoulder.

Of course, in these Neotropical forests there are dozens more species. The list for the Canopy Tower and its environs stands at about 550 species, and parties of birders staying for a few days usually record more than 200. However, much as listing is fun, and much as it is rewarding to pick out obscure birds in the undergrowth, the true value of this site lies in the superb views from the tower over the poorly known jungle canopy.

Right: Red-legged Honeycreepers are frequent members of mixed-species flocks in the canopy.

Monteverde

HABITAT Montane forest

KEY SPECIES Resplendent Quetzal, Three-wattled Bellbird, Coppery-headed Emerald, Violet Sabrewing, Emerald Toucanet, Black Guan, Prong-billed Barbet

TIME OF YEAR Good all year round, but weather conditions can be difficult between September and February

Few countries of the world are as accommodating for ecotourism as the pocket-sized Central American country of Costa Rica. With no less than 12 per cent of the land area set aside as national parks or biological reserves, the importance of natural heritage is ingrained in the culture. Add to that an excellent infrastructure, plus an enviable political stability within quite a volatile part of the world, and you have the perfect recipe for seeing superb wildlife in comfort and safety. With 850 species of birds recorded in 51,000 sq km, there is plenty of wildlife to enjoy.

Of all the many reserves in Costa Rica, perhaps the most famous is the Monteverde Cloud Forest, in the northern highlands. In many ways this area is a microcosm of the country, lying on the divide between the drier Pacific Slope and the wetter Caribbean Slope, and encompassing a mid-level altitudinal range from about 1,000 m up to 1,850 m. About 450 bird species have been recorded in the various different forest types in the area as a whole. Some stunted elfin forest grows on the highest ridges, while just below this the cloud forest takes over, festooned – almost overdressed – with epiphytes. Below this point, outside the reserve, the trees become taller, less laden, and the forest more open. All the different types of forest hold their own set of species.

With such a huge list, it is perhaps surprising that Monteverde should be associated with one bird above all the rest. What a bird it is, though. Few would doubt that the male Resplendent Quetzal should be placed high in a roll-call of the most beautiful birds in the world. With a bright crimson breast, shimmering iridescent grass-green plumage, short silky crest and extravagant tail streamers trailing up to 65 cm behind it, the Resplendent Quetzal makes an astonishing sight as it appears silently out of the foliage. This is no ordinary bird; it is the only one to give its name to the official currency of a country (neighbouring Guatemala), and it is perhaps no surprise that it was worshipped by both Aztecs and Mayans long before it came to be worshipped by birders. It really is the sort of bird that sets your spine tingling when you first set eyes upon it.

The Monteverde Cloud Forest Reserve and Tropical

Opposite: the Purple-throated Mountaingem is aggressive and dominant at Monteverde's hummingbird feeders.

Right: despite the huge richness of bird species, there's no doubting Monteverde's star bird – Resplendent Quetzal.

Opposite: Emerald Toucanet is a widespread species of Costa Rica's montane forests.

■ *Above: a calling Three-wattled Bellbird shows off its broad gape, which is ideally suited to feeding on fruit.*

Science Center has certainly made the most of this most famous inhabitant. The chance of seeing it attracts tourists in their thousands, which can at times make the reserve rather busy. Even before you visit, you can catch up with the birds breeding there by watching the 'Quetzalcam' on the internet, and that, no doubt, will make you even keener to come.

Another charismatic, if less famous, species at Monteverde is more noisy than showy. Throughout much of the year, between November and July, the loud *bock* of the Three-wattled Bellbird provides a backdrop to birding in the forest. The bird is hard to see, usually perching high in the canopy, but it is worth the effort, since the three grey wattles that hang down from around the bird's bill give the bizarre impression that the bellbird is eating spaghetti. Mainly chestnut-brown in plumage, but painted white on the head and upper breast, male Three-wattled Bellbirds would look rather smart but for these curious ornaments.

Both the Resplendent Quetzal and Three-wattled Bellbird have been thoroughly studied at Monteverde, and there are several interesting parallels. Both feed predominantly on fruit and both have a particular taste for members of the Lauraceae, especially wild avocados; their broad gapes make both species very important dispersers of these plants. Intriguingly, tracking studies have also shown that neither is entirely resident in the reserve, and that they are in fact altitudinal migrants. The bellbirds finish breeding in June, congregate on the nearby Pacific slope in July, and then spend a couple of months at sea-level in Nicaragua before returning to the Caribbean slopes to resume breeding. For their part, the quetzals are present at Monteverde only between January and July, and then migrate down the Pacific slope. They then transfer to the Atlantic slope before returning to the highlands again. These movements prove that the conservation of lowlands in Costa Rica, which has previously been somewhat neglected, is as important as preserving less fertile mountainous areas of the country.

Among the many other excellent forest birds at Monteverde are several other species which feed on fruit: the highly range-restricted Black Guan, the peculiar brownish-yellow Prong-billed Barbet, the Emerald Toucanet and the orange-bellied morph of the Collared Trogon. Other specialities include the Golden-browed Chlorophonia and Spangle-cheeked, Blue-and-gold and Black-and-yellow Tanagers. In order to see many of these, especially the Black Guan, the 3.6-sq-km Santa Elena Reserve, 6 km from the main park entrance, is often the best bet.

It is also well worth paying attention to the hummingbird feeders at Monteverde, since several very localized species occur in the area and can be appreciated at close quarters at the so-called 'Hummingbird Gallery', including one of Costa Rica's three mainland endemics, the Coppery-headed Emerald. With its plumage a mixture of green and copper, with much white in the tail, this species represents the smaller end of the hummingbird spectrum along with the Magenta-throated Woodstar, while Green-crowned Brilliant and Purple-throated Mountaingem are medium-sized, and lording over all of them is the spectacular Violet Sabrewing, with its deep flashy colours and fiery temperament.

This display of hummingbirds continues all day long and is an excellent distraction for the quieter middle period of the day. However, being at altitude, and relatively cool, Monteverde never really becomes completely still and sleepy. There is plenty here to keep birders satisfied, for almost all of the time.

San Blas

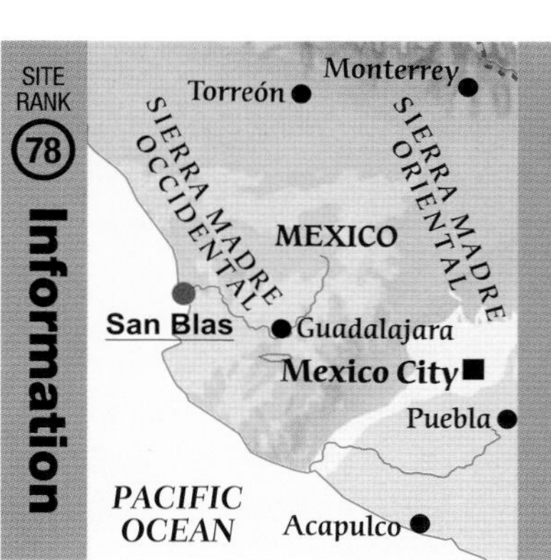

HABITAT Beach, mangroves, mudflats, offshore island, sewage lagoons, thorn scrub, deciduous forest, pine/oak forest

KEY SPECIES Boat-billed Heron, Elegant Quail, Bumblebee Hummingbird, Military Macaw, Citreoline Trogon, Black-throated Magpie-Jay, San Blas Jay

TIME OF YEAR October to April is best

Above right: a roadside stop near San Blas town is one of the best sites in the world for the rare and spectacular Military Macaw.

Below: the exotic-looking Black-throated Magpie Jay is a noisy and conspicuous species which is invariably seen in small parties.

If you travel overland through Mexico from the United States, there will come a time, as you go ever southward, when the northern elements fade into the distance and the balance of bird life tilts firmly in favour of the Neotropical. On the west side of Mexico this happens around the seaside town of San Blas, where such birds as woodcreepers and motmots begin to appear, and where you can almost feel the increase in diversity as you head towards the great riches of South America. This small Pacific coastal settlement, set on a quiet coastline well away from the main tourist centres of Acapulco and Cancun, is thus something of a biological frontier town, and it has become a firm favourite with those who are dipping their toe into Neotropical birding for the first time. With its rich diversity of habitats, including mangroves and pine/oak forest within easy reach, San Blas makes a superb and convenient base for a birding trip. There is an annual San Blas Christmas bird count, and the score from a single day almost always exceeds 250 species, and sometimes reaches 300.

One of the many advantages of this area is that plenty of west Mexican specialities can be found without leaving town. The settlement is charmingly under-developed, and there are plenty of well-wooded streets, gardens and scrubby areas, where you can find such species as the endemic sparrow-sized Mexican Parrotlet, the fruit-eating Citreoline Trogon with its banana-coloured breast, the brilliant two-toned Mexican Cacique and, perhaps the most exciting of all, the exquisite Black-throated Magpie-Jay with its extraordinary long, narrow tail and odd tuft on the head. The absurd gurgles of the Happy Wren compete with the equally upbeat whistles of the Sinaloa Wren, two Mexican endemic species that are so closely related that they react territorially to the others' song, and these

■ *Above: San Blas Jay is a social nester – one female lays eggs, but a female helper may well share incubation and up to 13 supernumeraries may help to feed the young.*

two provide a constant atmospheric backdrop to this easy-paced birding.

Just inland from town, especially around the village of Singayta, is some superb thorny, scrubby forest. This is a good place to look for a bird that is totemic of the area: the San Blas Jay. This distinctly smart, highly localized endemic is jet-black below and on the head and iridescent turquoise on the wings and back, a most tasteful combination, set off by a staring yellow eye. Alongside it the impressive supporting cast in this habitat may include groups of Elegant Quail and noisy Rufous-bellied Chachalaca, plus the smart, mainly green Russet-crowned Motmot. The Ivory-billed Woodcreeper also occurs here, in a drier habitat than would be tolerated by most of the rest of its family.

On the outskirts of San Blas town are significant stands of mangroves, and one of the many highlights of a visit here is a boat-trip upriver towards the freshwater springs of La Tovara, where people and Spectacled Caimans can eye each other up as they swim either side of a safety net. The gentle journey takes you through an amazing 'tunnel' of mangroves, where you can look out for a rare and poorly known mangrove specialist, the Rufous-necked Wood Rail. It is best to take this trip in the late afternoon, so that on your cruise you have a chance to catch up with a notable suite of twilight-loving herons: the Yellow-crowned Night Heron, the Boat-billed Heron – which catches its food by scooping – and the Bare-throated Tiger Heron, one of the few members of its group that is not impossibly secretive. The return journey in the dark may yield close views of Common Potoo, and you may also be treated to the antics of fish-catching bats.

There are plenty more excellent sites here. Just to the west of the town centre is a series of sewage ponds, where you can look for Least Grebes and where Northern Jacanas are impossible to miss. Nearby, the shrimp ponds are, not surprisingly, a reliable place to catch up with crustacean-eating Roseate Spoonbills, which swing their open bills from side to side in the shallow water. These sites, along with the tidal mudflats, play host to just about every heron, wader, gull and tern that can be recorded in the North American continent.

A North American visitor will, indeed, find plenty of familiar faces on a winter visit to the San Blas area. Many of the western members of the wood warbler family use Mexico as non-breeding quarters, including Black-throated Grey and MacGillivray's Warblers, as do several hummingbirds, including Black-chinned and Costa's, the former after a journey that may have started in Canada. For many visitors, catching up with these favourites is one of the highlights of a visit this far south.

Slightly further afield, a different set of birds can be seen in the mixed pine/oak woodlands around the Cerro de San Juan, about an hour's drive from San Blas. Set at a higher elevation, the specialities on these hills include the splendid Bumblebee Hummingbird, which avoids eviction by territorial hummingbirds by mimicking the flight of the bumblebee; at a mere 7.5 cm long, it is among the ten smallest birds in the world. The subtly patterned Grey Silky-flycatcher is also found here, alongside such striking characters as the Red-faced Warbler, Painted Whitestart and Red-headed Tanager, whose crimson colour just seems to glow. On the way back from these hills every birder makes a stop at the roadside Mirador del Aguila, one of the world's most reliable sites for Military Macaw. The birds, all green and blue, fly with majestic slow beats over the hill forest, a sight which, especially in the fading light, can make a fitting finale to a visit to this remarkably rich area of western Mexico.

Veracruz river of raptors

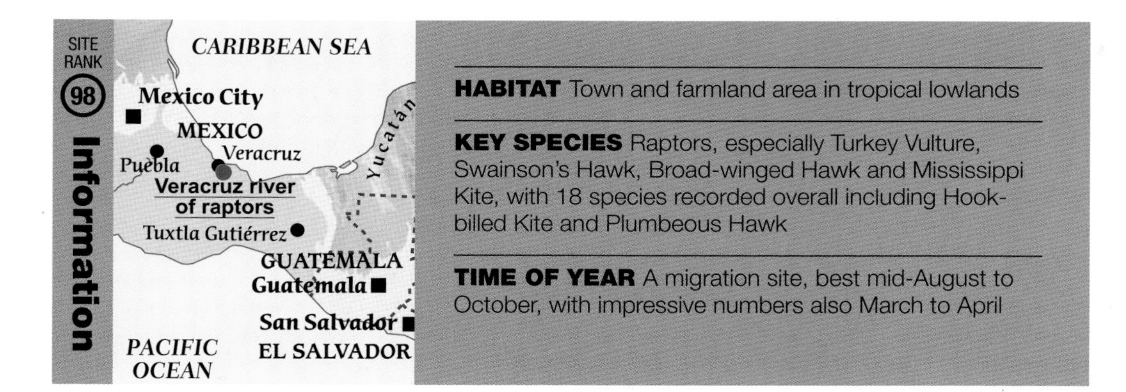

HABITAT Town and farmland area in tropical lowlands

KEY SPECIES Raptors, especially Turkey Vulture, Swainson's Hawk, Broad-winged Hawk and Mississippi Kite, with 18 species recorded overall including Hook-billed Kite and Plumbeous Hawk

TIME OF YEAR A migration site, best mid-August to October, with impressive numbers also March to April

The evocative name for this raptor migration viewpoint in eastern Mexico is no exaggeration. At times, when wide streams of birds fly over in continuous wheeling flocks, with no discernible beginning or end, almost from dawn until dusk, the effect really is of a vast aerial river of moving birds passing over, the circling birds being the river's eddies. Such a wondrous spectacle is truly special and can be seen from several tropical watchpoints, especially in Central America, but the Veracruz river of raptors has the distinction of being much the most impressive in terms of numbers of birds seen. More than five million birds, mainly North American breeders, pass over the site during each post-breeding migration, which is easily the highest flight count anywhere in the world. Incredibly, more than a million raptors have been counted passing this area in a single day.

Veracruz is situated at a latitude where the North American continent narrows to form the funnel that is the Mesoamerican Land Corridor. Furthermore, it lies on the east side of the mouth of the funnel, on the Gulf of Mexico, specifically on a narrow strip of coastal plain sandwiched between the gulf to the east and a spur of the Sierra Madre Oriental just 7 km to the west. Since, in this hot climate, the birds have no need for mountain updrafts, migrating birds avoid the highlands and are greatly concentrated into this gap, or 'bottleneck', between mountain and sea.

Furthermore, this river of raptors flows at the point where several major flyways meet. Birds travelling down via the Appalachian Mountains meet birds following the Rocky Mountain Chain, and these mix with migrants that have used the corridors in between. Together, these huge flows of raptors converge on Veracruz to create what is equivalent to a huge junction in the sky, the world's biggest bird motorway. South of this bottleneck, the paths of several species actually diverge, with Swainson's Hawks and Turkey Vultures travelling the Pacific

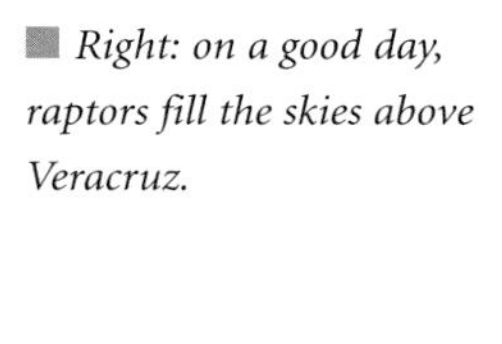

Right: on a good day, raptors fill the skies above Veracruz.

Above: among the 18 species of raptors recorded here are relative scarcities such as Zone-tailed Hawk, a species which is recorded passing at a rate of 140 individuals per season.

Slope, while Broad-winged Hawks take the Caribbean Slope. However, they all seem to pass through the Veracruz area.

Between September and November, the counts of raptors are simply enormous. Imagine, for example, seeing 707,798 Turkey Vultures in a single day, as happened on 17 October 2003, or 95,989 Mississippi Kites, as counted on 1 September 2002. Between 1999 and 2003, the mean counts for the major species were as follows: Turkey Vulture just over two million birds per season, Broad-winged Hawk 1,919,708, Swainson's Hawk 901,827, Mississippi Kite 192,132, American Kestrel 7,322, Sharp-shinned Hawk 3,870, Osprey 3,256 and Cooper's Hawk 2,716. Several of these species move through in rather short time-periods, so that on certain days thousands of birds of the same species can be seen. No wonder that this remarkable site, only recently discovered in its entirety, is beginning to draw raptor enthusiasts from all parts of the world.

There are several features of a semi-tropical site like this that differ from the more traditional hawk watch-points of the Northern Hemisphere, many of which are rather closer to the birds' breeding grounds. Apart from the very high numbers recorded, raptors at Veracruz tend to move through less fitfully than at other sites. Movement is less weather-dependent, since thermals are more reliable, and birds will sometimes fly in conditions that would normally be considered unsuitable. Owing to the increased solar radiation, 'streets' of thermals form easily, allowing the birds to glide from one thermal to another as they move in a southerly direction. Thirdly, because of the angle of the sun, the ground warms quickly in the morning, allowing migration to begin much earlier than at more temperate sites. All in all, through-migration at this latitude is much faster than further north or south.

Variety adds spice to every hawk watch, and at Veracruz, as anywhere else, there is always the prospect of seeing something less common. Altogether, of the 18 species recorded on the watch here, 11 are irregular, including an eclectic selection of birds from north and south. These include Common Black Hawk, Harris's Hawk, Snail Kite, Plumbeous Hawk and even Northern Goshawk. No doubt future years will see additional oddities turning up.

However, of greater interest are some true migrants that also arrive in smaller numbers, but regularly. These include several species that are largely sedentary in their breeding range, but which at these latitudes have turned into short-distance migrants. They include Hook-billed Kite (an average of 204 per season), Grey-lined Hawk (323) and Zone-tailed Hawk (140), all of them exciting for birders from further north.

At present the raptor counts are performed concurrently at two sites 11 km apart, one on a platform at the edge of a football field and one on the roof of a hotel, the latter with all facilities. It is perfectly possible for volunteers, or the simply curious, to witness these counts, and the organization involved, Pronatura Veracruz, has a strong education and media presence in the local community.

It's hard to believe that, 20 years ago, this, the world's largest movement of diurnal raptors, went largely unrecognized across these otherwise unremarkable skies.

Below: an astonishing total of more than 700,000 Turkey Vultures has been counted passing Veracruz in a single day.

Asa Wright Centre

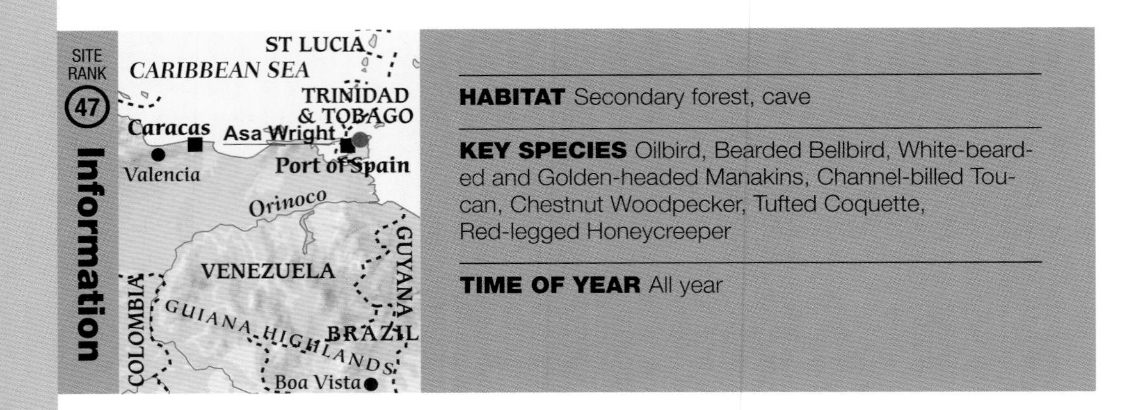

HABITAT Secondary forest, cave

KEY SPECIES Oilbird, Bearded Bellbird, White-bearded and Golden-headed Manakins, Channel-billed Toucan, Chestnut Woodpecker, Tufted Coquette, Red-legged Honeycreeper

TIME OF YEAR All year

The veranda of the main building in the Asa Wright Nature Centre, overlooking the Arima Valley in northern Trinidad, is one of the most famous in the ornithological world. Many of the world's greatest Neotropical birders have sipped their rum punches here, and several pivotal studies into the behaviour of tropical birds have been made within the 1.7 sq km of adjacent grounds and further down the valley, notably by William Beebe and David and Barbara Snow. This is not a bad record for what used to be a coffee, cocoa and citrus plantation. It was acquired from the sympathetic plantation owner, Asa Wright, in 1967, and has been a centre for observing and studying the natural history of the forest ever since.

It is perfectly possible to sit all day on the veranda, watching the birds go by. The view is breathtaking. Just in front are bird-tables and hanging feeders that attract a constant stream of colourful avian visitors, including the ubiquitous Bananaquit, Blue-grey, Palm and Silver-beaked Tanagers and Blue-crowned Motmot. Nearby, the hummingbird feeders receive visits from up to ten species, including the aggressive White-necked Jacobin and the tiny, but stunningly beautiful, Tufted Coquette. Meanwhile, in the treetops down the valley, Channel-billed Toucans feed, Blue-headed Parrots and Orange-winged Amazons pass the time of day, and you can find the magnificent Chestnut Woodpecker, a superb short-billed species with unusual brown plumage and golden crown. Birds using the airspace overhead include Double-toothed Kites and Zone-tailed Hawks, while a pair of Ornate Hawk-Eagles recently raised young a few hundred metres from the main building.

Captivating as this birding pageant is, the call from the forest below the centre inevitably proves irresistible. Here, among Trinidad's rich mix of Neotropical birds (160 species have been recorded in the grounds) you can catch up with some of the special characters, in the place where they were first studied in detail. The sites of the White-bearded and Golden-headed Manakin leks, for example, occupy exactly the same yardage of forest as they did in the 1960s, when David Snow postulated that, by eating fruit and thus fulfilling their nutritional needs easily, such birds have had the chance to evolve complicated social organization and display routines.

Right: pioneering studies of Golden-headed Manakin leks took place at the Asa Wright Nature Centre during the 1960s.

Right: the diminutive Tufted Coquette is one of Trinidad's most sought-after hummingbirds. This female lacks the gaudy head plumes of the male.

If you visit a manakin lek soon after dawn, you are likely to observe some of the remarkably ritualized behaviour that has made these small birds so famous. Leks, by way of explanation, are gatherings of communally displaying male birds; the idea is that, by visiting a lek a female can observe many males in display at once and so make an informed choice as to whom she fancies as her sexual partner. Genes are everything and, once she has copulated, the female manakin will receive no help in the upbringing of eggs and chicks. When a female arrives at the lek pandemonium, not surprisingly, breaks out, as each male (there may be up to 12) simultaneously leaps about its favoured perch in an effort to catch the female's eye; later, further perfectly choreographed rituals lead to the act of copulation. In order to make sure they don't miss a visit from a female, the males stay at their posts for 90 per cent of the hours of daylight, and throughout the year except for a brief break when moulting.

The two species of manakin at Asa Wright have different display routines. The White-bearded Manakin clears a patch of ground on the forest floor about 1 m in diameter, and displays by jumping from the ground to one of several vertical perches and back again, snapping its wings as it takes off and making a sound like a mini firecracker. When a female arrives, both birds

Right: Dunston Cave's famous Oilbird colony has been visited by thousands of birders over the years – a place on an organized tour should be booked well in advance.

jump between the favoured 'mating perch' and the ground, passing each other in mid-air until, finally, the female comes to rest and the male, landing above her, slides down the mating perch and on to her back. The Golden-headed Manakins, on the other hand, make their plays on horizontal perches 6–12 m above the forest floor. One of their routines is to lift their head and tail and 'moonwalk' backwards, Michael Jackson-style, along the stem, using very small and rapid steps. Both displays, if you are fortunate enough to see them, will fill you with a heady mix of wonder and amusement.

Another lekking species in these forests is also the noisiest, the Bearded Bellbird. It is not long after dawn that you will first hear the clamorous *tonk* of this bird, greeting the new day with a sound like a blacksmith striking a forge. These neat brown, black and white birds sing from privileged high perches, the rights to which may be viciously fought over by all the birds in the neighbourhood. It is ownership of one of these sought-after song posts that usually sways the female's choice.

Perhaps the most famous of all the inhabitants of the Arima Valley is, however, not a lek species at all. Within the grounds of the Asa Wright Centre is Dunston Cave, home to what is probably the most accessible, or at least the most visited colony of Oilbirds in the world. These oddities, assumed to be most closely related to the nightjars, are the only nocturnal frugivores in the avian world, plucking what they need while flying over the forest at night. While foraging they are guided by sight and smell, but back in their caves they turn to echolocation to find their way around in the total blackness. The Dunston Oilbirds, whose population numbers 120–150 individuals, sometimes make foraging trips of 120 km, which may take them over the sea to Trinidad's sister island, Tobago.

Oilbirds are quite vulnerable to disturbance at their caves, and access to them is strictly controlled. There are plans afoot, however, to set up a system of infra-red cameras that would allow the birds to be seen on monitors, and ultimately on the Internet. The monitors would, of course, be set up on that hallowed veranda.

■ *Right: the* tonk *call of the male Bearded Bellbird is far carrying and very useful for any birder wishing to find the species – it can be very difficult to locate these birds during the part of the year when the males are silent.*

Zapata Swamp

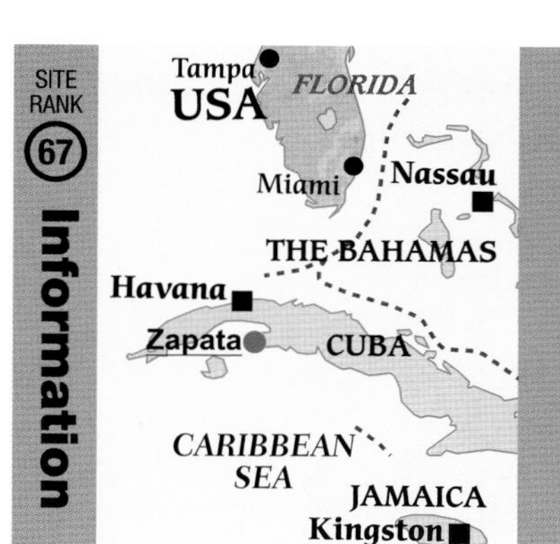

HABITAT Swamp, dry forest, mangroves, mudflats

KEY SPECIES Zapata Wren, Zapata Rail, Zapata Sparrow, Bee Hummingbird, Fernandina's Flicker, Gundlach's Hawk, Stygian Owl, Great Lizard-cuckoo, Blue-headed Quail-Dove, Red-shouldered Blackbird

TIME OF YEAR All year round

Zapata Swamp lies towards the western end of Cuba, not far from the capital, Havana. It occupies much of a large peninsula that juts out to the west of the infamous Bay of Pigs (Bahia de Cochinos), where an abortive attempt to depose president Fidel Castro in 1961 nearly plunged the world into war. The peninsula is 130 km long, and about half of it, 3,278 sq km, is designated as the Cienaga de Zapata National Park. Within this limited area is a unique ecosystem with great ornithological wealth, including two highly distinctive endemic birds.

The coastline of the peninsula is lined with mangrove forests and saline lagoons, which hold a variety of water birds including the colourful Roseate Spoonbill and the even more shockingly pink American Flamingo. Inland from this are two zones of dry forest, each 1–2 km wide and differing in floristic diversity, height and the number of palm trees (all of which increase away from the coast). Inside this zone is the swamp proper, much of which consists of bushes and impenetrable looking saw grass up to 2.4 m high. This last-named zone is where the two endemic species live.

There could hardly be a more difficult habitat in which to seek birds, especially secretive ones, than the saw grass swamp. As well as being very tall and thick and painful to the skin, it is seasonally inundated, and therefore any visitors will soon be convinced that they have virtually no chance of seeing the ultra-shy Zapata Rail. They may hear its odd bubbling call, but it is said that hardly any of the resident birders here have ever seen one, let alone visitors to the area. For the record, the Zapata Rail is slightly smaller than a Common Moorhen, with grey underparts, olive-brown unstreaked upperparts, and red legs. No juvenile has ever been seen.

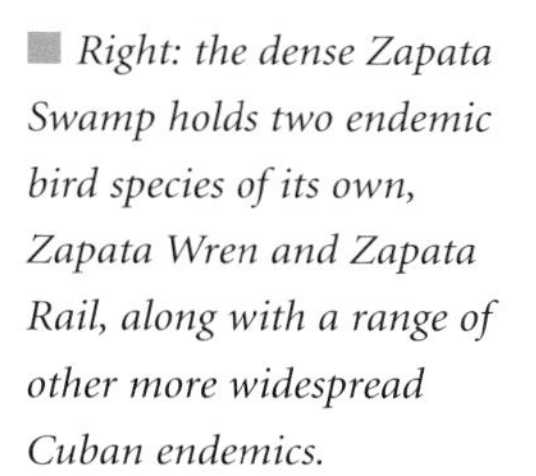
Right: the dense Zapata Swamp holds two endemic bird species of its own, Zapata Wren and Zapata Rail, along with a range of other more widespread Cuban endemics.

■ *Right: Gundlach's Hawk is a specialist hunter of birds.*

Visitors are likely to have more success with the other endemic, the Zapata Wren, because it will at least deign to sing from the tops of bushes; its outbursts seem to go on for as long as one of Fidel Castro's speeches! This delightful, tail-heavy wren with densely black-barred upperparts is known to scratch the ground in order to locate food. In common with the Zapata Rail it is not just an endemic species, but is unusual enough to be assigned its own genus.

Although these two species are endemic to the swamp itself, several other Cuban specialities occurring here are now very hard to find anywhere else. The Zapata Sparrow, despite its name, is actually found in three disparate areas of Cuba, each with its own subspecies, but the version found here is the most numerous and easy to find. It is an attractive sparrow with largely unmarked plumage but with a strong lemon-yellow wash to the underparts. Another great rarity is Fernandina's Flicker, a densely-barred woodpecker with a buff-coloured head. The world population of this species could be down to fewer than 300 individuals, of which nearly half are found in the forested areas of the Zapata Swamp, mainly in areas with abundant Sabal Palms. The birds usually excavate their nest-holes in these distinctive 'punk-topped' trees.

However much you admire these star birds at Zapata, it is likely that one other species will actually steal the show. This is the charismatic Bee Hummingbird, the smallest bird in the world. This mite is quite ludicrous-

■ *Below: in common with several of Zapata Swamp's birds, Fernandina's Flicker is becoming exceedingly rare. Its world population may be a low as 300 individuals.*

ly tiny – however much you are expecting minuteness, the reality is just incredible – and you are likely to suffer any number of false alarms while looking for it, since many flying insects here are just as big, or bigger. Only 5.7 cm long, the males of this species are spectacular to look at, with dark green back and brilliant iridescent fiery red crown and throat, the latter with long lateral plumes. They are slightly smaller than the females, and have an endearing habit of singing their squeaky song from high in a tree-top, as if this were some demonstration of power. There actually must be something in Cuba's air that encourages small size, because living on this same island is one of the very smallest bats in the world, and also the smallest frog in the Northern Hemisphere.

One advantage of being small is to be more or less immune from bird predators, and the Bee Hummingbird would barely register as a snack for the large Gundlach's Hawk, another of the swamp's star rarities. This fine predator, Cuba's endemic equivalent to North America's Cooper's Hawk, is a bird specialist. The females primarily hunt for larger prey such as Cuban Amazons, pigeons, and even Cuban Crows, the local endemic corvid which, incidentally, makes sounds like an inebriated parrot. The smaller male, meanwhile, hunts more in forested areas, catching thrushes and doves, including the swamp's glittering array of attractive ground doves. This hunter was once considered to be on the verge of extinction, but was subsequently found to be quite widespread, just elusive.

The story of Gundlach's Hawk bucks the trend in the fortunes of Cuba's birds. Many of the unique species of the island are in decline and some, including the Bee Hummingbird, the flicker, the rail and the wren are critically endangered. The swamp itself is subjected to burning in the dry season, together with some illegal logging and agricultural encroachment. The protected status of some parts of the swamp is confused and inadequate and is also compromised by Cuba's dire economic state. There have not yet been complete surveys of the endemic taxa, either, so it is difficult to know exactly how perilous the situation is.

The chances are that things are dire. This needs rectifying urgently, because Zapata Swamp is not just a jewel in Cuba's crown, but also one of the world's environmental treasures.

Below: endemic to Cuba, the Bee Hummingbird is officially the world's smallest bird, although several other hummers push it close.

North America

Although North America has only about 750 regularly occurring bird species, like Europe, those present are extraordinarily well known and well loved, and include some of the best studied in the world. There are also many birders here, and the hobby has a large and passionate following.

In particular North America is special in its ability to provide spectacle, from the teeming masses of auks and other seabirds around the Bering Sea to cranes and geese in the mid-west, and from innumerable shorebirds along the mudflats of the Gulf of Mexico to the remarkable leks of grouse on the prairies. Encompassing the range of latitudes from the Arctic to the Subtropical, North America also presents some of the best migration-watching in the world, from birds of prey pouring through at favoured routes to clouds of small songbirds dropping from the sky at certain watchpoints in the right seasons and conditions.

Working from north to south, the continent begins with a tundra zone which teems with waders and ducks in the summer but is all but evacuated in winter. To the south, the great forest belt of the taiga, largely grown with conifers and dotted with innumerable bogs and lakes, is equally seasonal, providing insect food for a variety of waders, thrushes, warblers and vireos, as well as many other species. Further south there are impressive deciduous forests in the north-east and breathtaking tall, mossy forests in the north-west, where some of the tallest trees in the world grow. The mid-west is best known for the Great Plains, one of the world's largest areas of grassland, while further south the dry conditions give rise to several different types of deserts, all with their own birds; the best known of these is the Sonoran, with its giant cacti and the world's smallest owl. In the far west, California provides its own special habitat in the form of a type of scrub known as chaparral, while in the east there are vast areas of cypress swamp, open conifer forests and, in the south of Florida, subtropical marshes and islands.

Although North America is short of unique families, it is very rich in such groups as New World warblers, icterids, tyrant flycatchers, grouse, auks, gulls and New World sparrows.

Right: Snow and Ross's Geese in New Mexico.

Bosque del Apache

HABITAT Wetlands, riparian woods, arid uplands

KEY SPECIES Sandhill Crane, Snow and Ross's Geese, Bald Eagle, Gambel's Quail, Chihuahuan Raven

TIME OF YEAR Most spectacular from early November to February, but excellent all year round

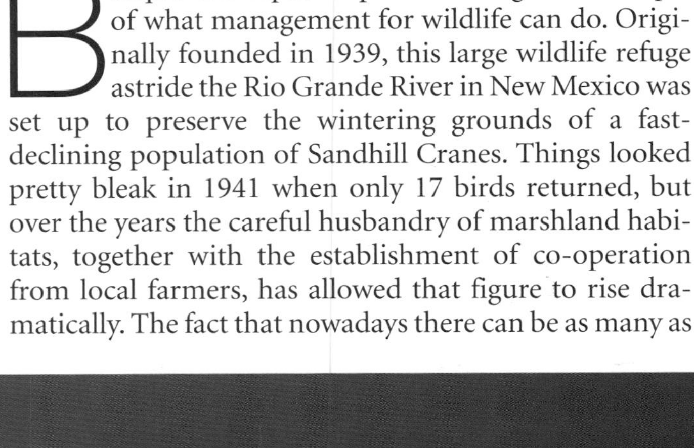

Bosque del Apache provides a great example of what management for wildlife can do. Originally founded in 1939, this large wildlife refuge astride the Rio Grande River in New Mexico was set up to preserve the wintering grounds of a fast-declining population of Sandhill Cranes. Things looked pretty bleak in 1941 when only 17 birds returned, but over the years the careful husbandry of marshland habitats, together with the establishment of co-operation from local farmers, has allowed that figure to rise dramatically. The fact that nowadays there can be as many as 18,000 cranes here in November, as well as many thousands of other waterbirds, speaks for itself. What the conservationists got right here, they got right in style.

Even then, it is still fair to assume that the Bosque del Apache of today might actually exceed those pioneers' wildest dreams. The reserve they created has a quite magical combination of stunning scenery, spectacular wildlife, convenience for visitors and a cast of birds sizzling with star quality. Birders and more casual visitors can sit alongside one another and be equally overwhelmed. It is that sort of place.

Bosque del Apache's flagship species is still the Sandhill Crane. Tall and majestic, it can be seen very easily here between November and February. The birds feed mainly on corn, which is grown by local farmers along with a cash crop, alfalfa. The cranes can usually be seen scattered about in family parties or larger groups, but at the end of the day they all commute to wetter areas to roost in safety. Then it is possible to see large flocks flying across the sky in neat 'V' and 'W' formations, all the while uttering their loud, wild-sounding bugling calls.

So special and appreciated are these cranes that, every November, when they arrive, the local town, Socorro and

Below: dawn rises on the wintering Sandhill Cranes of Bosque del Apache.

■ *Above: a snowstorm of Snow Geese is one of the great sights of North American birding; there's probably a Ross's Goose in there somewhere as well!*

the wildlife refuge get together to run the annual Festival of the Cranes. Over the selected weekend (the 20th annual Festival ran in 2007) there is a programme of lectures and tours, together with exhibitions of wildlife art and other activities. It is one of the most popular birding gatherings in the whole of North America.

Yet the cranes, amazingly, don't entirely steal the show here. Sharing the wetlands with them are huge numbers of wildfowl, and these include about 30,000 Snow Geese and 1,000 Ross's Geese that together are locally and logically termed 'white geese'. These, like the cranes, have a daily routine of commuting to and from a safe roosting site at dusk and dawn. In particular, to witness their regular morning 'flyout' as they leave their roost at sunrise is one of the greatest wildlife experiences in the whole of North America.

After being moderately dispersed overnight, almost all the geese on the reserve congregate in the pre-dawn darkness on a single large lake close to a viewing platform known as the 'flight deck', not far from the visitor centre. Quite suddenly, as the increasing light intensity trips a trigger, there will be a roar of wings and almost all of the geese will take off as one, calling loudly to one another with harsh, slightly hysterical honks. They may circle for a while, but most eventually break out for the north. For a few sweet moments the dark fields and lightening sky are overwhelmed by this blizzard of white birds, which can often be seen against the red sun of dawn. Within a few minutes the last calls fade away and the show is over for another morning, remaining only in the memories of the witnesses and their digital cameras.

One of the good things about the early morning flyout is that it leaves a full day to explore the rest of the core reserve. The wetland and farm sections are served by a one-way loop road, open at daybreak, vehicular access to which costs no more than $3 for a day. Along its 19 km are numerous viewing decks, hides and trails, all of which invite investigation. Many of the decks are fitted with telescopes through which visitors can admire the birds and work on the trickier identification puzzles, such as separating the very closely related Snow and Ross's Geese. The

loop is, in fact, divided into two. In winter most visitors will want to go along the Farm Loop, which leads north to the main feeding areas of the cranes and geese. The Marsh Loop is good for breeding birds in the summer, including various ducks, herons and rails.

The presence of so many birds inevitably attracts predators. Some human visitors have been fortunate enough to witness what must be quite a regular event, a huge cloud of birds flushed by a Coyote wandering in from the nearby hills. More common, but still impressive, are the large numbers of wintering raptors here: these can include 20 or more Bald Eagles and dozens of Red-tailed Hawks, with the odd Ferruginous Hawk thrown in. These birds sit in the cottonwoods and other trees beside the wetland, and wait.

When observing all these wetland birds it is easy to forget that Bosque del Apache is actually on a high-elevation (1,500 m) arid plain. However, a quick glance around the nearby countryside quickly gives the context. Gambel's Quail can be seen close to the visitor centre, and many other scrub-desert-loving birds breed in those parts of the wildlife refuge away from the river. These include Greater Roadrunner, Curve-billed Thrasher, Verdin and Pyrrhuloxia, to name just a few. You can also test your identification skills on the ravens here. Both Northern Raven and Chihuahuan Raven breed on the reserve, the latter distinguishable only when the wind ruffles its neck to reveal the white-based feathers. However, such subtleties are easily ignored here. Most visitors just come away overwhelmed.

Above: up to 20 Bald Eagles can be seen in the area in winter.

Right: check the area around the visitor centre for the desert-loving Gambel's Quail.

Monterey Bay

HABITAT Open ocean, inshore waters

KEY SPECIES Black-footed Albatross, Black, Ashy and Least Storm Petrels, Rhinoceros and Cassin's Akulets, Craveri's, Ancient and Xantus's Murrelets

TIME OF YEAR Autumn from August to October is best for a wide variety of species, but there is something of interest to see all year

Above right: despite the gentle undulations on land, Monterey Bay lies above an enormous underwater canyon, from which nutrients rise from the seabed to provide food for large numbers of seabirds.

Seabirds can be among the hardest of groups for birders to see, since many of them live exclusively far out in the ocean and rarely, if ever, come close to shore. At Monterey Bay in California, however, owing to a quirk of geography, just a few kilometres from dry land there lies the rim of a deep underwater canyon that shunts cold water, together with a rich broth of nutrients, up towards the surface. This upwelling is so rich in food that seabirds from far and wide simply cannot help themselves and come well within range of medium-sized vessels crowded with seabirders on day trips. As a result, Monterey Bay is North America's most famous pelagic destination.

Seabirds appear off Monterey throughout the year. The autumn period, between August and October, usually brings the richest crop of species, and this is when most trips are run. However, several specialities occur throughout the year, including the famous Black-footed Albatrosses, and some, such as the rare Laysan Albatross, are more frequent in winter. Of course, pelagic sea-birding is highly unpredictable, so it usually requires several trips to get sightings of all your desired species.

Some of the birds seen at Monterey travel vast distances to get there. Buller's Shearwater, for example, which is a scarce visitor in the autumn, comes from its breeding grounds in New Zealand, and the rare Short-tailed Albatross, now a very rare visitor, comes from Japan, on the opposite side of the Pacific Ocean. Meanwhile the two other albatrosses, the Black-footed and Laysan, both breed in the Hawaiian islands in mid-Pacific, more than 3,000 km away. Radio-tracking studies have shown that albatrosses sometimes make vast journeys from their

Right: one of Monterey's star species is the Black-footed Albatross.

Right: Heermann's Gull, something of a California speciality, is common on the Pacific shoreline.

breeding sites to fetch food for their young and, should you see one of these birds in the winter, it is incredible to believe that it could merely be engaged on what is effectively a shopping trip on a grand scale.

Other birds encountered on the pelagic trips have travelled much less far, including several of the region's rare and endemic storm petrels. The greyish-coloured Ashy Storm Petrel, for example, breeds mainly on the Farallon Islands and Channel Islands off California, and in late September and October large concentrations of this species gather off the northern rim of Monterey's submarine canyon. It is estimated that 90 per cent of the world population of this bird (9,000 individuals) may be involved in these gatherings, along with several other storm petrel species. Two of these, Black and Least Storm Petrels, breed mainly in the Gulf of California to the south and are much less numerous and reliable in their appearance than the Ashy. Nevertheless this large gathering of storm petrels, which at times includes Wilson's, Wedge-rumped and Fork-tailed Storm Petrels as well, is one of the major highlights of birding in Monterey Bay.

In a worldwide context, the other important group of birds seen on these northern Californian pelagics is the auks, even though their presence fails to register the same whoops of delight that greet every albatross sighting. Among these are the highly localized species, Xantus's and Craveri's Murrelets. These two black-and-white midgets are difficult to tell apart, and the situation is further complicated by the fact that Xantus's comes in two forms, with differing amounts of white around the eye and slightly different bill length; the forms are hotly tipped to be split into two species, with the northern form being known as Scripps's Murrelet. Craveri's Murrelet breeds mainly in the Gulf of California, while Xantus's breeds on the Pacific coast from California to Mexico. The population of neither species exceeds 10,000 birds.

Although the rare murrelets first appear in the autumn, the best time for auks is probably the winter. This is when enormous numbers of Cassin's Auklets, a small smoky-grey crustacean-eating auk with a short bill, begin to appear off these coasts. It is estimated that about one million birds can be found off California in winter. Not that most pelagic watchers would know it; Cassin's Auklets are notoriously shy, and flee over the water whenever a boat appears on the horizon. Another common winter visitor is the Ancient Murrelet, a smart black, white and grey auklet with a tiny yellow bill, while fish-eating Tufted Puffins are scarce.

Of course, many other seabirds can be seen along with these scarcer species and groups. The commonest bird seen is often the Sooty Shearwater, present here all year round, while Pink-footed Shearwaters can be common in autumn and Short-tailed Shearwaters in the winter. Waders are represented by Grey and Red-necked Phalaropes, and Western and Heermann's Gulls – both quite localized species in their own right – are abundant close to the shore. In late summer Sabine's Gulls pass through, and about the same time there may be large movements of Arctic Terns, pursued all the way by Long-tailed Skuas. In the winter, divers and grebes may also be in the mix.

Not surprisingly, rarities are recorded among the seabirds from time to time. These may include such birds as Red-billed Tropicbird, Masked Booby and Bulwer's Petrel, but of course, with the usual unpredictability of all things to do with birding, almost anything could turn up.

Prairie Potholes

SITE RANK 70 **Information**

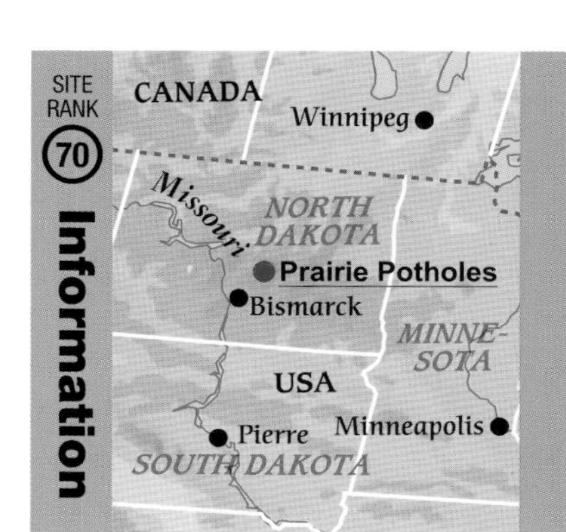

HABITAT Wetlands, grassland

KEY SPECIES Sprague's Pipit, grassland sparrows including Baird's, Le Conte's and Clay-coloured, wildfowl including Cinnamon Teal, Bufflehead and Hooded Merganser, Western and Clark's Grebes, Sharp-tailed Grouse, American White Pelican

TIME OF YEAR All year round, but the specialities are mainly found in the spring and summer

Above right: the male Wilson's Phalarope is smaller and less colourful than its mate. Wilson's is the only member of the phalarope family that breeds exclusively in North America; the other two species breed at higher latitudes and have circumpolar breeding distributions.

If the prairies are the wheat basket of North America, it could be argued that the Prairie Potholes region, comprising parts of south-central Canada, North and South Dakota, Wisconsin, Minnesota and Iowa, is the 'duck factory' of North America. It is estimated that about 45 million pairs of wildfowl breed each year in this flat, grassland area dotted by innumerable shallow lakes. That is half of all the ducks on the continent.

The characteristic flat, potted appearance of this region was caused by the action of glaciation. The numerous small depressions that were left behind nowadays form into lakes and marshes which are fed by snowmelt and rainwater. There are so many depressions that almost every type of wetland habitat occurs, from productive neutral to highly alkaline lakes, deep or very shallow water bodies and marshes, and from temporary ponds to permanent wetland systems. Furthermore, these habitats are surrounded by grasslands, which add considerably to the birding interest of the area and play host to some of the region's most threatened species.

Several of the best birding areas of the Potholes region are found in the state of North Dakota. Long Lake National Wildlife Refuge, for example, is a typically rich area, with a long list of over 300 species recorded. The common breeding ducks include Gadwall, Mallard, Northern Pintail, Northern Shoveler, Blue-winged Teal, Ruddy Duck, Canvasback and Redhead, while Green-winged and Cinnamon Teal, American Wigeon, Wood Duck, Ring-necked Duck, Lesser Scaup, Bufflehead and Hooded Merganser are less common or rare. Each species, of course, has its own ecological requirements: the Blue-winged Teal and Canvasback are drawn to the

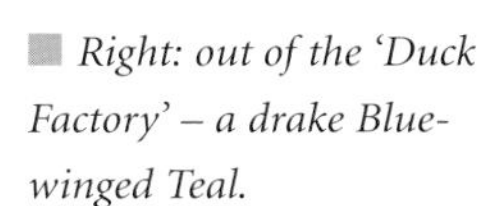

Right: out of the 'Duck Factory' – a drake Blue-winged Teal.

Above: the famous rushing display of the Western Grebe is a common sight in the spring.

marshy areas for breeding, while the Gadwall prefers alkaline areas, and both Ring-necked Duck and Wood Duck occur on the refuge's tree-lined wetlands. This remarkable diversity of ducks is quite typical of the region as a whole.

Grebes abound in many of North Dakota's refuges, with most of North America's species breeding in the state. Pied-billed Grebes are numerous in the sumptuous, thick marshes, while the open water provides a platform for the Western Grebe's stupendous courtship displays. Famous for 'walking on water', Western Grebes gather into pairs and make rushes across the surface, the whole body supported by the rapidly pattering feet; meanwhile, the participants keep their wings half open and bow their heads so that their necks arch elegantly forward, somehow managing to give an impression of supreme ease and grace, just like a hard-working ballerina. It isn't always a male and female that perform the rush together; sometimes a male will appropriate another male and together they will perform in the hope of impressing onlookers. On occasion, a Western Grebe will rush with the closely related Clark's Grebe at its side. The two species, which were only relatively recently separated, have identical displays, but manage to keep genetically apart by having different advertising calls, differently coloured bills and different amounts of black on the face.

Another bird of the open waters is the American White Pelican. It is common in the region and indeed Chase Lake National Wildlife Refuge holds the largest colony in North America, with about 30,000 birds. Oddly, Chase Lake itself is strongly alkaline and holds no food, so the birds must commute to local rivers or lakes to feed, and may travel over 100 km to do so. Apparently, their most numerous local food item is the Tiger Salamander.

Alkaline lakes do provide some food, though, mainly in the form of small invertebrates such as shrimps and insect larvae. These attract specialists such as the strikingly coloured American Avocet, with its delicate orange-buff neck and uptilted bill, which feeds on creatures in shallow water with a swishing, side-to-side movement of the bill. Another common inhabitant, the equally tastefully coloured Wilson's Phalarope, forages by swimming instead, and spinning around in characteristic phalarope fashion to stir particles within reach. Meanwhile, on the bare, somewhat desolate shores of the lakes, the very rare Piping Plover breeds. This fast-declining shorebird has a curiously disjunct distribution; besides these inland prairie lakes, the main population breeds along the Atlantic coast of North America.

Many of the true rarities of the Prairie Potholes region, however, occur not on the potholes but on the prairies themselves, and one very popular area with birders, Lostwood National Wildlife Refuge, holds a very fine selection of these. Its most famous residents are two small, streaky-brown passerines, Sprague's Pipit and Baird's Sparrow, but it also holds one of the largest populations in North America of a streaky-brown game bird, the Sharp-tailed Grouse. There are 40 leks of this species within the 10,880-ha refuge, some with up to 40 participating males.

Sprague's Pipit has suffered a steep decline in recent years, and has been placed on the National Audubon

■ Right: there are no less than 40 leks of the Sharp-tailed Grouse at Lostwood National Wildlife Refuge.

■ Right: the usually skulking Baird's Sparrow proclaims its territory from the top of a bush.

Society's WatchList, as well as being in BirdLife International's Vulnerable category. A bird that favours recently burnt areas in native shortgrass prairie, it has suffered as the grasslands have been turned over to agriculture in this fertile land, and is decreasing at a rate of nearly 5 per cent per annum. Unless this trend is reversed, this formidable songster, which has been known to perform song-flights 100 m in the air for up to 40 minutes at a time, could disappear in all but these most favoured sites.

Baird's Sparrow is one of an army of tricky sparrows that live in these grasslands, and spend much of their time driving birders mad with their skulking tactics. Baird's Sparrow at least has a decent song, a high-pitched jingle, while the Grasshopper Sparrow reels like its eponymous insect and the Clay-coloured Sparrow, a bird of more scrubby areas, makes an equally tuneless buzz. Interestingly, several of the grassland sparrows sing particularly well at night, including the Grasshopper Sparrow, the marsh-loving Le Conte's Sparrow and, if its name is to be believed, the moderately conspicuous Vesper Sparrow.

These sparrows are both a birder's dream and nightmare. Many are sadly now scarce (nearly one-fifth of all North America's sparrows are on the WatchList), they offer rather small, subtle differences in plumage, and they are often very elusive. Nevertheless, for many enthusiasts, the sight of a Baird's Sparrow perched in view for half a second is as memorable as the hordes of waterbirds that teem on the lakes of this marvellous area.

Chiricahua Mountains

HABITAT Semi-desert scrub and grass, mixed oak/pine chaparral, canyon and highland woodland

KEY SPECIES Hummingbirds including Magnificent and Lucifer, Elegant Trogon, Mexican Chickadee, Painted Whitestart, Arizona Woodpecker, Juniper Titmouse, Olive and Red-faced Warblers

TIME OF YEAR Good all year round

Right: one of the Chiricahua's star birds is the Elegant Trogon; the green head and upperparts and extensive red underparts identify this bird as an adult male.

As almost any North American birder will tell you, south-eastern Arizona is just about the best region in the whole of the United States for birding. It holds more breeding species than any other comparable area of the country and consistently scores one of the highest species totals during the Christmas Bird Count. Why go anywhere else?

Arguably the 'best single location' in this 'best general region' is the Chiricahua Mountains, one of the so-called 'sky island' ranges found in the very southern edge of the state. These rugged mountains, volcanic in origin, rise up from the surrounding desert plain to over 2,975 m. Within their bounds are some sites that, when they come up in conversation, turn birders watery-eyed and wistful, such as the celebrated Cave Creek Canyon, Rustler Park and the Chiricahua National Monument. No wonder: all in all, more than 300 species have been recorded in the Chiricahuas over the years, including dozens of great rarities.

The reason for that exceptional richness lies in the happy location of the range. Lying only 29 km from the Mexican border, the Chiricahuas lie sweetly at the convergence of four major biomes: the Sonoran and Chihuahuan Deserts, the Rocky Mountain chain and the Sierra Madre Range. The latter lies primarily in Mexico, and this northern extension enables a good many Mexican species to 'dip the toes' of their breeding ranges into the United States. The Mexican Chickadee, for instance, occurs in the USA only in these mountains. Rarities apart, though,the combination and mixing of these zones is what makes the Chiricahuas such a memorable birding destination.

The lower parts of the range contain arid semi-desert scrub and grassland, home to such typical Sonoran birds as Elf Owl, Greater Roadrunner, Gilded Flicker and Gila Woodpecker. Above about 1,000 m, though, the Chihuahuan birds take over, with the Chihuahuan Rave flying above the dry ground and the Scaled Quail hiding away in the brush. Despite its arid appearance, this area receives quite a lot of rain between mid-July and September, when many birders visit, and the canyons and valleys can have a miraculously lush appearance.

The famous Cave Creek Canyon lies in the transition zone, where the arid Chihuahuan grasslands are replaced firstly by thorny scrub and then by a variety of trees or shrubs with a Sierran flavour, mainly oaks and juniper. These woodlands are favoured by such Mexican

Above: nowhere in North America can rival south-east Arizona for hummingbirds – the Lucifer Hummingbird is one of an impressive list of 15 species that has been recorded in the Chiricahuas in a single season.

species as the bold black, white and red Painted Whitestart, the streaky Sulphur-bellied Flycatcher, Whiskered Screech Owl, Arizona Woodpecker, Montezuma Quail and the bird for which Cave Creek Canyon is renowned, the Elegant Trogon. This forest beauty, with its long, squared-off tail marked with bars below, tends to be found in Arizona Sycamores beside streams, and most birders strike lucky when looking for it. Meanwhile, these Mexican specials live alongside other more wide-ranging oak-juniper species such as Western Scrub Jay, Juniper Titmouse and Hutton's Vireo.

At higher elevations more coniferous trees appear, including Ponderosa Pine and Douglas Fir, representatives from the Rocky Mountains. The Rocky Mountain trees carry birds such as Pygmy Nuthatch, Band-tailed Pigeon, Steller's Jay and Northern Pygmy Owl down with them from further north. They mix with southern species such as Apache Pine and produce an eclectic mix of northern and southern birds. The pygmy owl, for example, shares the woods with the much rarer Flammulated Owl (the Chiricahua Mountains in general are excellent for owls, with eight or more species possible) and species that are rare or restricted in North America appear, including Grace's Warbler and the spectacular Red-faced Warbler, Mexican Chickadee, Greater Pewee, Hepatic Tanager, Yellow-eyed Junco and the peculiar Olive Warbler, considered sufficiently different from the rest of the American wood warblers to convince some to place it in its own family. One of these distinguishing features is that the nestlings defecate on the side of the nest, instead of having their faecal sacs removed by the adults. Such are the delicate minutiae of taxonomy!

Yet amazingly, this wondrous roll-call of birds is often a side-show, in birders' minds, to the Chiricahua Mountains' real star attraction – the hummingbirds. Most parts of the United States are graced by only one or two species of hummingbird, but in south-east Arizona it is possible to see ten in a day, and 15 in a season. The best time to look for them is the so-called 'second spring', which follows the beginning of Arizona's rainy season in mid-July. The rains produce a greening of the area and a flush of wildflowers, which attract hummingbirds in large numbers. These include not only the local breeding species but also early migrants from the north, such as Rufous and Calliope Hummingbirds, and post-breeding visitors from Mexico, such as the Lucifer Hummingbird. Birds in the latter category may also breed

■ *Right: the unmistakable Red-faced Warbler is unusual among New World warblers in building its nest on the ground.*

■ *Below: Painted Whitestart is a summer visitor to the oak and juniper forests.*

sporadically, and indeed the second nest of the Berylline Hummingbird ever found in North America was discovered at the Chiricahua National Monument.

Although the hummers are after real flowers, often the very best place to see them can be at a line of hummingbird feeders put up to attract them, for example those at the Coronado National Forest Station and around Ramsey Canyon and Sierra Vista in the neighbouring 'sky island' of the Huachuca Mountains. Such places can be buzzing with these tiny and often irascible birds. Regular breeding species in the Chiricahuas generally include Anna's, Rufous, Black-chinned, Broad-tailed and Lucifer, and in recent years the rare Violet-crowned has been added to the list for Cave Creek Canyon. This immediate area is the best site in the region for early migrant Calliope Hummingbirds, and also one of the few places where the rare early migrant Allen's Hummingbird appears, providing a good test of identification skills. Of course, if you are on a dedicated hummingbird trip (these are run commercially in the area), you may turn up one of the southern rarities, such as White-eared Hummingbird or Plain-capped Starthroat.

Ironically, one hummer you won't see is the otherwise familiar Ruby-throated, eastern North America's only regular species and the one that most birders in the USA were brought up upon. This bird was recorded in Arizona for the first time in 2005, and a repeat is considered highly unlikely.

Rio Grande Valley

HABITAT Coastal, palm woodland, riparian woodland, subtropical thorn woodland, wetlands, arid scrub

KEY SPECIES Green and Brown Jays, Altamira and Audubon's Orioles, Plain Chachalaca, Hook-billed Kite, Buff-bellied Hummingbird, Great Kiskadee, White-collared Seedeater

TIME OF YEAR All year round, but most strays from Mexico occur during the winter months

Below: one of the commonest of the Rio Grande specialities, the Green Jay will come to bird tables and picnic tables to feed on scraps.

The Rio Grande forms the south-eastern section of the border between the United States and Mexico, the two countries staring at each other across its muddy waters. Were it, say, 300 km to the north, it would probably not be especially interesting to birders, since its avifauna would not differ much from what was present further north. As it happens, though, a considerable number of species from the south creep as breeding species into the borders of the USA for about 100 km along the Rio Grande, giving the area an enviable list of reliable specialities that birders cannot see elsewhere in the country. Furthermore, a good many southern species also make odd and sporadic appearances here from time to time. No doubt American birders are a patriotic lot, but when thinking of life-lists they cannot seem to resist nipping down to their extreme southern border to see birds absconding from Mexico.

The Rio Grande Valley, which is really a delta, can be divided into three main sections, the Lower, Middle and Upper. The Lower and Middle are on the fertile plain, which has been largely lost to agriculture and develop-

ment over the last 100 years, leaving very few patches of native vegetation intact. The Upper Section, on the other hand, is far wilder and less populated, but is also much drier, with a smaller selection of birds.

A good example of the almost wholesale destruction of the Lower Valley can be seen at Sabal Palm Grove Sanctuary. It is a wonderful, rich and lush woodland of old-growth native palms, but is a miserly 0.13 sq km (with a separate block of 0.06 sq km situated 1.5 km away) amidst a sea of cultivated fields 6 km south of Brownsville. Sabal Palms were once a dominant part of the flora here, and yet are now reduced to fragments in nature reserves. As to birds, however, this sanctuary is a good place to get to know some of the species that are common here but extremely rare elsewhere in the United States. These include the exceedingly loudmouthed Plain Chachalaca, the White-tipped Dove, the smart, stripy Long-billed Thrasher, the extravagantly colourful Green Jay and Altamira Oriole, and a trio of large tyrant flycatchers, the brilliantly coloured Great Kiskadee and Couch's and Tropical Kingbirds. It is also known as a spot for winter visits from the sprightly Buff-bellied Hummingbird.

■ *Below: Plain Chachalaca is the only Cracid to occur in the United States; it lives in noisy groups.*

Further inland, the patches of land given over to nature reserves are slightly more generous, and in this area can be found two of the most famous birding sites in the entire United States: Santa Ana National Wildlife Refuge and Bentsen State Park. The former is the larger site, some 8.45 sq km encompassing large areas of riparian forest, with lakes and scrubland; the latter covers only 2.35 sq km, but contains a superb remnant of a rare habitat in the United States known as subtropical thorn woodland. The dense growth on the fertile alluvial land of both reserves, interspersed with old oxbow lakes known locally as resacas, contributes to its rich and unusual avifauna. Just about all the specialities of the Rio Grande Valley have been recorded in these reserves, along with rarities from Mexico such as Clay-coloured Thrush, Rose-throated Becard and Crimson-collared Grosbeak. The rare Hook-billed Kite, which has the distinction of being a specialist feeder on arboreal snails, can regularly be seen at both sites, while other reliable specialities include Grey-lined Hawk, the skulking Olive Sparrow and the rather dull-looking and elusive Northern Beardless Tyrannulet. The extensive wetlands at Santa Ana, in particular, are also good for the diminutive Least Grebe, and such exotic delights as Black-bellied Whistling Duck. It is one of the few places in the United States where you can see three species of kingfisher: the loud, hefty Ringed and the smaller, more self–effacing Green consorting with the familiar, continent-wide staple, the Belted Kingfisher.

Moving inland to the Upper Valley, the endless cultivation at last begins to melt away, and the countryside is wilder and more pleasing, even for list-obsessed birders. Indeed, somewhat unexpectedly, the influence of the south-west soon comes into play and, as the ground becomes drier and drier, such birds as Scaled Quail, Pyrrhuloxia and Verdin appear, characters out of the deserts of California, Arizona and New Mexico. However, with respect, most birders are not after these; they are heading for the area around Falcon Dam, where some of America's rarest breeding birds are to be found.

In the Upper Rio Grande, the fortunes of the great rarities are followed avidly by birders. They plead for stake-outs for such exceptionally capricious birds as the almost legendary White-collared Seedeater, a minute black, white and yellowish bird with a short, very thick bill. Apparently, this bird is something of a tease of tour leaders, being effortless to find one year and impossible the next; its last refuge in the USA is near the town of San Ygnacio. Another big target is the Muscovy Duck, a bird that has benefited from a nest-box scheme in Mexico and now has a foothold on the northern side of the river, but is hard to track down. It is slightly easier to find the other great rarities here, Audubon's Oriole and Brown Jay. Audubon's Oriole, although shy, will sometimes visit

■ *Right: an Altamira Oriole peeks out from its impressive pendulous nest.*

■ *Below: the striking profile of the Hook-billed Kite is due to the fact that its bill is adapted to feed on snails.*

feeding stations, where its yellow plumage, large size and long tail make such a contrast to most orioles familiar to US birders. The Brown Jay began nesting in this area only in 1974 and remains very rare. It is extremely dull to look at (just muddy brown), but of some biological interest; an inflatable sac in its breast probably helps with temperature regulation, but makes odd clicking sounds when the bird is vocalizing.

These great breeding rarities are a source of excitement to birders, and it is perfectly possible that more will add to their number in due course, especially under the conditions of global warming. If this happens, the Rio Grande Valley will maintain its position as one of the most popular and exciting birding areas in the whole of the United States.

Olympic Peninsula

SITE RANK 49 **Information**

HABITAT Temperate rainforest, upland coniferous forest, mountains, marine shoreline

KEY SPECIES Spotted Owl, Marbled Murrelet, Blue Grouse, Hermit Warbler, Black Brant, Black Oystercatcher, Surfbird, Varied Thrush, Pine Grosbeak, Grey Jay, Tufted Puffin, Rhinoceros Auklet

TIME OF YEAR All year round

Above right: the first Marbled Murrelet nest was only discovered in 1974. Remarkably, this otherwise pelagic species lays its eggs high up on the branches of a forest tree.

The Pacific coast of North America holds a superb selection of birds that can be appreciated in some of the wildest and most freshly wholesome scenery on earth. Few places are better for enjoying this combination than the Olympic Peninsula in Washington State, which lies about as far north as it is possible to go in the lower 48 states. As a bonus, the site also happens to be exceptionally important for two species that have acquired international celebrity status, one for being a conservation icon, and the other for having what is perhaps the oddest and most unexpected nest site of any species in the world.

This large block of land, about 100 sq km, which splinters away from the main coastline west of the city of Seattle and nestles against the southern coast of Vancouver Island, encompasses a variety of habitats, including temperate rainforest, seashore and upland coniferous forest. Of these, the most important habitat in global terms is the old-growth temperate rainforest, a habitat which is in rapid decline elsewhere and holds some birds that are considered threatened. The forest itself is truly magnificent with huge, towering trees, their branches bedecked with swathes of moss and lichen, creating a damp, rather dark atmosphere which, especially when combined with the frequent fogs, is rather surreal. This forest, which fringes

Right: a female Chestnut-backed Chickadee (right) receives food from its mate.

Above: a displaying Blue Grouse; these coastal birds have higher pitched hoots than their interior counterparts.

the coast, is sustained by the very high rainfall of the area, up to nearly 4 m annually. The abundance of the growth, with its high biomass, is also helped by the mild winters that are typical of the area. The forest is dominated by conifers: Sitka Spruce, Douglas Fir, Western Hemlock and the mighty Western Red Cedar, which can live for more than 1,000 years and grow to nearly 50 m in height. Any birdwatcher entering this arena soon feels dwarfed.

It was in forests like these, in 1974, that a great ornithological mystery was finally resolved, with the discovery of the very first verified nest of the Marbled Murrelet. The Marbled Murrelet is a small auk, which spends much of its life within 5 km of the shore, diving for fish and krill and behaving like any normal seabird. Despite being pretty numerous from California to Alaska, however, nobody had ever found a nest up until this point, despite searching all the likely coves, cliffs and rocky islands. So, where is the most unlikely place to look for the nest of such a bird? On the branches of a tree would be a good bet, and that's exactly where they found it, deep in these foggy forests. In a small depression among moss, on a thick branch up to 40 m above ground, this odd little seabird raises its single chick, serenaded by such unlikely bedfellows as the Varied Thrush, Hermit Warbler and Chestnut-backed Chickadee.

The other famous resident of the Olympic Peninsula is the Spotted Owl, a bird that has become a symbol of conflict between economic and conservation interests. In common with the Marbled Murrelet, the northern race of the Spotted Owl occurs in old-growth forests within the Olympic National Park and Olympic National Forest, and has something of an ecological predilection for the very richest parts, where trees are more than 200 years old. Here, cooled by the forest's special microclimate and biological richness, the owl lives an idyllic life munching the Northern Flying Squirrels and Dusky-footed Woodrats which abound here.

Outside protected areas, however, the owl shares its preferences with logging companies, for whom this old-growth forest is gold dust and a means of economic survival in an industry that is otherwise struggling. The two needs are pretty incompatible, since it has been shown that owls only reoccupy selected areas after about 40 years' re-growth, and at present there are no real winners in the

conflict. Several times the federal government has put forth action plans to protect the owl and to maintain the future for the loggers, but as yet there has been no all-party agreement. The Spotted Owl's fate is still in the balance.

Meanwhile, unhelpfully, the closely-related Barred Owl, a relatively common species across North America, has begun to spread into the Olympic Peninsula, adding to the pressures on the Spotted Owl. Barred Owls sometimes hybridize with Spotted Owls, but more often than not they simply drive the more specialized birds away and take over their territories, leading to the owners' demise. It is ironic that a close relative is putting further pressure on the Spotted Owl, alongside outside influences.

Naturally, both the coastal forests and their counterparts inland on the highlands of Olympic National Park hold an excellent selection of other species. These include the splendid Blue Grouse, whose slightly breathy hooting can be heard emanating from the forest depths. Birds in these coastal populations have higher-pitched hoots than their interior counterparts, and they also have a habit of calling from high perches, up to 25 m up, rather than on the ground. In display they inflate a warty patch of yellow skin on the breast sides, effectively by holding their breath. Other good species here include Northern Pygmy Owl, Pine Grosbeak and Grey Jay.

Away from the forests, the seashore habitats on the peninsula are also excellent for birds. Black Oystercatchers nest all along the rocky coasts, while rocky islands around the peninsula hold significant colonies of Brandt's Cormorant and Pelagic Shag, Tufted Puffin and both Rhinoceros and Crested Auklets. At Dungeness National Wildlife Refuge, in the north-west, there are important intertidal habitats providing refuge for migrating and wintering geese and shorebirds, including some 15,000 Brent Geese of the subspecies *nigricans* known as Black Brant. Two special shorebirds of the West Coast, the Surfbird and the Black Turnstone, occur at this site and more widely in the winter and on passage.

Above: Spotted Owl is an iconic species which is fussy in its habitat requirements. It has been used by conservationists to test the federal government's commitment to nature conservation.

Right: the Varied Thrush's song is a vital component of the atmosphere of north-western North America's forests. It sings an eerie succession of single-pitched whistles with long gaps in between, each whistle of different pitch.

Snake River

HABITAT Cliffs along river adjacent to sagebrush and steppe

KEY SPECIES Fifteen species of raptors and owls, including Golden and Bald Eagles, Prairie Falcon and Swainson's and Ferruginous Hawks

TIME OF YEAR All year round, but March to June is best for breeding raptors

It is possible, although difficult to prove, that a 130-km stretch of the Snake River, in the state of Idaho, could house the most concentrated breeding population of birds of prey in the world. Certainly, there is nothing quite like it in North America, and any site hoping to take the title would have to compete with the 800 or so nests that are occupied here each year by nine species of raptors and seven species of owls.

The reason behind the extraordinary richness of this rugged area of the Great Basin is the happy juxtaposition of superb breeding habitat and equally excellent feeding grounds. The Snake River canyon cuts through a plain covered by sagebrush and scrub that supports an exceptionally dense population of small mammals, including the Piute Ground Squirrel and Black-tailed Jackrabbit, as well as Pocket Gophers, Kangaroo Rats and Deer Mice. The walls of the canyon itself were eroded by the floodwaters of an ancient water-body, Lake Bonneville, which burst its banks some 15,000 years ago, creating ledges, cracks and crevices in the basalt and sandstone cliffs. These cliffs, which are up to 250 m high, could have been tailor-made for raptors, as they provide plenty of safe sites for these essentially nervous birds. Thus, the combination of secure nest-sites and plenty of food has, over the years, proved irresistible to these predators.

The most important species in the Snake River Birds of Prey National Conservation Area is the Prairie Falcon. Up to 200 pairs may nest each year in the canyon, at a density of one pair per 0.65 sq km, and nests have been recorded just 25 m apart; nowhere else is there a comparable density of breeding pairs of this North American endemic, with its world population of only about 6,000 pairs. There are also about 50 nests of the scarce Ferrug-

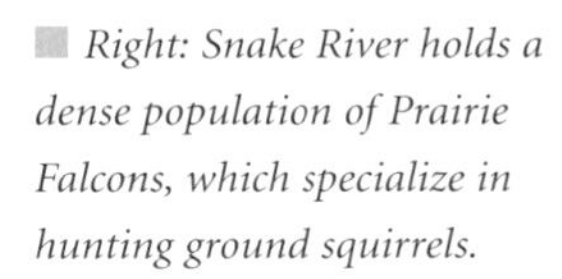
Right: Snake River holds a dense population of Prairie Falcons, which specialize in hunting ground squirrels.

Above: Golden Eagle is a relatively numerous breeding species in the gorge.

inous Hawk, plus good numbers of Swainson's and Red-tailed Hawks, Golden Eagle and American Kestrel. Other day-flying raptors that also breed, but less abundantly, are Northern Harrier, Osprey and Turkey Vulture.

The ecology of the Prairie Falcon gives a fascinating insight into the finely-tuned relationship between prey and predator. The main prey species of the falcon is the Piute Ground Squirrel, which is a small rodent adapted to the dry conditions of the nearby plains. The squirrels are highly seasonal in their life-cycle, appearing above ground only during the first half of the year, when there is an abundant growth of green shoots on which they can feed. They emerge in January, mate the following month and bring forth litters in March; then, for the next three months, adults and young stuff themselves with food, laying down fat reserves for a long period of torpor lasting between July and December. Meanwhile, the Prairie Falcons time their breeding season so that their eggs hatch in April, just as the ground squirrels are abundant and fattening up satisfactorily. It is estimated that the Prairie Falcons of the Snake River consume 50,000 Piute Ground Squirrels every year. However, once the glut of ground squirrels has passed, most falcons leave the area and spread out over a broad swathe of western North America, turning their predatory attentions to Pocket Gophers and various species of small bird.

For their part, the Golden Eagles of the Snake River are more partial to somewhat larger prey, Black-tailed Jack-Rabbits. Indeed, their population fluctuates with the cyclical abundance of their prey. Ferruginous Hawks, on the other hand, take quite a range of prey, from young jack-rabbits to Pocket Gophers and Kangaroo Rats. They will sometimes wait outside the burrows of Pocket Gophers, ready to snatch them as soon as they emerge.

The list of day-flying raptors is enhanced during the winter by the appearance of several other species, including Rough-legged Hawk, Bald Eagle, Merlin and Sharp-shinned and Cooper's Hawks. Some of these species also pass through on migration. In fact, the river is a major flyway for many species of birds, not just birds of prey, but also wildfowl and passerines. Adjoining the Conservation Area is Deer Flat National Wildlife Refuge, where 150,000 ducks and 15,000 geese have been recorded in winter. Mallard, Green-winged Teal, Northern Pintail, Common Merganser and Common Goldeneye are the most abundant species, and some of these provide a different sort of food for raptors wandering from the canyon.

The surrounding sagebrush and scrub is an important habitat for a number of specialized birds, including Sage and Brewer's Sparrows, Sage Thrasher and Long-billed Curlew.

Burrowing Owls also breed on the plains in healthy numbers, one of the seven owl species to nest in the area. In this part of their range these owls tend to appropriate a disused burrow of a Ground Squirrel or Kangaroo Rat for nesting, rather than digging their own, preying on those same animals, as well as a few insects.

Another ground-nesting owl is the Short-eared Owl. Instead of using burrows, this species lays its eggs on the ground, among tall grass or shrubs. In contrast to the Burrowing Owl it rarely, if ever, hunts on the ground, instead flying with wavering, jinking progress a few metres up, checking for movement below, plunging talons-first when it spots something. Its hunting action is similar to that of the Barn Owl, yet another species of the Snake River area, nesting in buildings or tree holes.

In the wooded areas there are four more species of owl: the Great Horned, Long-eared, Northern Saw-whet and Western Screech Owl. Together they complete the remarkable roll-call of the predatory birds of Idaho's Snake River Canyon.

Above: the distinctive Ferruginous Hawk tends to feed on large prey, such as jackrabbits and snakes.

Right: with so many ground-dwelling mammals making burrows, the Burrowing Owl has plenty of nest sites to choose from.

Yellowstone

HABITAT Forest (mainly coniferous) at various altitudes; meadows; sagebrush; lakes and rivers

KEY SPECIES Trumpeter Swan, Great Grey Owl, American White Pelican, Bald Eagle, Black Rosy Finch, woodpeckers

TIME OF YEAR Best in June and July; many roads in the park are closed from October to April

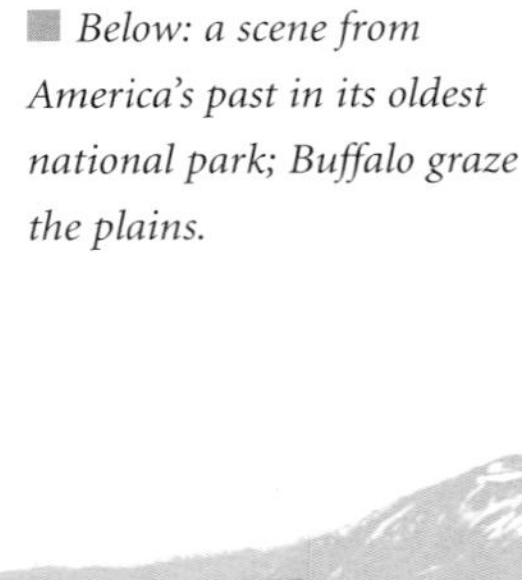

Yellowstone may well be the most famous national park in the world; it is certainly the oldest, having been declared on 1 March 1872. That date should be remembered gratefully by every birdwatcher and conservationist the world over. What began here affects us all and has safeguarded thousands of natural places worldwide.

To most people, Yellowstone National Park conjures up images of big landscapes, large roaming wild animals, and geysers. All of these impressions are accurate. The park is certainly huge, no less than 8,987 sq km in extent, and these days scientists tend to expand the definition and talk about the 'Greater Yellowstone Ecosystem', a network of national parks and national forests centred on north-west Wyoming and contiguous to Yellowstone. If you take in Grand Teton National Park and neighbouring national forests you have about 80,000 sq km of wilderness, roughly delineated by the distribution of Grizzly Bears, which constitutes what is one of the largest areas of pristine habitat in the temperate Northern Hemisphere. On the whole, the area is so undisturbed that scientists use it as a control for studies of how wildlife is affected by human land use.

There are plenty of large animals in the park: Elk, Moose, Bison and Grizzly Bears roam everywhere and, since the 1990s, reintroductions have established a new population of Wolves. The geysers are just a part of an impressive range of geothermal features – the 10,000 here constitute half the number recorded in the world. Yellowstone lies on top of what is succinctly described as a 'supervolcano'. It is still active and is due to blow 'soon', a catastrophe that could well spell the end of human life on earth.

You can take your mind off such cataclysms, however, while observing the ecological differences between American Three-toed and Black-backed Woodpeckers (the former likes denser forests) or Red-naped and Williamson's Sapsuckers (the two prefer subtly different mixes of trees, with the latter more closely tied to conifers). Despite being famous for its vistas and

■ *Below: a scene from America's past in its oldest national park; Buffalo graze the plains.*

scenery, 80 per cent of the national park area is covered by forest, mainly conifers such as Lodgepole Pine. Birders should not come here expecting to swell their life-lists so much as appreciate some great birds in as natural a setting as you can find anywhere.

There are plenty of birds here – 320 species have been recorded – but you will have to work hard for them, and cover a considerable sweep of area and habitats in doing so, sometimes hiking on trails away from the crowds. If you do this, you will soon gain a very good idea of the avifauna of these Rocky Mountain forests. This is one of the best places in North America to see the Great Grey Owl, while the Nearctic race of Tengmalm's Owl, known as Boreal Owl, being more nocturnal, is a trickier task. The conifers hold Blue Grouse, Northern Goshawk, Grey and Steller's Jays, Mountain Chickadee, Swainson's Thrush, Olive-sided and Hammond's Flycatchers, Pine Grosbeak, Cassin's Finch and Red Crossbill, while other types of forest and forest glade provide habitat for Broad-tailed and Calliope Hummingbirds, Dusky Flycatcher, MacGillivray's Warbler and Western Tanager. It will take you quite some time to find them all.

Another key habitat, although accounting for only 5 per cent of the national park area, is wetland. There are plenty of rivers in the park, many of them rushing downhill at high speed, providing habitat for the American Dipper and Harlequin Duck (the Harlequin Duck maintains a population of between 16 and 24 pairs). The park has 290 waterfalls, the highest being 94 m, and the fast-flowing water has cut a number of canyons (including that of the Yellowstone River, which is the park's own version of the Grand Canyon) which provide good breeding habitat for the White-throated Swift and raptors such as the Peregrine Falcon. Yellowstone Lake is the largest high-altitude lake in

Below: Yellowstone is home to a healthy population of Bald Eagles.

North America, with an area of 352 sq km, and there are numerous other lakes and pools, supporting strong populations of Barrow's Goldeneye and Great Northern Diver. One of the area's most famous bird inhabitants is the rare Trumpeter Swan, which often nests in the area of Seven Mile Bridge. From 30 pairs in the late 1990s recent years have seen a decline to two or three pairs; this superb swan, the largest native flying bird of North America, is only just clinging on. Flocks do still pass through on migration; 700 were once recorded.

The Molly Islands in Yellowstone Lake are an important breeding area for waterbirds, especially the American White Pelican. Counts in 2005 amassed 219 successful nests of this species on the two islands, one rocky and one sandy, together with 69 pairs of Double-crested Cormorant and 31 pairs of California Gull. These nests are sometimes raided by Bald Eagles, which thrive here; there were 34 active nests of these predators in 2005, along with 50 or so Osprey nests.

The remaining habitats of the area include meadows, which support a population of Sandhill Cranes, plus large acreages of montane sagebrush, home of such specialists as Sage Grouse, Sage Thrasher and Brewer's Sparrow (all of which easiest to find in Grand Teton National Park). Since the area encompasses altitudes ranging up to 3,462 m, there are some 'islands' of montane tundra. These are limited in their avifauna, but support a strong population of the rare North American endemic Black Rosy Finch. This species, which is found only in isolated pockets within the central Rocky Mountains, spends much time around the edges of ice-patches, feeding on insects that have been blown up from lower altitudes and have become disabled by the cold substrate.

The Black Rosy Finch could one day become the rarest bird in the whole area. Its mountaintop populations, with no refuge in the northern taiga, are exceptionally vulnerable to the global rise in temperature that is predicted in the coming years, and could be prone to extinction.

Above: male Williamson's Sapsucker has a uniform black back. This species is more closely associated with conifer forests than its relative the Red-naped Sapsucker.

Right: Trumpeter Swan is a scarce bird at Yellowstone.

Pribilof Islands

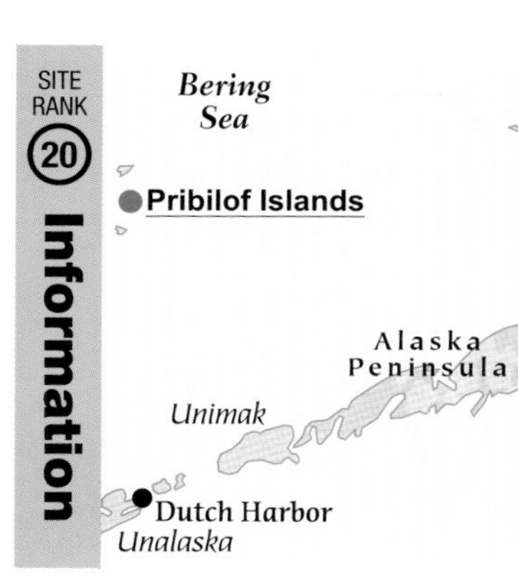

HABITAT Continental shelf islands, sea-cliffs, ocean

KEY SPECIES Red-legged Kittiwake, auks including Least and Parakeet Auklets, Pigeon Guillemot and Tufted and Horned Puffins, Harlequin Duck, Red-faced Shag, Rock Sandpiper, vagrants from Asia

TIME OF YEAR April to August is best

Right: the Tufted Puffin usually breeds on clifftops and grassy slopes.

Alaska is home to the greatest diversity of seabirds in the northern hemisphere, and no fewer than 90 per cent of all the seabirds breeding in the United States are found in this single state. Not surprisingly, therefore, it attracts birders from all over North America and, with eight species endemic to the Bering Sea area thrown in for good measure, it acts as a magnet for visitors from all over the world.

Where exactly is the best place to go, though? Alaska is a big place, with nearly 40,000 km of coast and innumerable islands, including the long Aleutian Chain that sprawls off the map and almost slams into the coast of Russia. Fortunately, there is an easy answer: the Pribilofs, a small group of inhabited islands which are situated 482 km west of the Alaskan mainland, and three hours by plane from the capital, Anchorage. With some three million breeding seabirds, these islands provide a superb spectacle that can be reached in relative comfort.

The Pribilof Islands are isolated from others in the Bering Sea, lying some 386 km north of the Aleutian

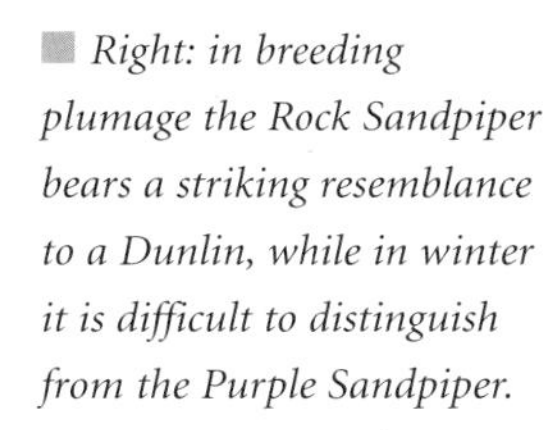

Right: in breeding plumage the Rock Sandpiper bears a striking resemblance to a Dunlin, while in winter it is difficult to distinguish from the Purple Sandpiper.

Above: a true speciality of the Bering Sea, the Red-legged Kittiwake feeds over deeper water than the widespread Black-legged Kittiwake.

Chain. They make up a small archipelago of just five basaltic volcanic islands: two large ones, St Paul and St George, and three small ones, Otter, Walrus and Sea Lion Rock, the latter three all being close to St Paul. Most birders stay on St Paul, which is the largest island, with an area of 104 sq km, and has a small town of the same name, with some visitor accommodation. St George, 70 sq km, is actually a better place for seabirds, with more cliffs, but it is harder to reach (it is almost 100 km from St Paul) and does not have any extra species to offer. The islands have a maritime climate, which usually means wind, rain and cool temperatures, often with fog. They lie within the shallow, food-rich waters of the continental shelf, and are also close to the shelf break of the eastern Bering Sea, where upwellings bring nutrients from the sea bed and make the whole area outstandingly biologically productive. Thus the colonies of birds here are among the largest in the world.

The cast list of breeding seabirds is superb. After all, there are very few other places in the world where you can see nine highly distinctive members of the auk family (Alcids) all together. The commonest are Least Auklets, which are not much bigger than sparrows. These tiny seabirds, the smallest of their family, specialize in foraging for plankton, and carry food for their young in a gular (throat) pouch which may hold 600 tiny bodies after a foraging trip. These auks have short, dark bills, white staring eyes, a single small white plume behind each eye and largely dark upperparts; the underparts can be mottled dark or pale. Least Auklets are quite shallow divers, usually working downstream of an upwelling, fielding drifting crustaceans such as copepods and euphasiids. Crested Auklets, however, which are also numerous here, are much more powerful and deeper divers, capable of working the roaring currents at upwellings, picking off a diverse selection of animals trapped in the flow, including some fish and squid as well as plankton. These birds are much larger than Least Auklets, with all-dark, sooty plumage, a short waxy orange bill and a slightly absurd-looking front-drooping crest. The third auklet that breeds in the Pribilofs is the Parakeet Auklet, which is similar to the Crested, but has white underparts and lacks the crest; its unusual uptilted bill enables it to eat jellyfish tentacles along with plankton.

Two puffins and three guillemots also occur on the islands. Brünnich's and Common Guillemots are both abundant. They are largely dependent on free-swimming fish, although the Brünnich's dives the deeper (a depth of 210 m has been recorded), and has a slightly more varied diet. These two very similar species may share the same cliff ledges, although the Brünnich's Guillemot, which has a diagnostic white streak along its gape line, is more of an Arctic species. The Pigeon Guillemot, mostly black with a large white patch on the wing, avoids narrow ledges, nesting in crevices or burrows instead. It has a wide diet which is obtained close to shore, among shallow rocky waters. The two puffins are also cavity nesters. The commoner Horned Puffin has white underparts, a yellow and orange bill and a peculiar fleshy protuberance (the 'horn') above its eye, while the Tufted Puffin is all black, with a white face, a pale staring eye and a straw-coloured wispy, curly 'wig' at the back of its head. Both puffins are fish-eaters, although neither dives as deep as the Brünnich's or Common Guillemots. The Tufted Puffin breeds mainly on grassy slopes, while the Horned Puffin tends to use more rocky areas.

Finally, one more alcid, the smart grey, black and white Ancient Murrelet, has also recently begun breeding on the Pribilofs. It visits its breeding sites only at night, making it the most difficult of the set to see.

Another, unrelated species that is active at night is also one of the most sought-after species in the Pribilofs. The Red-legged Kittiwake is something of a speciality to these waters, with 80 per cent of the world population breeding on St George (some 100,000 pairs). The more widespread Black-legged Kittiwake occurs on the islands, too, giving birders the perfect opportunity to

Right: breeding on ledges rather than in burrows, the Horned Puffin is ecologically separated from the larger Tufted Puffin. It also has a broader diet.

observe the rarer bird's darker mantle, short red legs and shorter bill. The Red-legged Kittiwake also has a larger eye than its relative, indicative of its habit of feeding at night on marine organisms, such as squid, that drift closer to the surface in the darkness. These birds feed over deeper water than Black-legged Kittiwakes, and it is thought that their smaller clutch size, one egg compared to two or three, is related to their need to travel further to find food for the young.

Another localized speciality found here is the Red-faced Shag. Confined to the Bering Sea, it is closely related to the Pelagic Shag, which also occurs, but has a large patch of red skin on the face, surrounding the eye. Both species search for fish and invertebrates around submerged rocks.

Other seabirds occurring on the Pribilofs include the widespread Northern Fulmar (most are of the Arctic 'blue' morph), various gulls and the splendid Harlequin Duck. On the tundra Rock Sandpiper is a speciality, along with Least Sandpiper and Red-necked Phalarope, while Snow Buntings and Grey-crowned Rosy Finches make up the main passerine count.

If you need yet another reason to visit the Pribilofs, they also happen to be an excellent place in spring for rare migrants from both east and west. While species from the North American mainland can be rare visitors – St Paul had its first record of Western Wood Pewee in 2006, for instance – it is the Old World wanderers that tend to excite American birders. These include regular Wood and Common Sandpipers, ducks such as Tufted Duck and Smew, plus passerines such as Brambling and Eyebrowed Thrush; mainly standard fare to European or Asian birdwatchers, but as thrilling to North American enthusiasts as the teeming seabirds.

Right: Least Auklets are notable for their odd trilling calls.

Cape May

SITE RANK 80 **Information**

HABITAT Seashore, coastal lagoons, freshwater marshes and meadows, woodland, scrub

KEY SPECIES Impressive migration of landbirds and waterbirds of many species, Black Skimmer, Piping Plover, Clapper Rail, Yellow-breasted Chat, Blue Grosbeak

TIME OF YEAR All year round, but best during the of migration periods [April to May and September to October]

Above right: Cape May is one of the most famous coastal migrations watchpoints in the whole of North America.

Below: Black Skimmer fishing. The lower mandible is a third longer than the upper, and is used to skim the surface of the water. If a fish is struck, the bill snaps shut.

What can you say about a single site in temperate North America with a list of 388 species? It must be good, obviously, and it must also attract a lot of migrants. That, however, is only half the story of Cape May, New Jersey, which is one of the finest sites for birding in the whole of North America. This place has everything: excellent breeding birds, including several rarities; exciting landbird migration; spectacular numbers of shorebirds; great sea-watching; major concentrations of migrating hawks; there are even some interesting wintering birds to reward those who brave the cold at that season. Cape May has a bird observatory with dedicated staff offering on-site help for visitors, and there is a continuing programme of bird walks and seminars to cater for all ornithological tastes. Cap that with ease of access and good facilities, and you can see why Cape May is firmly placed at the hub of East Coast birding.

Perhaps the best way to appreciate Cape May as a birding centre is to follow typical events through the year. The first sign of spring occurs in February, when there is a large build-up of divers (mainly Red-throated Divers) off Cape May Point, on the north side of Delaware Bay, together with a few ducks such as Long-

■ *Right: Piping Plover is a rare and threatened species that breeds on Cape May's beaches.*

■ *Below right: Blue Grosbeak, uncommon and at the limit of its range this far north, breeds in scrubby areas.*

■ *Below: May is a good month to find a migrant Prothonotary Warbler.*

tailed Ducks, Black and Surf Scoters and Buffleheads. Late in the month the first migrant raptors, mainly Northern Harriers, pass through over the area, in small, scattered movements. By March the sea can be almost covered with divers, with several hundred in view at once, including a few Great Northern Divers; a visit out to sea on a fishing boat can also be good, with possible sightings of Black-legged Kittiwakes and Northern Gannets. The changing season is indicated by the arrival of the first migrant Ospreys over the sea, while the first Great Egrets and Black-crowned Night Herons appear in the creeks and marshes.

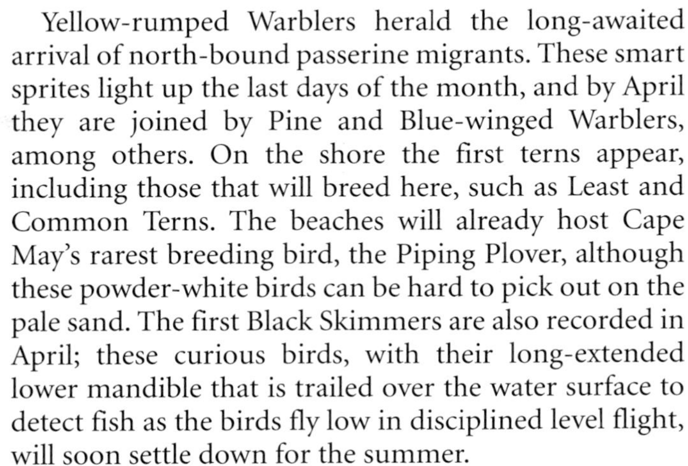

Yellow-rumped Warblers herald the long-awaited arrival of north-bound passerine migrants. These smart sprites light up the last days of the month, and by April they are joined by Pine and Blue-winged Warblers, among others. On the shore the first terns appear, including those that will breed here, such as Least and Common Terns. The beaches will already host Cape May's rarest breeding bird, the Piping Plover, although these powder-white birds can be hard to pick out on the pale sand. The first Black Skimmers are also recorded in April; these curious birds, with their long-extended lower mandible that is trailed over the water surface to detect fish as the birds fly low in disciplined level flight, will soon settle down for the summer.

May is a classic month for landbirds, not just because of the large numbers of migrant individuals, but also because the birds are in their spring finery and, if they are males, will be singing. This is a great time to get to grips with songs and, if you possess this skill, you may be able to pick up prizes such as Swainson's, Yellow-throated and Prothonotary Warblers, as well as a host of vireos, thrushes and flycatchers. Uncommon species such as Orchard Oriole and Summer Tanager are most likely to be seen at Cape May in May. The best conditions for seeing large numbers of north-bound birds in spring will be overcast days following south-westerly winds.

Much as the landbirds make Cape May famous, the shorebird numbers are also impressive in spring, as many thousands of birds rest here before the final push to their breeding grounds. It is estimated that more than

Right: it usually takes south-westerly winds during the migration seasons to bring the uncommon Summer Tanager to Cape May.

one million shorebirds pass through in May alone, including large numberes of species such as Red Knot, Short-billed Dowitcher and Grey Plover. On the beaches, shorebirds such as Sanderling and Ruddy Turnstone tuck into what is something of a speciality gourmet meal for this location, the glut of eggs of Horseshoe Crabs, which beach here at night in their millions.

June is a good time to appreciate the breeding birds of the area, such as on the marshes, where a good mix includes Tricoloured and Little Blue Herons, Least and American Bitterns, Marsh Wren, Virginia Rail and Sora on the fresher marshes around the migratory bird refuge (South Cape May Meadows), and Clapper Rail on the saltier areas. The woodlands around Higbee Beach Wildlife Management Area offer an eclectic mix, including everything from Yellow-breasted Chat and Blue Grosbeak (for which Cape May is noted) to Wild Turkey and American Woodcock.

July and August see the beginning of the main autumn migration. The first southward travellers to reappear are often shorebirds, some of which have completed rapid breeding duties and have left their mates up in the tundra to rear young alone. Owing to the short Arctic season, however, the shorebird season peaks at the end of August, at least a month before the main landbird passage. Nonetheless, early outbound land migrants such as Blue-grey Gnatcatchers and orioles are common in August and have largely passed through by mid-September.

September, though, is the busiest season of the year at Cape May, both for the birds themselves and for visitors. If a cold front passes, particularly one bringing north-westerly winds, the birds can arrive in large waves known as 'fallouts', and the favoured spots may play host to large crowds of birders as well as birds. In the autumn there have been estimates of 100,000 warblers at the Higbee Beach area alone in a single morning, which is enough to fulfil any birder's dreams. Any wind from south-west to north-east may funnel birds to the coast and down to Cape May, where the birds suddenly face the 10-km crossing of Delaware Bay if they are to continue southward. Along with the dozens of expected species, rarities are inevitable, and Cape May has an outrageous list of these, including birds from all corners of North America.

The emphasis shifts slightly in October, with fewer warblers but more sparrows, thrushes and other odds and ends refuelling in the woods and scrub. Wildfowl begin to build up, too, but what is really going strong is the hawk-watching. It seems unfair that Cape May is great for hawks as well as everything else, but over 20,000 have been seen in a single day, and 40,000 in a fall season is not unusual. There is a special hawk-watching platform at Cape May Point State Park, which is staffed by experts to help you along. Although most of the birds recorded are Sharp-shinned Hawks, there is an excellent array of other species. In particular, the remarkable passage of Ospreys (2,110 in the 2006 season), Peregrine Falcons (1,113) and Merlins (1,884) is probably the largest in the world.

Naturally, things tail off in November and December, but there are still good birds to see, such as Purple Sandpipers at the point and Short-eared Owls in the marshes. At Cape May, there is something interesting to see at any time of year.

Hawk Mountain

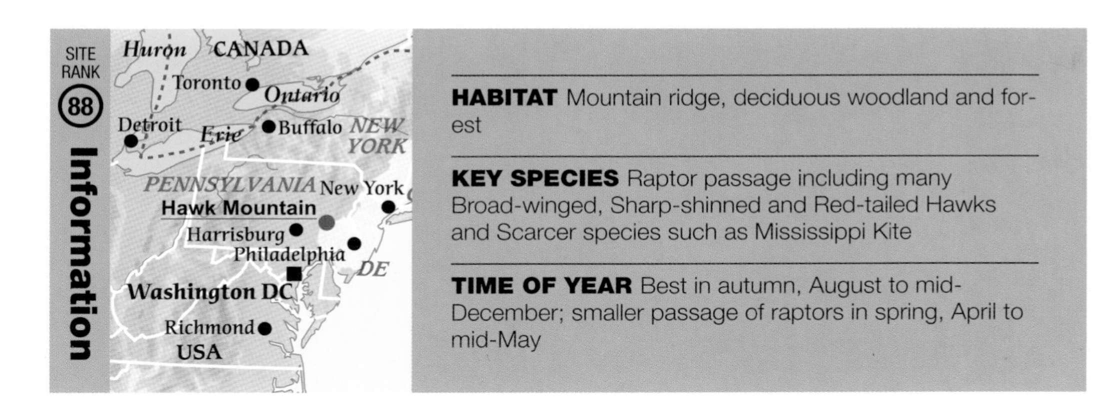

HABITAT Mountain ridge, deciduous woodland and forest

KEY SPECIES Raptor passage including many Broad-winged, Sharp-shinned and Red-tailed Hawks and Scarcer species such as Mississippi Kite

TIME OF YEAR Best in autumn, August to mid-December; smaller passage of raptors in spring, April to mid-May

The Kittatinny Ridge is a finger of high ground running for 485 km north to south within the Appalachian Mountain range, on the eastern side of North America. Positioned along this ridge lies Hawk Mountain, one of North America's most famous wildlife sites. Geography, combined with the fact that migrating raptors are basically lazy, makes this one of the best places on the continent, and indeed in the world, to watch large numbers of eagles, hawks, vultures and falcons passing by.

Raptors tend to be comparatively heavy birds, for whom migration is more of a drain than for most. Many are simply unable to cover large distances if they have to maintain energy-sapping flapping flight for any length of time. Instead, these birds tend to hitch a ride on suitable air-currents, either thermals – spirals of air heated by the ground, circulating upwards – or updrafts caused by wind passing over high ground. During the autumn, thermals are not dependable, so the birds tend to travel on a narrow front following mountain ridges; thus, in the east of North America, there is an Appalachian Flyway, a sort of major route for migrants. Hawk Mountain lies along this route, and is especially favoured because it lies on the eastern side of the mountain range, and the prevailing wind is north-west, typically funnelling the birds right past it.

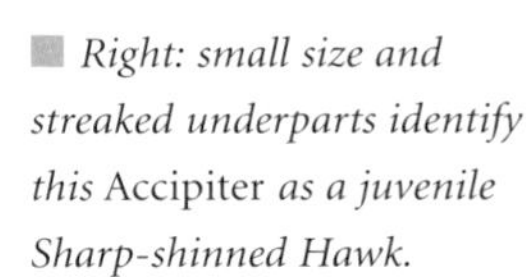

Right: small size and streaked underparts identify this Accipiter *as a juvenile Sharp-shinned Hawk.*

■ *Above: Broad-winged Hawk is recorded in large numbers at Hawk Mountain each autumn.*

Hawk-watching has been going on here since 1934, when a certain Rosalie Edge, a pioneering conservationist from New York City, bought the site and surrounding area (the sanctuary now covers 1,050 ha). This means that there are counts covering some 69 years (allowing for a three-year break during World War II), the world's longest record of raptor counts, and an invaluable reference on how populations have changed. Many species have decreased (the Red-shouldered Hawk, for example) but two, now frequently seen, are comparatively new arrivals, as their respective ranges have advanced northwards in the last few decades. These are the Turkey and American Black Vultures; the most recent ten-year average for these species is 206 and 46 birds per season respectively, but 2005 was a record year for the latter, with 114 birds seen.

The most numerous species at Hawk Mountain is the Broad-winged Hawk (about 6,000 birds per fall season), followed by the Sharp-shinned Hawk (5,000) and the Red-tailed Hawk (3,700). Slightly less numerous but regular species include cut Osprey, American Kestrel, Bald Eagle, Northern Harrier and Cooper's Hawk, the latter usually turning up in singletons and providing an excellent opportunity for comparison with the Sharp-shinned. The keener birders are inevitably enticed by the prospect of rarities, of which only one or two a day might be seen: these include Golden Eagle (94 birds seen per year), Merlin (127), Peregrine Falcon (43) and Northern Goshawk (78). Ten or more species a day is a routine total, ensuring that up to 60,000 people visit Hawk Mountain every year, mostly armed with binoculars.

The migration period lasts from the beginning of August until mid-December, with a few records dribbling in up until mid-January. Different species are numerous at different times: Broad-winged Hawks peak

■ *Above: once a great rarity, Turkey Vulture is now seen regularly at Hawk Mountain, reflecting its overall increase in range and numbers in the north-east of North America.*

in mid-September, when several hundred may pass in a day, but their passage has virtually petered out by the beginning of October. Red-tailed Hawks, on the other hand, peak much later, in mid-October to early November. Sharp-shinned Hawks have a much longer season, and in fact are seen almost every day until the end of November. Although rarities may be seen at any time, early November (for such species as Golden Eagle and Northern Goshawk) is the best bet.

Hawk Mountain is a convenient place to bird. It is easy to find and has all the facilities one expects in the United States, including a spacious visitor centre and bookshop. In the autumn migration season there are designated hawk-spotters who are stationed on the North and South Lookouts and will call the hawks out as they come over, explaining the differences between them. A trail fee of a few dollars provides you with a map of the site and a chart illustrating the possible birds and, as a further nice touch, the spotters will also hold up wooden silhouettes of each species.

For a birder it is important to get the weather right. Although a day rarely goes by without something being seen, there can be very sharp contrasts even between consecutive days. To take an example: on 12 September 2006, 7,584 birds passed by, mainly Broad-winged Hawks, but also eight other species, whereas on the following three days not a single bird was seen, as the site was fogbound. Therefore birders should, if possible, time visits to coincide with a couple of days after a cold weather front has passed through, leaving north-west winds in its wake. If this happens, constant streams of birds will appear in front the North Lookout, many of which can be seen soaring upwards in groups (known as 'kettles'); at times there will be wheeling birds in every part of the sky.

In the spring the raptor passage is nowhere near as impressive as in the autumn. The prevailing winds are easterly and keep the travelling birds to a more westerly route. Nevertheless, patient counters regularly watch between late March and the middle of May, and have recorded such totals as 540 Broad-winged Hawks and 170 Sharp-shinned Hawks. The spring also brings visitors a slender chance of seeing a Mississippi Kite, which is rare in these parts.

Spring is also a time to appreciate the other birds that occur at Hawk Mountain. About 150 species breed at the sanctuary each year, mostly associated with the mature oaks, maple and birches. These include an interesting mixture of northern species such as Hermit Thrush and southern species such as Worm-eating Warbler.

High Island

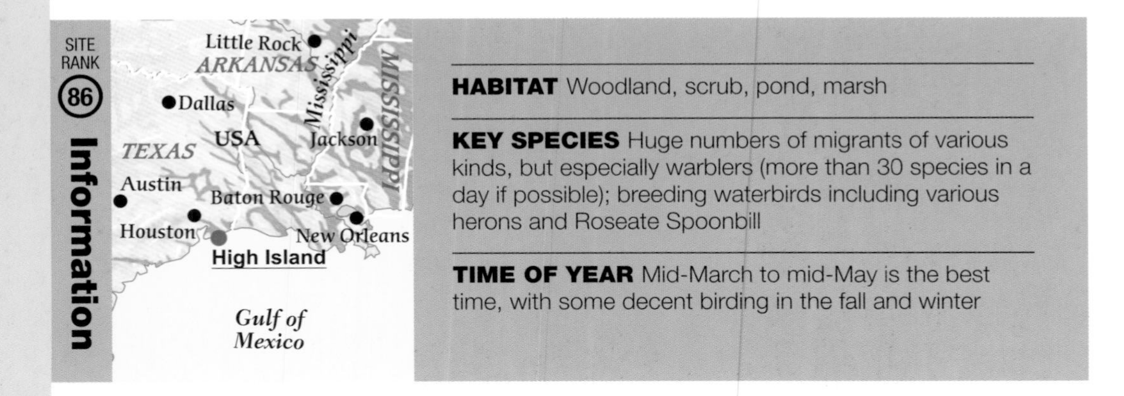

HABITAT Woodland, scrub, pond, marsh

KEY SPECIES Huge numbers of migrants of various kinds, but especially warblers (more than 30 species in a day if possible); breeding waterbirds including various herons and Roseate Spoonbill

TIME OF YEAR Mid-March to mid-May is the best time, with some decent birding in the fall and winter

Whoever gave High Island its name certainly had a sense of humour, because this small wooded rise on the Texas coast is neither an island nor is it very high. In fact it is the skin of a subterranean salt dome that lies close to the land's surface and lifts the ground very slightly above the marshy plain upon which it lies. This creates little more than a pimple on the flat coastline, but it provides the only opportunity for miles around for trees to take root and grow.

The ironic name is not very funny, but High Island has a big reputation for putting smiles on birders' faces. Many is the visitor who has come away from this modest-looking place blabbering that it has provided the best single birding experience of his or her entire life. High Island is consistently in everybody's top ten birding sites in the USA. Yet if you visit in the summer you will be completely bewildered by such accolades, since the best you will see is plenty of waterbirds and a lot of Northern Cardinals and Blue Jays. Spring is the time to come.

High Island's reputation is built on its magnetic attraction for northbound migrants. It lies just over 1 km from the Gulf Coast of Texas, on a major migration route for birds that have wintered in Mexico or South America and are heading north to breed in the forests of eastern North America. In spring, millions of travelling birds are funnelled through to the northern tip of the Yucatan Peninsula in Mexico and from there they set out on a 1,000-km flight across the Gulf of Mexico to Texas. In the right conditions many thousands of birds will be undertaking this journey simultaneously.

If you are a bird approaching from a height, you will see High Island as a small patch of dark green lushness amid a vast sea of grassy marsh. After your 18-hour flight, you could be forgiven for wanting to drop down and take a rest and a feed the moment you sight land; and since High Island is the only joint in town, it is here that you will end up. Thus, on a typical spring day, there will always be some migrants foraging in the woodlands here.

However, these 'normal' conditions are not what really excite birders. At times during the migration season, strong cold fronts cross the Texas coast and enter the Gulf during the daytime, and when this happens they bring strong northerly winds and rain behind them. Suddenly, things change for the worse for the migrants, who are still flying over the water. Approaching the end of their marathon journey, with reserves already depleted, they hit strong headwinds and rain and their routine flight becomes an emergency. The journey becomes a struggle for survival, and many perish. Those that do make it, however, often make landfall in a state of exhaustion, and it is here that High Island comes into its own for both birders and birds. The birds gratefully flop down into shelter and rich feeding grounds, and the birders thrill to a phenomenon known as a 'fall-out'.

Although fall-outs are bad for the birds themselves, there is no doubt that they are among the most exciting events for a birder. The name is apt, because birds may literally fall out of the sky in a veritable shower, and

■ Right: 'fallout' conditions can bring migrant Grey Catbirds to High Island, often in large numbers.

■ Opposite: if birds are exhausted after their journey, normally shy species can give fantastic views to birders. This is a male Rose-breasted Grosbeak.

bushes that were empty a few minutes before can almost instantaneously become full of colourful sprites. You can see as many birds together in a single view as you would encounter in thousands of square kilometres of their breeding habitat, and of course there will also be dozens of species all showing at the same time. It is not uncommon in spring for 30 species of American wood warblers to be on High Island simultaneously; you could see Grey-cheeked, Swainson's and Hermit Thrushes, plus Veery, feeding alongside each other; you can see 20 or 30 Scarlet Tanagers or Rose-breasted Grosbeaks in the same tree, and even more Grey Catbirds in the

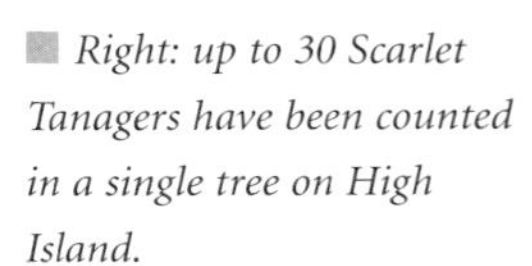

■ Right: up to 30 Scarlet Tanagers have been counted in a single tree on High Island.

■ *Above: as a sideshow to the migration-watching, more than 80 pairs of Roseate Spoonbills breed at Smith Oaks.*

same patch of scrub. All the while, the birds are too exhausted to care about what danger birders might be to them, so they go about their business within a few metres of enthralled watchers.

A quirk of High Island is that, since the trip from Mexico takes a night and half a day, the main arrivals of birds here are not at dawn, as at most migrant traps in the world, but usually from midday onwards. A birder arriving too early will receive amused looks from the volunteer rangers.

The season at High Island begins in mid-March and ends in mid-May. During that time, the volunteer guides are always present in the main wooded areas, Boy Scout Woods (Lewis Smith Woods) and Smith Oaks Bird Sanctuary, both of which are owned and run by the Houston Audubon Society. The former site consists of groves of hackberry and oak trees, and it has an excellent boardwalk and trail that cater for wheelchair users. Birds can perform to the gallery here, because there literally are galleries, tiered rows of seats that overlook certain hot-spots; this may be the only place in the world with such an arrangement for general birding. Meanwhile, Smith Oaks is larger and the trees are older. This sanctuary also has a pond with a truly splendid waterbird colony on an island; in 2006 there were 190 pairs of Great Egrets here, plus 97 pairs of Snowy Egrets, 67 of Cattle Egrets, 35 of Tricoloured Herons, one of Little Blue Heron, three of Black-crowned Night Herons, 69 of Neotropic Cormorants and no fewer than 86 of Roseate Spoonbills. They make great viewing when the birding is slow.

The season at High Island begins with Black-and-white Warblers in mid-March and ends with Mourning Warblers in mid-May. April is the peak month but autumn can be excellent, although it is never as spectacular or as showy as in the spring. High Island, therefore, is very much a seasonal hot-spot – but what a hot-spot it is.

Everglades

SITE RANK 16 **Information**

HABITAT Freshwater marshland, swamp, estuary, mangroves

KEY SPECIES Snail Kite, Limpkin, Wood Stork, Anhinga, Reddish Egret, Short-tailed Hawk, Mottled Duck, White-crowned Pigeon, Mangrove Cuckoo, Seaside Sparrow

TIME OF YEAR All year round

Above right: although it looks still, all the water in the Everglades runs slowly south towards the Gulf of Mexico at a speed of about 30 m per day.

Right: the snail-eating Limpkin is an Everglades speciality.

Southern Florida could be thought of as the toe of North America dipping into the edge of the tropics. As such, it famously draws vast numbers of people from further north to bask in its balmy climate. For birders, the attraction of Florida is quite different; as a subtropical outpost the state holds a range of species found nowhere else on the continent, thus making it one of North America's most enduring birding destinations.

On the south-western corner of the peninsula of Florida lies a huge wetland complex of marshes, swamps and mangroves known as the Everglades. Even without its enticing geographic location, this would be a special place. Its sheer size, over 6,000 sq km, and its relatively undisturbed nature, with only one main road and innumerable inaccessible corners, make it essentially unique. The ornithological treasures of the Everglades – its eclectic mix of northern and southern species breeding together in large numbers – cannot be replicated anywhere else in the world.

The Everglade ecosystem is actually a very complex one. Although much of the area looks like a huge flat marsh grown with saw grass, it is perhaps more accurately thought of as a vast, sluggish river. In the summer, rainwater falls in the Kissimmee River Basin in central Florida and fills up the 1,890-sq-km Lake Okeechobee. The excess water flows in a wide band down through the Everglades, travelling at no more than 30 m a day, and eventually empties into the Gulf of Mexico. Here there are different habitats, including hammocks (small rises in the land overgrown with hardwoods), and limestone ridges interrupt the flow, but the system is essentially one of water circulation. Furthermore, as winter progresses, the flooded land dries out and concentrates the fish and other aquatic creatures into the deeper parts of the wetland, creating a network of different habitats.

Much of this world heritage site now comprises the Everglades National Park, one of Florida's many popular tourist attractions. With most of the park pretty much inaccessible, birders are confined to sections along the access road, which runs from Florida City to the resort of Flamingo, as well as several areas on the north

side along Route 41. Most of these sections are fitted with boardwalks, trails, towers and other facilities; some routes, such as the celebrated Anhinga Trail, are famous birding destinations in their own right. All of the localized species can be seen from one trail or another.

Not surprisingly, it is the wetland species that are the most conspicuous. The Anhinga Trail will introduce you to the most obvious ones, many of which allow a ridiculously close approach and make photography a breeze. Not least of these, of course, is the Anhinga, a cormorant-like bird that is adept at controlling its buoyancy and can thus swim half submerged, with only its head and neck showing. The Anhinga, in contrast to cormorants and herons, routinely catches fish by impaling them with its bill; it will also take more exotic fare, such as young alligators and smaller turtles. Equally obvious is the gawky Purple Gallinule, a large, brightly coloured rail that uses its long toes to wander over marshy vegetation, including lily-pads. You will also be unable to miss the many long-legged wading birds, such as the White Ibis and herons of all sizes, from the tiny Green Heron to the Great Blue Heron. Some of the Great Blues in the Everglades are actually, perversely, white; this rare morph, sometimes dubbed the Great White Heron, is something of a local speciality. So is the Wood Stork, North America's only stork, a huge fish-eating species that usually nests high up in groves of cypress trees. In common with several local species, it breeds in the winter, as water levels dip and

■ *Above: although spectacular, the rather inelegant plunge-dive of the Brown Pelican has been unkindly likened to throwing a stack of washing onto the water.*

■ *Opposite top: the Reddish Egret is famous for its balletic chases after skittish fish.*

■ *Opposite: walking on floating vegetation is no problem for the Purple Gallinule, thanks to its huge feet.*

make fish more readily available in the diminishing pools.

Two more localized species of the saw grass sloughs share a highly unusual diet – apple snails. These are the Snail Kite, a blue-grey coloured bird of prey that quarters low over the marshes and swoops down to snatch its prey, and the white-spotted brown Limpkin, a somewhat ibis-like bird which obtains its snails from the water as it wades around: two quite unrelated species using different hunting methods to the same end.

The Everglades are generally a good place for birds of prey. Ospreys are commonly seen diving down to catch fish; Swallow-tailed Kites, becoming rarer and rarer in the rest of the southern USA, are still numerous here, snatching insects in mid-air in expert aerial manoeuvres. With effort you might spot a real tropical rarity, the Short-tailed Hawk. Often soaring higher than the very common Red-tailed Hawks and Turkey Vultures, this *Buteo* frequently dives down to snatch birds from the treetops.

On the seaward side of the freshwater marshes, mangroves hold sway and support a good range of unusual species, as well as providing nest sites for many of the herons and egrets. Two species in particular attract the attention of birders here, the White-crowned Pigeon and the Mangrove Cuckoo. The Snake Bight Trail, near West Lake, is a good place to find them, although the effort is often torturous. Both species, especially the cuckoo, which seems to be drawn to the impenetrable thickets, can be pretty hard to track down, and in order to look you have to brave the area's infamous mosquitoes, which delight in making your life intolerable. Whether seeing a dark pigeon with a snowy white cap makes the ordeal worthwhile is a moot point.

At the edge of the sea the estuarine flats hold yet another guild of birds. Members include a number of distinctive species, including Black Skimmer, Brown Pelican and Reddish Egret. This latter species, which also occurs along the Gulf Coast, is a real fidget when it comes to feeding in the shallows; it often runs after fish, or lifts its wings to make shade, tending to shun the patient approach adopted by most large herons. Knowing that some Great Blue Herons in Florida are white, it probably won't surprise you that some Reddish Egrets, which are normally salmon-coloured on the head and neck, are also pure white.

Talking of pink birds, south Florida also holds a small flock of American Flamingos, the only ones on the North American continent. They are often found near their eponymous visitor centre in the Everglades. In years long past flamingos used to visit Florida from the Bahamas, but the origins of the current crop are suspect. They may have escaped from zoos – after all, half the birds in Miami seem to be escaped cagebirds, so why not the flamingos? Try not to confuse them with the equally pink Roseate Spoonbills; driven beyond sensible judgment by the mosquitoes, it could happen.

Alaka'i wilderness

SITE RANK 6 **Information**

HABITAT Montane forest and bog

KEY SPECIES [Kauai] Elepaio, Puaiohi, Kauai Amakihi, Anianiau, Akekee, Akikiki, Iiwi, Apapane

TIME OF YEAR All year round

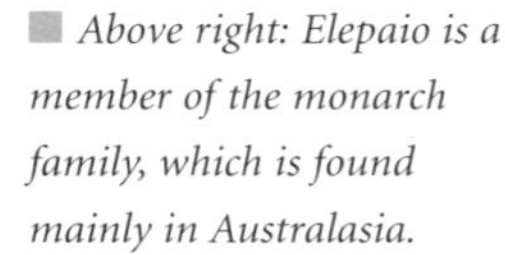

Above right: Elepaio is a member of the monarch family, which is found mainly in Australasia.

The 50th state of the USA, Hawaii, is situated more than 3,000 km from the nearest landmass, in the middle of the North Pacific Ocean, making it the most isolated archipelago in the world. Not surprisingly for such a remote place, the birdlife is very special indeed, containing, among the land dwellers, almost nothing but endemic forms. These forms include one completely endemic family, the Hawaiian honeycreepers, which present perhaps the most complete example of avian adaptive radiation in the world. From a single colonization, the honeycreepers' finch-like ancestor evolved into a family of birds showing profound differences in bill shape, colour, vocalization and ecology – indeed, a much more impressive variety of forms than the better known Darwin's finches of the Galapagos. Today it is still possible to see some of these wonderful birds, but every meeting is also tinged with regret. For Hawaii, as well as presenting a wondrous cradle of evolution, has also become a desperate battleground where species are fighting for their very existence.

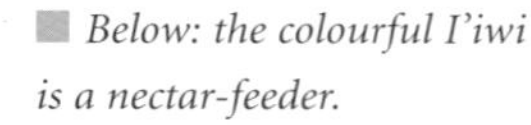

Below: the colourful I'iwi is a nectar-feeder.

Nowhere is the war fought more fiercely than in the highlands of Kaua'i, the westernmost of the main islands. This is one of the wettest places in the world; indeed, Mount Waialeale experiences rain for 360 days a year, adding up to an annual total of 10 m. On the west side of this island lies Koke'e National Park and the adjacent Alaka'i Wilderness Area, a large expanse of alpine forest that still retains high proportion of its original flora and provides a last refuge for much of the island's birdlife. It is a remote and lonely place, with long trails that go through thick forest and traverse ravines; there is also the so-called Alaka'i Swamp, a bog with spongy ground and almost impenetrable vegetation. It resembles, and indeed is, something of a last frontier.

The dedicated birder will be able to find up to six native bird species here, while the fanatic may find two more with a great deal of effort. By taking the popular Pihea Ridge trail any visitor should soon come across several species of 'dreps', the affectionate term used by local birders for Hawaiian honeycreepers, the Drepanididae. The commonest are the Apapane, the Kaua'i Amakihi, the Anianiau and the Akekee. The Apapane is a delightful nectarivorous tit-sized songbird, a bright red species with a long, sharp bill. It can be seen almost everywhere, often guzzling its favourite red o'hia blooms (this eucalypt being one of the dominant plants of the forest). This is the most abundant of all the dreps, being found on most of the main islands. The Amakihi, by contrast, is a small, yellow-and-green species with a short, decurved bill; it gleans insects from branches and trunks, as well as also taking some nectar. This is another fairly common species, the only one that also occurs in the lower parts of the island. The Anianiau is similar to the Amakihi but much brighter yellow all over, and with a very narrow bill that is used to probe for nectar into the o'hia blooms; in common with the other nectar-feeding dreps it has a brush-like tongue that is also effective for catching tiny insects. Besides foraging in o'hia trees, Anianiaus also nest in them, as does the Akekee, yet another small yellowish bird which, in contrast to the others, appears to shun nectar in favour of an invertebrate diet, and has a comparatively short bill.

Above: parts of Hawaii are among the wettest places on earth.

Right: the Kaua'i Amakihi is mainly insectivorous; it forages by gleaning foliage.

Another common bird here is the Elepaio, yellowish-brown with two pale wing-bars, another native and a member of the monarch family. Less common, but still likely to be seen, is everybody's favourite drep, the stunning Iiwi. This gorgeous bird is brilliant red in colour, with black wings and white spots on the inner secondaries; its bill is sharply decurved, betraying its nectar-feeding habit. Although common on some other islands, this species appears to be declining here.

By being a bit more adventurous, the birder can also seek out two great rarities. These are the Puaiohi, a rather dull-plumaged thrush, mainly found in the swamp area, and the Akikiki, a drep that is the Hawaiian equivalent of a nuthatch, finding its invertebrate food by foraging unobtrusively along trunks. This sharp-billed species, grey-brown above and whitish below, appears to be in severe decline and is now very difficult to find.

Soon the Akikiki will probably be impossible to find. Its decline seems to be remorseless and its range is contracting towards the east of the site, which has ominous parallels with several other species from this part

Above: an Apapane guzzles from its favourite O'hia blooms.

of Kaua'i that have now become extinct – unless somewhere, somehow, deep in this inhospitable terrain, they still cling on to survival. The sad fact is that this very site, where nowadays visitors can be stunned by the sight of the Iiwi and Apapane, has lost more birds to extinction in the last 25 years than anywhere else on earth. The Greater Akioloa was last seen in 1967 and was the forerunner of several species lost in quick succession: the last confirmed Kauai Oo, a male, was seen in 1985; the last Ou in 1989; the last Kamao in 1993 and the last [Kauai] Nukupuu in 1998. Almost all of them were last seen around the area of the Alaka'i Swamp.

The reasons for this sorry state of affairs include most of the usual blights on restricted island species. Deforestation and development reduce the amount of available habitat. Introduced plants compromise the native vegetation. Introduced animals – in particular here, pigs – further denude and disrupt the habitat. Introduced birds (there are more species of them in Hawaii than anywhere else on earth, and they reach even these montane forests) out-compete the natives for food. Introduced predators, such as rats, attack nests.

In Hawaii, however, there are other threats, too, which are even more insidious. In the archipelago as a rule, the main wave of local bird extinctions began with the introduction of mosquitoes. Carried by introduced birds, avian malaria and pox killed thousands of dreps and other native birds, and at a stroke confined them all to regions above 1,219 m, where mosquitoes could not reach. At the same time, introduced ants and wasps have predated young birds or competed for nectar; the local birds have had no defence against these alien creatures.

Worryingly, there are signs that mosquitoes are beginning to reach the higher altitudes of Kaua'i. If they do, it is difficult to see how some of the native birds can survive. If you would like to see some of these special birds, you should probably visit soon.

Niagara Falls

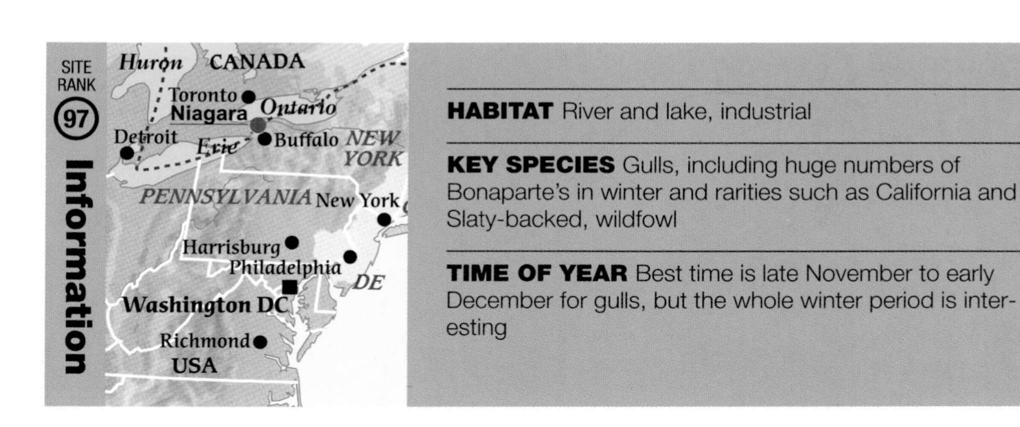

HABITAT River and lake, industrial

KEY SPECIES Gulls, including huge numbers of Bonaparte's in winter and rarities such as California and Slaty-backed, wildfowl

TIME OF YEAR Best time is late November to early December for gulls, but the whole winter period is interesting

To the average visitor Niagara Falls conjures up words such as 'awesome', 'amazing' and 'majestic'. To bird enthusiasts, it can be summed up in one word – gulls. The Niagara River, which runs south to north for 56 km from Lake Erie to Lake Ontario, is one of the best places in the world for observing these birds. In the early winter there may be 100,000 gulls along this fast-flowing river, and at least 20 different species have been recorded here (the exact number depends upon which taxonomy you follow). Amazingly, 14 species have been seen in a single day, which is surely a world record. So, if your thing is tertial steps, gonydeal angles and fourth-winter plumage conundrums, you won't find anywhere else that offers such complete bliss. You can spend hours – days, even – staring through a telescope and examining the minutiae of the plumages before you.

The gulls start gathering in September, when the river's most numerous species, Bonaparte's Gull, begins to arrive after breeding by lakes in the forest-tundra of northern Canada and Alaska. These delicate gulls, which are unusual among larids for their habit of nesting in trees, join the Ring-billed, American Herring and Great Black-backed Gulls that are here more or less year-round, milling around the falls, rocks, river and jetties. A careful look at this time of the year may also reveal the odd Sabine's Gull, perhaps a juvenile fresh from the tundra of eastern Canada and Greenland. This

Below: the larger, Arctic-breeding 'white-winged' gulls are seen in greatest numbers during January. This is a first-winter Glaucous Gull.

■ *Right: Little Gull is a rare but regular visitor to the falls. Note the distinctive underwing pattern of this winter-plumaged adult.*

■ *Right: numbers of Bonaparte's Gulls peak in early December, when 20,000 may be present; this is a first-winter.*

species winters far out in the ocean off the coast of Africa, and passes Niagara for only a brief window lasting until early November. It appears to prefer the area around the falls themselves.

The build-up continues in October, when the first Franklin's Gulls arrive from the Great Plains. No more than a few of these birds stray this far east, since they should be on their way to winter on the Pacific Coast from Mexico to Chile. In some ways it is remarkable that any Franklin's Gulls survive, as these birds build floating nests in deep marsh vegetation which are forever on the verge of sinking. This species excites gull experts because it is the only one that undergoes two complete moults a year, as opposed to one complete moult in autumn and a partial moult in spring.

November marks the beginning of the main season, where numbers and variety climb sharply upwards. The Little Gull, a Eurasian species almost unknown in North America before colonizing the Great Lakes area in 1962, disperses this way en route to the Atlantic Coast, and may be seen in single-figure numbers. The Black-legged Kittiwake may also arrive; at this time of the year this species should be out on the open Atlantic Ocean, but some evidently take a wrong turn after breeding on the north-east coast of Canada. The sought-after 'white-winged' gulls begin to appear at the end of November; these are so-named because they lack any black on their wing-tips. They include the Iceland Gull which, despite its name, will actually have wandered in from its breeding grounds in Eastern Canada, and the huge Glaucous Gull, a highly predatory species of the Canadian tundra, which has been known to swallow adult alcids whole. Another species, Thayer's Gull of the Canadian High Arctic, is an honorary member of the 'white-winged' group because although it actually has some black on the wing-tips, this only appears in adult plumage. All of these species rarely appear in numbers above single figures.

The species mentioned so far are predictable, but November also inevitably brings great rarities and, with them, equally great challenges for birders. How, for example, do you find a single Black-headed Gull among 20,000 Bonaparte's? Somehow enthusiasts do, and they manage this feat every year. They also pick out Lesser Black-backed Gulls, also from Europe, and every year a

Above: gull enthusiasts may take time to notice that there's a waterfall at Niagara, too.

California Gull seems to arrive here from its breeding grounds in the Prairies. These days a confirmed gull-watcher, or a participant on one of the various specialist gull tours that operate here, would feel disappointed not to see these three species.

Every few years sees something exceptional turn up to add to the Niagara list. Laughing Gulls from the Atlantic coast, a Mew Gull from the Canadian north-west, a Common Gull from Europe, Ross's and Ivory Gulls from the Arctic pack-ice: all of these have been logged. Even a Slaty-backed Gull from east Asia has, incredibly, found its way here and, equally incredibly, was picked out by some gull genius. No doubt the list will be added to in the future; there are plenty of possibilities in North America alone.

The Niagara River covers quite a large area, but the best places for gulls are considered to be the International Control facility above the falls (this regulates the flow of water; it is also an excellent place for ducks), the immediate falls area, the Sir Adam Beck Power Plant on the Canadian side, where gulls come to feed in the outflow, and the Hydroelectric Power Plant on the American side between Niagara Falls and Lewiston. In addition, evenings bring a mass movement of Bonaparte's Gulls past the settlement of Niagara-on-the-Lake (on the Canadian side) out into Lake Ontario, where the birds roost. If you have spent the day straining your eyes examining gull plumages, this amazing fly-past of thousands of white birds in fading light adds extra magic to your experience.

The number of Bonaparte's Gulls peaks in early December, when 20,000 or more may be present. After this, the variety dwindles and the only birds increasing in abundance are the large, hardy species such as the white-winged gulls, which may reach double figures in January. By then, most of the gull enthusiasts are working their local patches.

Why is it that there so many gulls here? One reason is that there is plenty of food, with fish stocks being continually replenished by the fast-flowing waters, and lots of human activity to provide scraps. Another important reason is that the water rarely freezes here, even when all around the countryside is gripped in ice. However, scientists are not convinced that these factors tell the whole story; perhaps there is another reason, still waiting to be found.

Point Pelee

HABITAT Deciduous woodland, marsh, swamp, lakeshore

KEY SPECIES Migrants, especially warblers including Cerulean, Blackburnian, Prothonotary and Golden-winged

TIME OF YEAR Most popular month is May, but migration is strong from March to June and from September to November

There is not a birder in North America whose pulse doesn't quicken at the very mention of the name of Point Pelee. This tiny national park, the smallest in Canada, is universally regarded as the finest site for migrant birds in inland North America. It attracts up to 400,000 human visitors each year, most of them hoping to witness one of the famous 'waves' of warblers, vireos, sparrows and tanagers that turn up here throughout a season. Over the years a quite incredible total of 372 species has been recorded here, including many rarities. Warblers are a particular feature and it is standard for visitors to see 20 species of warbler in a day, and on a few occasions the remarkable total of 34 has been achieved.

Point Pelee is exceptional because of its geographical position. It is a peninsula that juts some 10 km out from the northern shore of Lake Erie, in the southern part of the Great Lakes region; thus it is a first landfall for birds crossing the lake in the spring, and acts as a natural funnel concentrating southbound migrants in the autumn. It is the southernmost point in mainland Canada, with a relatively mild climate, and is ideally placed to host northbound migrants and southern species overshooting. Its rich forests and wetlands also provide excellent feeding and sheltering areas for many thousands of migrants in both seasons. Additionally, the small area (20 sq km) ensures that it is an easy place for birders to work.

Although the teeming hordes of birders tend not to arrive until May, the first migratory movements of the birds themselves actually take place in March, when wildfowl feature and such species as American Black Duck, Redhead and Canvasback are at their peak. At the same time, typical early migrants such as American Robins, Mourning Doves and Killdeers arrive, buoyed by warm weather fronts moving north. At this season the phenomenon of reverse migration can sometimes be witnessed at 'The Tip', Point Pelee's southernmost point. Birds such as Common Grackles and Red-winged Blackbirds appear to decide to retreat, and fly back south across the lake's waters.

Throughout April and May birds arrive in intermittent waves, some large, with birds touching down en masse, and some much less obvious. Large waves occur when a warm front from the south meets a cold front from the north, and the wind and rain cause the birds to make landfall at the first opportunity as they move north. When this happens there may seem to be birds in every tree and every bush, and even the beach areas around the point will be full of weary travellers.

The spring migration reaches its height in mid-May, when the greatest variety of birds can be seen. Warbler enthusiasts will be checking through the commoner species, such as Northern Parula, Chestnut-sided, Blackburnian, Magnolia, Bay-breasted, Golden-winged and Canada Warblers, for rarities such as Prairie and Hooded Warblers and the now endangered Cerulean Warbler. Another uncommon species is the delightful buttercup-yellow Prothonotary Warbler, which actually remains to breed in small numbers in the swamp areas, although it can be hard to find.

■ *Above: Point Pelee is renowned for its New World warblers, which peak in numbers and variety around mid-May. This is the aptly-named Chestnut-sided Warbler.*

■ *Opposite: the woods of Point Pelee are close to the northern limits of the Blue-grey Gnatcatcher's breeding range.*

■ *Right: the scarce and declining Cerulean Warbler usually occurs only in small numbers.*

Above: a bright orange throat and extensive white wing panel identify this bird as a male Blackburnian Warbler.

There are, in fact, several other species nesting in the Point Pelee area that are more typical of milder climates, besides the Prothonotary Warbler. The warming effect of the Great Lakes means that the climate here is unusually mild, and Point Pelee is within the so-called Carolinian Zone, dominated by broad-leaved rather than coniferous trees. Carolina Wrens, Blue-grey Gnatcatchers and Yellow-breasted Chats reach the northern limits of their breeding distribution in this area, and can be found around Point Pelee in June and July, along with more common breeding birds.

August marks the beginning of the main southward migration, although shorebird movement actually begins a month earlier. One of the first landbird migrants to reappear will be the sprightly Black-and-white Warbler which, in contrast to most of its relatives, is easy to identify at any season. Other warblers, now bereft of their colourful plumage and with their voices reduced to a few call-notes, will present a much sterner test of identification skills, although they may present in equal or even greater numbers than in spring. In contrast to the spring, however, there are fewer overwhelming waves, since autumn migration is a more leisurely affair than the unseemly northward rush to reach and claim territories in spring. Furthermore, the subtly marked birds can be well hidden by the copious yellow and brown autumn foliage so, perhaps not surprisingly, Pelee is less popular with birders at this time of year.

Nevertheless, there is still plenty to enjoy, and it isn't all a question of peering into bushes. Cold fronts in September may see large southerly movements of Broad-winged Hawks and other raptors into Pelee's airspace, and there are often spectacular movements of Blue Jays overhead; this is one of the few places where birders actually notice this otherwise routinely familiar bird. At the point, the later autumn is also an excellent time to observe visible migration as birds head south over the lake in the morning hours. Others seem to think about departure, become fidgety but eventually stay around instead, soon tucking into insects, or feeding on the fruits of the commonest tree in Pelee's forests, the Hackberry.

By October and November the landbird passage has passed its peak, and most of the warblers are gone. A few species, notably Ruby-crowned and Golden-crowned Kinglets and Black-capped Chickadees, may be at their commonest at this season, but they are definitely in the minority. For some reason, though, this is also a good time for great rarities to appear, including a bewildering number that have strayed from the western side of the North American continent, such as Mountain Bluebird or Western Tanager. However, finding rarities in November is really a pastime of the local birders. The main crowds, both of birds and human visitors, will return in the spring.

Churchill

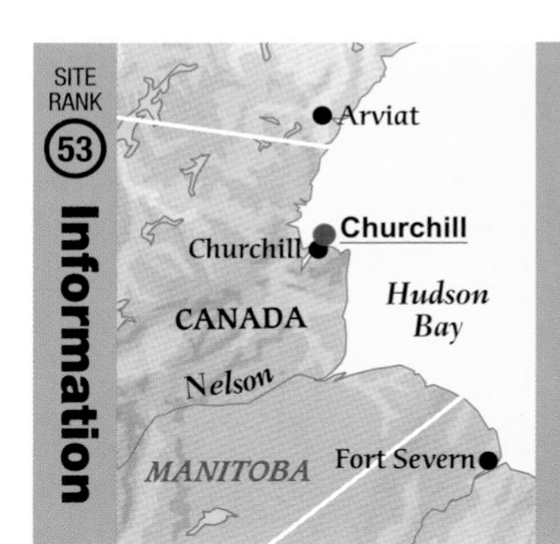

HABITAT Tundra, taiga forest, coast

KEY SPECIES Ross's Gull, Snowy Owl, 15 species of breeding waders including Hudsonian Godwit, Spruce Grouse, Smith's Longspur

TIME OF YEAR Late May through to July is the best time for spring migrants and breeding birds; late autumn for a few birds (including Rock Ptarmigan) and Polar Bears

Above right: the uncommon Smith's Longspur brings its colourful breeding regime to the summer tundra of Churchill.

Below: Ross's Gull is a rare spring migrant to Churchill, although it has bred there twice in the past.

When the Hudson Bay Railway finally rolled into Churchill in 1929, the builders could hardly have envisaged that their long, lonely track would one day carry tourists with binoculars. Covering the 1,600 km from the Manitoba state capital, Winnipeg, the railway was constructed to bring grain from the prairie wheat belt to the shores of Hudson Bay, and thence by sea to the rest of the world. Churchill, on the shores of the great bay, grew up as a frontier town, a remote outpost where no one but those in need of work would ever go. Yet these days it is far more famous as an eco-tourism destination, and the vast grain silos and container ships are merely the backdrop to some fantastic wildlife viewing.

There are few places in the world where it is easy to observe the birds of the true tundra, but Churchill is one of them. During the short summer growing season, which begins in late May and ends in August, the tundra turns a vibrant mix of subtle colours as the mosses, lichens, sedge-grasses and sphagnum are at last unveiled by melting snow. Dotted with innumerable lakes and bogs, the sun-warmed tundra supports unimaginably high populations of mosquitoes, midges, spiders and other invertebrates, and these in turn provide food for a wealth of Arctic breeding birds. In this bloom of productivity, as many bite-plagued birders will attest, there is plenty of everything to go around.

To visit the tundra at this time is a noisy, as well as a visually stunning experience, since one of the most

dominant groups of birds here, the waders, spend all the many hours of daylight performing their spirited song-flights. The sounds are incessant; birds tend to arrive unpaired and must acquire a mate as quickly as possible. Thus the sharp Arctic air rings with the whinnying sounds of Wilson's Snipe, the braying of the Stilt Sandpiper and the spluttering of the Semipalmated Sandpiper, among many others, each male trying to grab the attention of a nearby female.

Studies of shorebirds here and at other tundra sites have revealed that their breeding ecology is full of fascinating details. For example, while most waders are monogamous here, several are not. A female Spotted Sandpiper, for example, might copulate with up to four different males, lay maybe four clutches of eggs and try to enlist each male to incubate one of them – the ultimate in female emancipation! The male Pectoral Sandpiper, by contrast, often attempts to acquire several

■ *Opposite: after breeding at Churchill, the Hudsonian Godwit makes a 3,700 km journey to winter in South America; this is often undertaken as a single, non-stop flight.*

female mates, while Wilson's Snipe's overall behaviour appears to be a promiscuous free-for-all. When it comes to protecting the nest-site, several waders go in for the broken-wing distraction display, but Hudsonian Godwits instead resort to making a constant deafening noise, one of the loudest made by a wader and undoubtedly irritating to the sensitive ears of an Arctic Fox. Another aspect of wader ecology is that, in many species, only one parent remains to tend the chicks, while the other deserts the area; in most species, such as the Least Sandpiper, the attending parent is the male.

The waders don't have a monopoly on odd breeding systems. Recently another tundra bird, Smith's Longspur, has given up a few secrets of its own. These birds effectively live in small breeding groups, with females copulating with two or three males and the males copulating in turn with several females, a system known as polygynandry. When a female is most fertile she may copulate 350 times in a single week, and the males repay the favour by helping to feed her offspring at the nest.

Churchill lies in the transition zone between the open tundra and the dominant vegetation further south, the taiga; this is, in fact, the core habitat of the longspur. At Twin Lakes, within reach of town, are mixed spruce and larch forests that house more waders, including the Solitary Sandpiper, a wader that has the rare habit of laying its eggs above ground-level, in the abandoned nests of American Robins and other breeding birds such as Bohemian Waxwing and Grey Jay, which, incidentally, are both easy to see. The forest here also holds Spruce Grouse and, in some years, Great Grey Owls.

Although it has a fine range of breeding birds, Churchill's position at the western side of Hudson Bay and at the outflow of the large south-to-north flowing Churchill River places it on one of the major flyways used by true Arctic species on their way farther north. Thus, from early June, any day can bring flocks of such birds as White-rumped Sandpiper, Red Knot and Sanderling, which are simply passing through. This is also the favoured time for Churchill's most famous species, Ross's Gull. When this graceful, delicately-hued gull nested here in 1980 it was the first breeding record for the North American continent. However, there was only one further breeding attempt, and since the late 1980s this bird has only been seen passing through at Churchill, often only a few times each year. It is usually seen out on the bay, dipping to catch marine invertebrates among the melting icebergs, occasionally in the presence of a group of snow-white Beluga Whales.

The birding season carries on into July but, by early August, most visitors have left. When the tourists return, in October and November, it is for quite a different reason. Churchill is simply the best place in the world to see Polar Bears, and visitors come in their thousands to take a 'tundra buggy' for virtually guaranteed sightings of these superb animals. As a bonus, they almost always see the Aurora Borealis, too; Churchill is also apparently the best place in the world to see this phenomenon, which occurs on 300 nights a year. Not many birds are found around this time, but it is well worth looking out for the Snowy Owl, if you can tear yourself away from watching all these other wonders.

■ *Below: Churchill lies in the transition zone between tundra and taiga.*

Bibliography

BirdLife International. 2003. *Threatened Birds of the World.* Lynx Edicions, Barcelona.

Brooke, M. 2004. *Albatrosses and Petrels across the World.* Oxford University Press.

Brown, L.H., Urban, E.K., Newman, K., Keith, S., and Fry, C.H. (eds). *The Birds of Africa.* Vols 1-7. Academic Press, London/Christopher Helm, London.

Chipley, R.M., Fenwick, G.H., Parr, M.J. and Pashley, D.N. 2003. *The American Conservancy Guide to the 500 Most Important Bird Areas in the United States.* Random House, New York.

Cohen, C., Spottiswoode, C. and Rossouw, J. 2006. *Southern African Birdfinder.* Struik, Cape Town.

Cramp, S. and Simmons, K.E.L. (eds). 1977-83. *Handbook of the Birds of Europe, the Middle East and North Africa: The Birds of the Western Palearctic.* Vols 1-3. Oxford University Press.

Cramp, S. (ed.). 1985-92. *Handbook of the Birds of Europe, the Middle East and North Africa: The Birds of the Western Palearctic.* Vols 4-6. Oxford University Press.

Cramp, S. and Perrins, C.M. (eds). 1993-94. *Handbook of the Birds of Europe, the Middle East and North Africa: The Birds of the Western Palearctic.* Vols 7-9. Oxford University Press.

Davies, S.J.J.F. 2002. *Ratites and Tinamous.* Oxford University Press.

Del Hoyo, J., Elliott, A. and Sargatal, J. (eds). 1992-2001. *Handbook of the Birds of the World.* Vols 1-7. Lynx Edicions, Barcelona.

Del Hoyo, J., Elliott, A. and Christie, D.A. (eds). 2003-2007. *Handbook of the Birds of the World,* Vols 8-12. Lynx Edicions, Barcelona.

Ehrlich, P.R., Dobkin, D.S., Wheye, D. and Pimm, S.L. 1994. *The Birdwatcher's Handbook.* Oxford University Press.

Elphick, C., Dunning, J.B. (Jr) and Sibley, D. 2001. *The Sibley Guide to Bird Life and Behaviour.* Christopher Helm, London.

Ferguson-Lees, J and Christie, D.A. 2001. *Raptors of the World.* Christopher Helm, London.

Fjeldså, J. 2004. *The Grebes.* Oxford University Press.

Frith, C.B and Beehler, B.M. 1998. *The Birds of Paradise.* Oxford University Press.

Frith, C.B. and Frith, D.W. 2004. *The Bowerbirds.* Oxford University Press.

Gaston, A.J. 2004. *Seabirds: A Natural History.* T & AD Poyser.

Gill, F. and Wright, M. (IOC). 2006. *Birds of the World: Recommended English Names.* Christopher Helm, London.

Gorman, G. 2006. *Birding in Eastern Europe.* WildSounds, Norfolk, UK.

Green, I. and Moorhouse, N. 1995. *A Birdwatchers' Guide to Turkey.* Prion, Sandy, UK.

Harrap, S. and Redman, N. 2003. *Where to Watch Birds in Britain.* Christopher Helm, London.

Harrison, J. 1999. *A Field Guide to the Birds of Sri Lanka.* Oxford University Press.

Jaramillo, A. *Field Guide to the Birds of Chile.* Christopher Helm, London.

Kazmierczak, K. and Singh, R. 1998. *A Birdwatchers' Guide to India.* Prion, Sandy, UK.

Kaufmann, K. 1996. *Lives of North American Birds.* Houghton Mifflin, New York.

Marchant, S. *et al.* 1990-2003. *Handbook of Australian, New Zealand and Antarctic Birds (HANZAB).* Vols 1-5. OUP, Melbourne, Australia.

Marsh, K. 2005. *The Good Bird Guide.* Christopher Helm, London.

Mayr, E. 1978. *Birds of the South-west Pacific.* Tuttle, Tokyo.

Nielsen, L. 2007. *Birding Australia – A Directory for Birders.* Birding Australia.

Perrins, C. (ed.). 2003. *The New Encyclopedia of Birds.* Oxford University Press.

Pratt, H.D. 2005. *The Hawaiian Honeycreepers.* Oxford University Press.

Restall, R., Rodner, C. and Lentino, M. 2006. *Birds of Northern South America.* Vols 1-2. Christopher Helm, London.

Robinson, H. 2003. *Australia: An Ecotraveller's Guide.* Arris Books, Moreton-in-Marsh, UK.

Robson, C. 2000. *A Field Guide to the Birds of South-East Asia.* New Holland, London.

Rose, L. 1995. *Where to Watch Birds in Spain and Portugal.* Hamlyn, London.

Schodde, R. and Tidemann, S.C. 1986. *Reader's Digest Complete Book of Australian Birds.* Reader's Digest, Sydney, Australia.

Sinclair, I. and Langrand, O. 1998. *Birds of the Indian Ocean Islands.* Struik, Cape Town.

Snow, D.W. and Perrins, C.M. (eds). 1998. *The Birds of the Western Palearctic. Concise edition* Vols 1-2. Oxford University Press.

Turner, A. and Rose, C. 1994. *A Handbook to the Swallows and Martins of the World.* Christopher Helm, London.

Van den Berg, A. and Lafontaine, D. Wheatley, N. 1996. *Where to Watch Birds in Holland, Belgium and Northern France.* Hamlyn, London.

Wells, D.R. 1999-2007. *The Birds of the Thai-Malay Peninsula.* Vols 1-2. Christopher Helm, London

White, M. and Lehman, P. 2006. *National Geographic Guide to Birding Hotspots of the United States.* National Geographic Society, Washington.

Wheatley, N. 1995. *Where to Watch Birds in Africa.* Christopher Helm, London.

Wheatley, N. 1996. *Where to Watch Birds in Asia.* Christopher Helm, London.

Wheatley, N. 1998. *Where to Watch Birds in Australasia and Oceania.* Christopher Helm, London.

Wheatley, N. 2001. *Where to Watch Birds in Central America and the Caribbean* . Christopher Helm, London.

Wheatley, N. 2000. *Where to Watch Birds in Europe and Russia.* Christopher Helm, London.

Wheatley, N. 1994. *Where to Watch Birds in South America.* Christopher Helm, London.

Index

A

B

Map No.	Site	Rank	Page
1.	North Norfolk	55	14
2.	The Outer Hebrides	66	17
3.	The Camargue	61	20
4.	Organbidexka Col Libre	94	23
5.	Lac du Der-Chantecoq	89	25
6.	Waddensee	56	27
7.	Coto Doñana	39	30
8.	Extremadura	63	33
9.	Po Delta	95	36
10.	Varanger Peninsula	83	38
11.	Falsterbo	71	41
12.	Oulu	90	44
13.	Lake Myvatn	79	47
14.	Matsalu Bay	87	50
15.	Bialowieza Forest	57	53
16.	High Tatras	93	56
17.	Danube Delta	36	59
18.	Eilat	25	64
19.	Dubai	75	67
20.	Korgalzhyn	40	70
21.	Pontic Alps	35	73
22.	Keoladeo Ghana	7	76
23.	Goa	48	79
24.	Taman Negara	10	82
25.	Gomantong Caves	92	85
26.	Limithang Road	11	88
27.	Bogani Nani Wartabone	8	91
28.	Katanglad Mountains	14	94
29.	Sinharaja Forest	17	97
30.	Mai Po	29	100
31.	Liaoning	100	103
32.	Beidahe	41	105
33.	Wolong	74	108
34.	Lena Delta	59	111
35.	Ussuriland	62	114
36.	Arasaki	68	117
37.	Merzouga	72	122
38.	Banc D'Arguin	43	125
39.	Djoudj	65	128
40.	Gambia River	50	131
41.	Ivindo	19	134
42.	Bale Mountains	21	137
43.	Rift Valley Lakes	12	140
44.	Serengeti	31	143
45.	Caprivi Strip	54	146
46.	Bwindi	24	149
47.	Dzalanyama Forest	96	152
48.	Bangweulu Swamps	52	154
49.	Sir Lowry's Pass	84	157
50.	Mkhuze	46	160
51.	Andasibe	9	163
52.	Spiny Forest	1	166
53.	The Seychelles	30	169
54.	The Strzelecki Track	51	174
55.	Lamington	77	177
56.	New South Wales pelagic	85	179
57.	Kakadu	33	182
58.	Queensland Wet Tropics	42	184
59.	Rivière Bleue	23	187
60.	Stewart Island	22	190
61.	Tari Valley	2	193
62.	Taveuni	44	196
63.	South Georgia	34	202
64.	Manu	3	206
65.	Abra Patricia	13	209
66.	Santa Marta	26	212
67.	Galapagos Islands	5	214
68.	Tandayapa	64	217
69.	Pantanal	15	220
70.	Serra da Canastra	45	223
71.	Manaus	28	225
72.	Iguassu Falls	69	228
73.	Ushuaia	32	231
74.	La Escalera	4	234
75.	Los Llanos	27	236
76.	Lauca	18	239
77.	Tikal	37	244
78.	Canopy Tower	91	247
79.	Monteverde	73	250
80.	San Blas	78	253
81.	Veracruz river of raptors	98	255
82.	Asa Wright Centre	47	258
83.	Zapata Swamp	67	261
84.	Bosque del Apache	38	266
85.	Monterey Bay	76	269
86.	Prairie Potholes	70	271
87.	Chiricahua Mountains	82	274
88.	Rio Grande Valley	81	277
89.	Olympic Peninsula	49	280
90.	Snake River	99	283
91.	Yellowstone	58	286
92.	Pribilof Islands	20	289
93.	Cape May	80	292
94.	Hawk Mountain	88	295
95.	High Island	86	298
96.	Everglades	16	301
97.	Alaka'i wilderness	6	304
98.	Niagara Falls	97	307
99.	Point Pelee	60	310
100.	Churchill	53	313